教育部高等学校材料类专业教学指导委员会规划教材

国家级一流本科专业建设成果教材

材料科学与工程专业英语

孙大林 黄高山 赵 婕 等 编著

ENGLISH FOR MATERIALS SCIENCE AND ENGINEERING

·北京·

内 容 简 介

《材料科学与工程专业英语》为教育部高等学校材料类专业教学指导委员会规划教材，全书内容分为7章。第1章为材料科学与工程概述，第2章为金属材料与合金，第3章为陶瓷材料，第4章为聚合物材料，第5章为生物材料，第6章为纳米科技与纳米材料，第7章为材料表征技术。每章均包含多篇课文，重点介绍材料科学相关知识，同时注重英文的专业性和准确性。此外，每章还提供了延展性的阅读材料，提供与课文相关的背景知识，进一步拓展专业内容。在每篇课文和阅读材料之后，均配有词汇表和难点注释、练习题，可进一步帮助学生加强对专业英语知识的掌握。

全书选材前沿且知识丰富，有助于学生了解材料科学发展的新态势，并激发学生从事材料相关科学研究的兴趣和热情。

本书为材料类各专业本科生、研究生的教材，也可作为教师及材料领域研究人员的学习参考书。

图书在版编目（CIP）数据

材料科学与工程专业英语/孙大林等编著. —北京：化学工业出版社，2023.6（2024.10重印）
ISBN 978-7-122-43152-3

Ⅰ.①材… Ⅱ.①孙… Ⅲ.①材料科学-英语-高等学校-教材 Ⅳ.①TB3

中国国家版本馆 CIP 数据核字（2023）第 050233 号

责任编辑：陶艳玲　　　　　　　　　　　文字编辑：丁　瑞
责任校对：王鹏飞　　　　　　　　　　　装帧设计：史利平

出版发行：化学工业出版社（北京市东城区青年湖南街13号　邮政编码100011）
印　　装：北京天宇星印刷厂
787mm×1092mm　1/16　印张18½　字数459千字　2024年10月北京第1版第2次印刷

购书咨询：010-64518888　　　　　　　　售后服务：010-64518899
网　　址：http://www.cip.com.cn
凡购买本书，如有缺损质量问题，本社销售中心负责调换。

定　价：59.00元　　　　　　　　　　　　　　　　　　版权所有　违者必究

序

国以才立，业以才兴。

现今，国家的发展对高等教育的需要更加迫切，对科学知识和卓越人才的渴求更加强烈。教育部自2019年以来推进实施一流本科专业建设"双万计划"，提出做强一流本科、建设一流专业、培养一流人才，提高高校人才培养能力，以实现高等教育内涵式发展。

为进一步加强材料类专业教材建设，促进材料类专业人才培养水平的提升，教育部高等学校材料类专业教学指导委员会（以下简称材料教指委）成立了教材建设工作组，按照《普通高等学校教材管理办法》，开展材料类专业教材建设工作，切实提高材料类专业教材建设水平。复旦大学材料科学系在本科教学中坚持需求导向，强化专业特色，注重培育前沿交叉人才，针对工科专业学生培养的新形势，大力完善教学资源，启动了系列教材编撰工作。"材料科学与工程专业英语"课程教学团队教师在多年教学和科研经验的基础上，参阅大量国内外最新文献资料，对教学内容进行了全新编排，编写了本教材。本教材内容涵盖金属材料、陶瓷材料、聚合物材料、生物材料、纳米科技与纳米材料以及材料表征技术。团队教师开拓创新，对教材内容进行了改编和取舍，强化了内容的多样性与文字的可读性，在专业知识上有的放矢，紧扣材料科技最前沿。

《材料科学与工程专业英语》内容新颖，知识系统，逻辑严谨，时代感强，能有效加强学生英语读写能力，提升材料科学与工程专业知识水平，拓展学生的学科前沿视野。该教材不仅可对培养新工科一流人才发挥重要作用，更对发展面向国家需求的新时代一流本科教育具有重要的价值。

2018—2022年教育部高等学校材料类专业教学指导委员会秘书长

前　言

在科技创新驱动发展的国家政策背景下,对人才培养提出了新要求,学生的基本素质、科学知识及实践能力的培养受到空前重视。为进一步加强材料学科专业教材建设,促进材料类专业人才培养水平的提升,教育部高等学校材料类专业教学指导委员会(以下简称材料教指委)成立了教材建设工作组,按照《普通高等学校教材管理办法》,开展材料类专业规划教材建设工作,切实提高材料学科专业教材建设水平,本书有幸被列为规划教材。

英语水平是衡量大学生素质能力的一项重要指标。我国基础英语教育工作已发展成熟,但专业英语教学相对滞后,教材内容的选择与编排、教学内容的系统性及教学形式的多样性等问题亟须改善。近年来,材料科学研究与材料工程实践的内涵和外延发生了巨大变化,这些变化主要体现在以下几方面。

(1)材料科学与工程学科的地位日益突显。材料科技创新在国家重大需求、国防建设和生命健康等方面正发挥着重要作用。一种新材料的面世预示着一项革新技术的到来;同样,以需求为导向的科技发展,也推动了材料科技的创新。

(2)新材料成为新兴产业发展的源动力。高效能源材料的突破是实现"碳达峰""碳中和"目标的技术基础;光电材料是集成电路、大数据、人工智能等高技术产业的基石;生物和医用材料推动着医疗技术的变革,催生了事关人类健康和生活质量的高新技术产业。更为重要的是,学科交叉融合是当今科学技术发展的最突出特征,新旧材料的交替迭代速度加快,这对材料类专业的学生培养提出了新的要求。

(3)材料学科的国际影响力增强。我国的材料研究水平不断提高,材料学科的排名处于世界前列。为了彰显我国材料研究的影响力和话语权,要求材料专业的人才具有国际化视野,具备很好的国际合作和交流能力,这对材料专业英语教材的编写提出了更高要求。

(4)信息技术改变了教与学的方式。在信息技术高度发达的今天,教育理念的变革、教学形式的创新和教学方法的多样化,成为培养学生的必由之路。人才培养要提倡采用互联网+等新形式的教学模式,发挥教育资源信息化提供的便利优势。

为了积极顺应这些变化,在吸收已有教材优点的基础上,我们认为有必要对相关知识体系进行重新设计和更新,并按照教育部高等学校材料类专业教学指导委员会规划教材的要求,编写《材料科学与工程专业英语》这本新教材。

本教材特点:(1)知识面宽。涵盖了与材料学科相关的基础知识,内容包括材料科学与工程概论、金属材料、无机非金属材料、聚合物材料、复合材料、纳米结构材料及生物材料等,还特别增加了材料(原位)表征技术等内容;(2)内容新颖。各章的阅读材料参考了国外最新出版的专著和期刊论文。选材前沿且知识丰富,有助于学生了解材料科学发展的新态势。同时,特别注重选择国内材料方面的研究成果作为教材内容,以激发学生从事材料研究的兴趣和热情;(3)编写人员具备相应的实力。编写成员由具有材料、化学、物理等交叉专业背景的青年学者组成,他们具有海内外学习和工作经历,科研成果丰硕,对本专业人才英语能力的培养需求较为了解,对如何学好专业英语颇有心得体会。

内容与结构:教材分为 7 章(Chapter),共 29 个单元(Unit)。第 1 章为材料科学与工程概述(三个单元),第 2 章为金属材料与合金(五个单元),第 3 章为陶瓷材料(四个单元),第 4 章为聚合物材料(五个单元),第 5 章为生物材料(三个单元),第 6 章为纳米科技与纳米材料(五个单元),第 7 章为材料表征技术(四个单元)。大部分单元除课文外还包含一篇阅读材料(reading material),共计材料专业文章 53 篇(课文 29 篇,阅读材料 24 篇)。阅读材料提供与课文相应的背景知识或课文内容的拓展。在每篇课文和阅读材料之后,配有词汇表和难点注释,并提供相应的练习题以帮助学生加强对专业英语知识的掌握。

本教材第 1 章和第 5 章由赵婕编写,第 2 章由刘洋编写,第 3 章由吴仁兵编写,第 4 章由梁佳、赵岩、黄高山编写,第 6 章由黄高山、王飞编写,第 7 章由朱凡、周永宁编写。全书由孙大林、黄高山、赵婕、方方统稿。

本教材为教育部高等学校材料类专业教学指导委员会规划教材,适应于材料类各专业的本科生和研究生使用,也可作为教师及相关领域研究人员的学习参考书。本书在编写过程中得到复旦大学材料科学系的大力支持,在此表示感谢。由于水平所限,且教材涉及面较广,书中难免出现疏漏,希望广大读者不吝指正,以便使本教材在使用过程中进一步改进和完善。

编者
2023 年 6 月

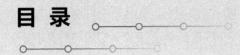

目录

Chapter 1 Introduction to Materials Science and Engineering

Unit 1.1 Text Materials science and engineering / 001
Reading Material Brief introduction to materials science / 005
Unit 1.2 Text Classification of materials / 007
Reading Material Biomaterials / 015
Unit 1.3 Text The structure of the atom / 017
Reading Material "Why didn't we think to do this earlier?" chemists thrilled by speedy atomic structures / 030

Chapter 2 Metallic Materials and Alloys

Unit 2.1 Text Introduction to metals and alloys / 034
Reading Material The hydrogen storage alloys for electrochemical applications / 039
Unit 2.2 Text Thermal equilibrium diagram / 042
Reading Material Failure analysis on leaked titanium tubes of seawater heat exchangers / 048
Unit 2.3 Text Characters of metallic materials— ductility, malleability, and corrosion / 052
Reading Material Revisiting the effect of molybdenum on pitting resistance of stainless steels / 058
Unit 2.4 Text Superalloys and nonferrous alloys / 063
Reading Material High-entropy alloys / 067

Unit 2.5　**Text**　Metal-matrix composites　/ 072
　　　　　Reading Material　Mechanical behavior of particle reinforced metal matrix composites　/ 077

Chapter 3　Ceramics

Unit 3.1　**Text**　Introduction to ceramics　/ 082
Unit 3.2　**Text**　Bioceramics　/ 089
　　　　　Reading Material　High-entropy ceramics　/ 095
Unit 3.3　**Text**　3D-printing technologies for ceramics　/ 098
Unit 3.4　**Text**　Ceramics solid electrolytes　/ 104

Chapter 4　Polymers

Unit 4.1　**Text**　Introduction to polymers　/ 109
　　　　　Reading Material　Polymer crystallinity　/ 116
Unit 4.2　**Text**　Methods for synthesis of polymers　/ 121
　　　　　Reading Material　Step-growth polymerization　/ 126
Unit 4.3　**Text**　Structures and properties of polymers　/ 132
　　　　　Reading Material　History of polymers and their typical synthesis approaches　/ 137
Unit 4.4　**Text**　Polymers for food packaging　/ 141
　　　　　Reading Material　Biopolymers and their potential as a packaging material　/ 148
Unit 4.5　**Text**　Photocontrol of fluid slugs in liquid crystal polymer microactuators　/ 154

Chapter 5　Biomaterials

Unit 5.1　**Text**　Biomaterials and biomaterials science　/ 162
　　　　　Reading Material　Make better, safer biomaterials　/ 165
Unit 5.2　**Text**　Traditional and new biomaterials　/ 169
　　　　　Reading Material　3D printing metallic implants: Technologies available and the future of the industry　/ 178
Unit 5.3　**Text**　Medical applications of biomaterials　/ 182
　　　　　Reading Material　Biomaterials for tissue repair　/ 190

Chapter 6 Nanoscience, Nanotechnology, and Nanomaterials

Unit 6.1 Text Introduction to nanoscience and nanotechnology / 195
Reading Material Carbon dots / 201
Unit 6.2 Text Classification of nanomaterials / 204
Reading Material 3D nanomaterials fabricated by rolled-up nanotechnology / 208
Unit 6.3 Text Synthesis and applications of nanomaterials / 212
Reading Material Nanomaterials for energy conversion and storage / 219
Unit 6.4 Text Intriguing physical properties of nanomaterials / 224
Reading Material Natural and artificial nanotechnologies / 231
Unit 6.5 Text Nanographenes and optoelectronic devices / 235
Reading Material Catalytic nanozymes / 241

Chapter 7 Material Characterization Methods

Unit 7.1 Text Introduction to spectroscopic methods / 247
Reading Material X-ray absorption spectroscopy / 254
Unit 7.2 Text Operando spectroscopy / 257
Unit 7.3 Text Introduction to microscopic methods / 262
Reading Material Microscopic aberrations / 270
Unit 7.4 Text Modern in situ characterization techniques / 273
Reading Material Ultrafast transmission electron microscope and diffraction / 280

参考文献

Chapter 1

Introduction to Materials Science and Engineering

【本章导读】

本章是材料科学与工程概论，分为三个单元，包括三篇课文和三篇阅读材料。三篇课文分别讲解了材料科学与工程的发展、材料的分类和材料的结构。三篇阅读材料分别为材料科学简介、生物材料简介和分子结构快速分析技术。

Unit 1.1

Text

Materials science and engineering

这篇课文从不同角度介绍了材料科学与工程的发展背景。简言之，人类社会对于材料的使用逐步经历了从石器到青铜器再到现代材料的过程。课文简述了材料科学与材料工程的区别：材料科学侧重于研究材料结构、特性、制备方法及其相互关系；材料工程的重点是利用材料的理论基础和应用知识，将其转化为社会需要或期望的产品。课文的最后详细介绍了材料加工、结构、性质和性能之间的关系。

In this chapter, we will introduce you to the field of materials science and engineering using different real-world examples. Materials science is the basis for most technological advances. The most important aspect of materials is that they make things happen. Historically, the development and progress of society have been closely linked to the ability of its members to produce and manipulate materials to meet their needs. Prehistoric humans were limited to materials that were naturally available, such as stone, wood, bone, and fur. Over time, they moved from the Stone Age to the more recent Copper (Bronze) and Iron Ages. Besides, it was discovered that the properties of materials could be altered by heat treatment and the addition of other substances. In this regard, the utilization of materials is a process of selection, i.e., from a given and rather limited number of materials, it is decided that they are best suited for a certain application based on their properties. It is only recently that scientists have begun to understand the relationships between the structural elements of materials and their properties. This knowledge, gained over the last 100 years or so, has given them the ability to extend the properties of materials to a large extent. As a result, tens of thousands of different materials have been developed quite specialized properties to satisfy the needs of our modern and complex society; these materials include metals, plastics, glass, and fibers.

Materials science and engineering is an interdisciplinary field concerned with the invention of new materials and the improvement of already known materials via a deeper understanding of the relationships between microstructure, composition, synthesis, and processing. It is sometimes helpful to subdivide the discipline of materials science and engineering into two sub-disciplines, materials science and materials engineering. Materials science focuses primarily on identifying basic knowledge about the internal structure, properties, and processing of materials. Materials engineering is mainly concerned with using the basic and applied knowledge of materials to enable their transformation into products needed or desired by the society. From a functional perspective, the role of a materials scientist is to develop or synthesize new materials, while a materials engineer is called upon to create new products or systems from existing materials, and/or to develop materials processing techniques.

One of the most fascinating aspects of materials science involves the study of the structure of materials. In brief, the structure of a material is usually related to the arrangement of its internal components. The structure involves the electrons within individual atoms and the interactions with the nucleus. At the atomic level, structure encompasses the organization of atoms or molecules in relation to each other. The next larger domain of structure, which contains large groups of atoms that are usually clustered together, is called microscopic, meaning that it can be directly observed using some type of microscopes. Finally, structural elements that can be observed with the naked eye are referred to as macroscopic. The concept subatomic of properties deserves elaboration. During using, all materials are exposed to external stimuli that cause some type of reactions. For example, a specimen under stress will deform, or a polished metal surface will reflect light. A property is a material characteristic in terms of the type and degree of response to a particular stimulus.

In addition to the structure and properties, the science and engineering of materials involves two other important components: processing and performance. The term "processing" refers to how materials are formed into useful components. In terms of the relationships between these four components, the structure of a material depends on how it is processed. In addition, the performance of a material depends on its properties. Thus, the relationship between processing, structure, properties and performance is shown as in the schematic diagram in **Figure 1.1.1**. In this text, we draw attention to the relationships between these four components in terms of the design, manufacture and the use of materials.

(**Selected from:** Callister Jr W D, Rethwisch D G. Fundamentals of Materials Science and Engineering: An Integrated Approach [M]. John Wiley & Sons, 2020.)

New Words and Expressions

manipulate *v.* 操作，使用
prehistoric *adj.* 史前的
alter *v.* 改变

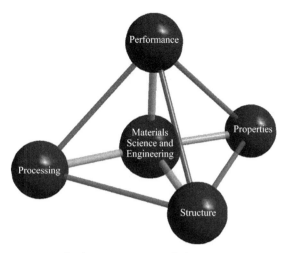

Figure 1.1.1　The four components of the discipline of materials science and engineering and their interrelationships.

heat treatment　热处理
in this regard　在这方面
utilization　*n.* 利用，使用
i. e.　也就是，即
interdisciplinary　*adj.* 跨学科的，各学科间的
synthesis　*v.* （化学物质的）合成
processing　*n.* 加工，处理
subdivide　*v.* 细分
arrangement　*n.* 整理，排列
subatomic　*adj.* 亚原子的，原子内的
encompass　*v.* 包含，环绕
elaboration　*n.* 阐述
specimen　*n.* 样品
polish　*v.* 磨光，改进
characteristic　*n.* 特征
stimulus　*n.* 刺激（物）
manufacture　*v.* 制造；*n.* 制造

Notes

(1) In this regard, the utilization of materials is a process of selection, i. e., from a given and rather limited number of materials, it is decided that they are best suited for a certain application based on their properties.

—in this regard 是一个常用的表达，意为在这个方面。
—i. e. 是拉丁语 id est 的缩写，意思为"即……"。
—参考译文：在这方面，材料的应用是一个选择过程，即从给定且有限的材料中，根据它们的特性决定其最适合的某种应用。

(2) It is only recently that scientists have begun to understand the relationship between the structural elements of materials and their properties.

—此句为强调句,It is ... that 是强调句的格式。
—参考译文:直到最近,科学家们才开始理解材料的结构单元与其特性之间的关系。

(3) This knowledge, gained over the last 100 years or so, has given them the ability to shape the properties of materials to a large extent.

—gained over the last 100 years or so 为插入语,丰富了句意。
—参考译文:过去100多年里获得的这种知识,使他们有能力在很大程度上塑造材料的性质。

(4) As a result, tens of thousands of different materials have developed quite specialized properties to meet the needs of our modern, complex society; these materials include metals, plastics, glass, and fibers.

—参考译文:因此,数以万计的材料已经发展出特定特性,以满足现代多元社会的需求;这些材料包括金属、塑料、玻璃和纤维。

(5) Materials science and engineering is an interdisciplinary field concerned with the invention of new materials and the improvement of already known materials through a deeper understanding of the relationships between microstructure, composition, synthesis, and processing.

—此句子较长,阅读理解时应该明确句子的结构成分,并进行合理的翻译。
—参考译文:材料科学与工程是一个跨学科领域,它通过更深入地理解微观结构、成分、合成和加工之间的关系来发明新材料和改进已知材料。

(6) From a functional perspective, the role of a materials scientist is to develop or synthesize new materials, while a materials engineer is called upon to create new products or systems from existing materials, and/or to develop materials processing techniques.

—from ... perspective 译为从……角度来看/从……角度去思考,是学术英语中一个常用的表达。call upon 译为号召,在这里形象地表达材料工程师所要做的事情。
—参考译文:从功能角度来看,材料科学家的任务是开发或合成新材料,而材料工程师则是基于现有材料开发新产品、新系统和新技术。

(7) At the atomic level, structure encompasses the organization of atoms or molecules in relation to each other.

—参考译文:在原子层面上,结构包含了原子或分子之间的相互组织关系。

Exercises

1. Question for discussion

(1) What is the main difference between materials science and materials engineering?

(2) Which kinds of specialized materials were developed to meet our modern, complex

society?

(3) What is microscopic structure?

(4) Please illustrate four important components of the science and engineering of materials.

2. Translate the following into Chinese

(1) Historically, the development and progress of society have been closely linked to the ability of its members to produce and manipulate materials to meet their needs.

(2) In addition, it was discovered that the properties of materials could be altered by heat treatment and the addition of other substances.

(3) It is sometimes useful to subdivide the discipline of materials science and engineering into two sub-disciplines, materials science and materials engineering.

(4) Materials science focuses on finding basic knowledge about the internal structure, properties, and processing of materials.

3. Translate the following into English

(1) 子学科　　　　　(2) 铁器时代
(3) 物质　　　　　　(4) 纤维
(5) 领域　　　　　　(6) 外部刺激
(7) 簇　　　　　　　(8) 反射光
(9) 除了结构和特性外，材料科学与工程还涉及另外两个重要组成部分：加工和性能。
(10) 就这四要素间的关系而言，材料的结构取决于其加工方式。
(11) 材料科学是大多数技术进步的基础。
(12) 我们要注重四要素在设计、制造和材料使用方面的关系。
(13) 性质是材料对特定刺激的响应类型及程度的本征属性。

Reading Material
Brief introduction to materials science

这篇课文介绍了材料科学领域的一个重大突破。19世纪后期，美国科学家Josiah Willard Gibbs阐述了材料物理特性与热力学性质间的关系，从而推动了现代新材料的发展。

The material of choice of a given era is often a defining point. Phrases such as Stone Age, Bronze Age, Iron Age, and Steel Age are historic, if arbitrary examples. Originally deriving from the manufacture of ceramics and its putative derivative metallurgy, materials science is one of the oldest forms of engineering and applied science. Modern materials science evolved directly from metallurgy, which itself evolved from mining and (likely) ceramics and earlier from the use of fire. A major breakthrough in the understanding of materials occurred in the late 19th century, when the American scientist Josiah Willard Gibbs demonstrated that the thermodynamic properties related to atomic structure in various phases are related to the physical properties of a material. Important elements of modern materials science were products of the Space Race; the understanding and engineering of the metallic alloys, and silica and carbon materials, used in building space vehicles enabling the exploration of space. Materials science has driven, and been driven by, the development of revolutionary technologies such as rubbers, plastics, semiconductors, and biomaterials.

Before the 1960s (and in some cases decades after), many eventual materials science departments were metallurgy or ceramics engineering departments, reflecting the 19th and early 20th century emphasis on metals and ceramics. The growth of materials science in the United States was catalyzed in part by the Advanced Research Projects Agency, which funded a series of university-hosted laboratories in the early 1960s, "to expand the national program of basic research and training in the materials sciences". The field has since broadened to include every class of materials, including ceramics, polymers, semiconductors, magnetic materials, biomaterials, and nanomaterials, generally classified into three distinct groups: ceramics, metals, and polymers. The prominent change in materials science during the recent decades is active usage of computer simulations to find new materials, predict properties and understand phenomena.

(**Selected from:** Wikipedia. Materials science (2023-06-26) [2023-07-29]. https://en. wikipedia. org/wiki/Materials_science)

New Words and Expressions

era *n.* 时代，纪元
arbitrary *adj.* 任意的
putative *adj.* 假定的
derivative *n.* 衍生物；*adj.* 衍生的
mining *n.* 采矿业
breakthrough *n.* 突破，重大进展
thermodynamic *adj.* 热力学的
silica *n.* 二氧化硅，硅土
revolutionary *adj.* 革命性的
semiconductor *n.* 半导体
prominent *adj.* 重要的，突出的

Notes

(1) Originally deriving from the manufacture of ceramics and its putative derivative metallurgy, materials science is one of the oldest forms of engineering and applied science.

—one of ＋复数名词 表示"……之一"。
—参考译文：材料科学最初源于陶瓷的制造及其衍生的冶金学，是最古老的工程和应用科学之一。

(2) A major breakthrough in the understanding of materials occurred in the late 19th century, when the American scientist Josiah Willard Gibbs demonstrated that the thermodynamic properties related to atomic structure in various phases are related to the

physical properties of a material.

— 此句为 when 引导的时间状语从句，从句在前主句在后。
— in the understanding of... 对……的认识。
— 参考译文：对材料认识的重大突破发生在 19 世纪后期，当时美国科学家 Josiah Willard Gibbs 证明了与各物相原子结构相关的热力学性质和材料的物理性质具有相关性。

(3) The growth of materials science in the United States was catalyzed in part by the Advanced Research Projects Agency, which funded a series of university-hosted laboratories in the early 1960s, "to expand the national program of basic research and training in the materials sciences".

— 此句为 which 引导的定语从句，which 后的从句用于解释和补充 the Advanced Research Projects Agency。
— 参考译文：美国高级研究计划局（Advanced Research Projects Agency）一定程度上推动了美国材料科学的发展，该局在 20 世纪 60 年代初期资助了一系列依托大学建设的实验室，"以扩展材料科学基础研究和（人才）培养的国家计划"。

Unit 1.2

Text

Classification of materials

这篇课文介绍了工程材料的主要分类：金属、陶瓷和聚合物，并列出了这几类材料的特性及分类标准。此外，课文还对复合材料的性质及应用领域作了详细阐述，并对用于高科技领域先进材料的性能进行了介绍。

For convenience, most engineering materials are classified into three main or basic categories: metallic materials, polymeric materials, and ceramic materials. This scheme is based primarily on chemical composition and atomic structure, and most materials fall into this or that separated group. In addition, there are composites, which are engineered combinations of two or more different materials. A brief explanation of the classification and representative characteristics of these materials follows. Another category is advanced materials—materials used for high-tech applications such as biomaterials and nano-engineered materials, which will be discussed in detail in other chapters.

Metallic materials

These materials consist of one or more metallic elements (such as iron, aluminum, copper, titanium, gold, and nickel) and usually relatively small amounts of non-metallic elements (such as carbon, nitrogen, and oxygen). Metals have a crystal structure in which the atoms are arranged in an orderly manner. In general, metals are good conductors of heat and electricity. Many metals are relatively strong and ductile at room temperature, and many retain good strength even at high temperatures.

Metals and alloys are usually divided into two categories: ferrous metals and alloys that contain large amounts of iron, such as steel and cast iron; and nonferrous metals and alloys that contain no or only relatively small amounts of iron. Examples of non-ferrous metals are aluminum, copper, zinc, titanium, and nickel. The reason for the distinction between ferrous and non-ferrous alloys is that steel and cast iron are used and produced in significantly higher quantities than other alloys. Alloys and pure metals are used in many industries, including aerospace, biomedical, semiconductor, electronics, energy, civil structures, and transportation. **Figure 1.2.1** shows a picture of rocket Chang'e 5 made primarily of metal alloys and superalloys.

Figure 1.2.1   The rocket in the picture (Chang'e 5) is mainly made of metal alloys. It uses the latest high-temperature, heat-resistant, and high-strength metal alloys.

Ceramic materials

Ceramics are compounds between metallic and non-metallic elements; they are most commonly oxides, nitrides, and carbides. Ceramic materials can be crystalline, amorphous, or a mixture of both. For example, common ceramic materials include alumina (or aluminum oxide, Al_2O_3), silica (or silicon dioxide, SiO_2), silicon carbide (SiC), silicon nitride (Si_3N_4), in addition to some called traditional ceramics—made up of clay minerals (i.e. porcelain) as well as cement and glass. Most ceramic materials are characterized by high hardness and high temperature strength, but exhibit extreme brittleness (lack of ductility) and fracture very easily. Newer ceramics are designed to have better fracture resistance; these materials are used in cookware, tableware, and even automotive engine parts. The advantages of ceramic materials for engineering applications include lightweight, high strength, hardness, good heat and wear resistance, friction reduction, and insulation. Insulating properties, along with the high heat and wear resistance of many traditional ceramics, make them useful in heat treatment and furnace linings for metals such as molten steel. Advanced ceramic components are finding their way into applications normally dedicated to metals, such as power resistors, fuse protectors, and battery sealing connectors for electric vehicles (**Figure 1.2.2**).

The applications for ceramic materials are truly limitless, as they can be used in aerospace, metal fabrication, biomedical, automotive, and many other industries. The two main disadvantages of these materials are (1) they are difficult to machine into finished products and therefore expensive, and (2) they are brittle and have a low fracture toughness compared to metals. If the technology for developing high toughness ceramics is further developed, there could be a tremendous upsurge in engineering applications for these materials.

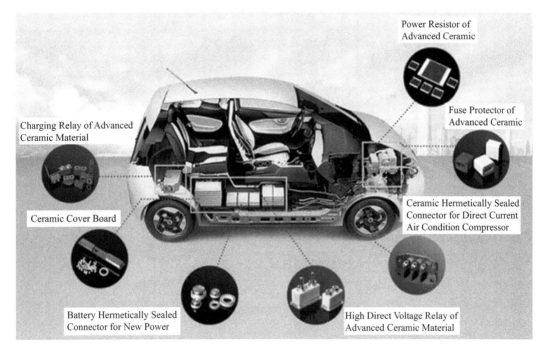

Figure 1.2.2 Examples of a newly developed generation of engineering ceramic materials for advanced electric vehicles.

Polymeric materials

Polymers include the familiar plastic and rubber materials. Many of these are organic compounds that are chemically based on carbon, hydrogen, and other non-metallic elements (i. e., O, N and Si). Most polymeric materials consist of long molecular chains or networks, usually with a backbone of carbon atoms. In general, polymeric materials have low densities and relatively low softening or decomposition temperatures. The strength and ductility of polymeric materials vary widely, from low strength and high deformability (rubber bands) to high strength, low deformability, and high durability (vulcanized rubber used in tires). Polymers have thousands of applications, from bulletproof vests, compact discs (CDs), cords, and liquid crystal displays (LCDs) to clothing and coffee mugs. Due to the nature of their internal structure, most polymeric materials are poor electrical conductors. Some of these materials are good insulators and are used in electrical insulation applications. Some frequently noted applications for polymer materials are packaging materials, automotive tires, and new flexible display fabrics (**Figure 1.2.3**). These applications show the diverse utility and importance of polymers in our everyday lives.

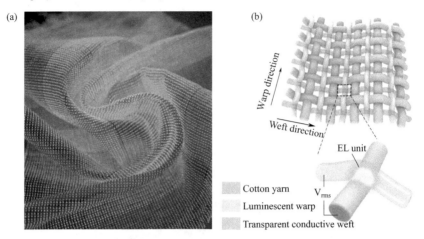

Figure 1.2.3 (a) A new type of flexible display fabric made of polymer composite fibers interwoven by warp and weft. (b) The luminescent warp and transparent conductive weft form an electroluminescent unit.

Composite materials

The main consideration in developing composites is to mix the properties of different materials. Composites consist of two (or more) separate materials from the previously discussed categories of metals, ceramics, and polymers. These components maintain their properties, while the overall composite will have properties different from their respective ones. Most composites consist of selected fillers or reinforcing materials and compatible resin binders to produce the desired properties and performance. Typically, these components do not dissolve into each other and they can be physically identified from each other through the interface. A large number of composite types are represented by different combinations of metals, ceramics, and polymers. In addition, some naturally occurring materials are also composites — for example, wood and bone. However, most of what we consider in our discussion are synthetic (or human-made) composites.

Composites have replaced many metal components, especially in the aerospace (**Figure 1.2.4**), avionics, automotive, civil structures, and sports equipment industries. One of the most common and familiar composite materials is glass fiber, in which small glass fibers are embedded in a polymeric material (usually epoxy resin or polyester). Glass fibers are relatively strong and stiff (but also brittle), while polymers are more flexible. Some advanced composites have similar stiffness and strength to some structural metal alloys but are significantly less dense and therefore have a lower total component weight. These properties make advanced composites extremely attractive in situations where the weight of the component is critical. In general, like ceramic materials, the main drawback of most composites is their brittleness and low fracture toughness. In some cases, some of these drawbacks can be improved by proper selection of the matrix material.

 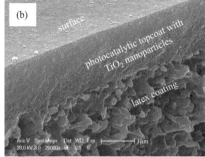

Figure 1.2.4 (a) The dust-removal composite coating materials on the solar cell surface of the "Zhurong" Mars rover. (b) The TiO_2-based photocatalytic self-cleaning coating on architectural latex coatings.

Biomedical materials

Biomaterials are employed in parts of the body that are implanted to replace diseased or damaged body parts. Our bones and dental parts are made of a naturally occurring ceramic called hydroxyapatite. Some artificial organs, bone replacement parts, cardiovascular stents, orthodontic stents, and other parts are made from different plastics, titanium alloys, and non-magnetic stainless steel. These materials must not produce toxic substances and must be compatible with body tissues (i.e., cannot cause adverse biological reactions). All the previously mentioned materials—metals, ceramics, polymers, composites, and semiconductors—can be processed as biomaterials.

Nanomaterials

One new class of materials with fascinating properties and great technological promise is nanomaterials. Nanomaterials can be any of the four basic types—metals, ceramics, polymers, and composites. However, unlike these other materials, nanomaterials are not distinguished by their chemical properties, but by their size. Nanomaterials are typically defined as those with characteristic length scales (i.e., particle diameter, grain size, layer thickness, etc.) of less than 100 nm (1 nm = 10^{-9} m). Nanomaterials can be metals, polymers, ceramics, electronics, or composites and in various shapes (**Figure 1.2.5**).

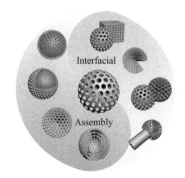

Figure 1.2.5 Fabrication of nanomaterials of various shapes by interfacial and self-assembly methods.

(**Selected from:** Smith W F, Hashemi J, Presuel-Moreno F. Foundations of Materials Science and Engineering [M]. McGraw-hill, New York, 2006.

China National Space Administration, Chang'e-5 Launch Highlights [DB/OL], (2020-11-24) [2020-11-24]. https://www.cnsa.gov.cn/n6758823/n6758842/c6810575/content.html.

Zhao T, Zhang X, Lin R, et al. Surface-Confined Winding Assembly of Mesoporous Nanorods [J]. Journal of the American Chemical Society, 2020, 142(48): 20359-20367.)

New Words and Expressions

scheme *n*. 方案；*v*. 策划

primarily *adv*. 主要地，首要地

composite *adj*. 合成的；*n*. 复合材料；*v*. 合成

category *n*. 类别，种类

high-tech *adj*. 高科技的

ductile *adj*. 柔软，易延展的

retain *v*. 保持

ferrous *adj*. 亚铁的，含铁的

zinc *n*. 锌

titanium *n*. 钛

nickel *n*. 镍

cast iron 铸铁，锻铁

aerospace *n*. 航天；*adj*. 航空和航天的

nitride *n*. 氮化物；*v*. 使氮化，渗氮于

carbide *n*. 碳化物

brittleness *n*. 脆性

cookware *n*. 炊具

tableware *n*. 餐具

automotive *adj*. 自动的

lining *n*. 内层

power resistor 功率电阻器

battery sealing connector 电池密封接头

machine *n*. 机器

upsurge *n*. 高涨；*v*. 涌起

decomposition *n*. 分解

deformability *n*. 可变形性

rubber band 橡皮筋

tire *n*. 轮胎

cord *n*. 电线

coffee mug 咖啡杯

warp and weft 经纬

respective *adj*. 分别的

filler *n*. 填料

compatible *adj*. 兼容的

resin *n*. 树脂；*v*. 涂树脂

binder *n*. 黏结剂

dissolve *v*. 溶解

avionics *n*. 航空电子学
embed *v*. 嵌入
epoxy *n*. 环氧树脂
polyester *n*. 聚酯
matrix *n*. 矩阵，基质
hydroxyapatite *n*. 羟基磷灰石
cardiovascular *adj*. 心血管的
stent *n*. 支架
orthodontic *adj*. 齿列矫正的
adverse *adj*. 不利的，有害的
fabrication *n*. 组装

Notes

（1）A brief explanation of the classification and representative characteristics of these materials follows.

——follow 在后面进行介绍。

——参考译文：下面（将）简要说明这些材料的分类和本征特性。

（2）Many metals are relatively strong and ductile at room temperature, and many retain good strength even at high temperatures.

——此句使用了几个关于性能表述的形容词。

——参考译文：大部分金属在室温下相对坚硬且具有延展性，有些甚至在高温下也能保持良好的强度。

（3）For example, common ceramic materials include alumina (or aluminum oxide, Al_2O_3), silica (or silicon dioxide, SiO_2), silicon carbide (SiC), silicon nitride (Si_3N_4), in addition to some called traditional ceramics—made up of clay minerals (i.e. porcelain) as well as cement and glass.

——参考译文：例如，常见的陶瓷材料包括氧化铝（Al_2O_3）、二氧化硅（SiO_2）、碳化硅（SiC）、氮化硅（Si_3N_4），此外还有一些被称为传统陶瓷的材料——由黏土矿物（即瓷器）以及水泥和玻璃制成。

（4）Advanced ceramic components are finding their way into applications normally dedicated to metals, such as power resistors, fuse protectors, and battery sealing connectors for electric vehicles.

——find way to 在非生命名词作主语时，通常为一种被动的形象表达。

——意译为：先进的陶瓷材料正进入通常专属于金属材料的应用领域，例如新能源汽车上的功率电阻器、保险丝保护器和电池密封连接器。

（5）In addition, some naturally occurring materials are also composites—for example, wood and bone.

——参考译文：此外，一些天然存在的材料也是复合材料——例如木材和骨骼。

(6) Biomaterials are employed in parts of the body that are implanted to replace diseased or damaged body parts.

—employed 在此句中相当于 used。
—参考译文：生物材料被植入体内以替换患病或受损的身体部件。

(7) Nanomaterials can be any of the four basic types—metals, ceramics, polymers, and composites.

—be of 后接名词等同于形容词的表达。
—参考译文：纳米材料可以是四种基本类型中的任何一种——金属、陶瓷、聚合物和复合材料。

Exercises

1. Question for discussion

(1) Please list the applications of biomaterials.
(2) What kinds of materials can be used as biomaterials?
(3) Please identify the size of nanomaterials.
(4) What are the merits and demerits of composite materials?
(5) Please list the applications of polymer materials.
(6) What are the advantages of metallic materials?

2. Translate the following into Chinese

(1) For convenience, most engineering materials are classified into three main or basic categories: metallic materials, polymeric materials, and ceramic materials.

(2) These materials consist of one or more metallic elements (such as iron, aluminum, copper, titanium, gold, and nickel) and usually relatively small amounts of non-metallic elements (such as carbon, nitrogen, and oxygen).

(3) Metals and alloys are usually divided into two categories: ferrous metals and alloys that contain large amounts of iron, such as steel and cast iron; and nonferrous metals and alloys that contain no or only relatively small amounts of iron.

(4) If the technology for developing high toughness ceramics is further developed, there could be a tremendous upsurge in engineering applications for these materials.

(5) Typically, these components do not dissolve into each other and they can be physically identified from each other through the interface.

3. Translate the following into English

(1) 纳米工程 (2) 元素
(3) 有序推进 (4) 一般来说
(5) 耐热 (6) 易延展的
(7) 断裂 (8) 熔断器
(9) 硫化 (10) 织物
(11) 金属具有原子有序排列的晶体结构。

（12）一般来说，金属是热和电的良导体。

（13）钢和铸铁的使用量和生产量明显高于其他合金。

（14）陶瓷材料的应用确实是无限的，因为它们可以用于航空航天、金属制造、生物医学、汽车和许多其他行业。

（15）由于其内部结构的性质，大多数聚合物材料是电的不良导体。

Reading Material

Biomaterials

这篇课文介绍了生物材料的发展历程及其在环保领域发挥的重要作用。课文还对生物材料的加工与使用进行了简要的阐述。

Biotic material or biological derived material is any material that originates from living organisms. Most such materials contain carbon and are capable of decay. The earliest life on Earth arose at least 3.5 billion years ago. Earlier physical evidences of life include graphite, a biogenic substance, in 3.7 billion-year-old metasedimentary rocks discovered in southwestern Greenland, as well as, "remains of biotic life" found in 4.1 billion-year-old rocks in Western Australia. Earth's biodiversity has expanded continually except when interrupted by mass extinctions. Although scholars estimate that over 99 percent of all species of life (over five billion) that ever lived on Earth are extinct, there are still an estimated 10-14 million extant species, of which about 1.2 million have been documented and over 86% have not yet been described. Examples of biotic materials are wood, straw, humus, manure, bark, crude oil, cotton, spider silk, chitin, fibrin, and bone.

The use of biotic materials, and processed biotic materials (bio-based material) as alternative natural materials, over synthetics is popular with those who are environmentally conscious because such materials are usually biodegradable, renewable, and the processing is commonly understood and has minimal environmental impact. However, not all biotic materials are used in an environmentally friendly way, such as those that require high levels of processing, are harvested unsustainably, or are used to produce carbon emissions.

When the source of the recently living material has little importance to the product produced, such as in the production of biofuels, biotic material is simply called biomass. Many fuel sources may have biological sources, and may be divided roughly into fossil fuels, and biofuel.

In soil science, biotic material is often referred to as *organic matter*. Biotic materials in soil include glomalin, Dopplerite and humic acid. Some biotic material may not be considered to be organic matter if it is low in organic compounds, such as a clam's shell, which is an essential component of the living organism, but contains little organic carbon.

(**Selected from:** Wikipedia, Biotic Material, Wikimedia Foundation [DB/OL], (2023-03-02) [2023-03-02]. https://en.wikipedia.org/wiki/Biotic_material.)

New Words and Expressions

biotic material　生物材料

graphite　*n.*　石墨

metasedimentary rock　变质碎屑岩
biodiversity　*n.* 生物多样性
straw　*n.* 麦秆，吸管
humus　*n.* 腐殖质，腐殖土
manure　*n.* 肥料，粪肥
crude oil　原油
spider silk　蜘蛛丝
chitin　*n.* 壳质，几丁质，角素，甲壳素
unsustainably　*adv.* 不可持续地，无法确定地

Notes

（1）Earlier physical evidences of life include graphite, abiogenic substance, in 3.7 billion-year-old metasedimentary rocks discovered in southwestern Greenland, as well as, "remains of biotic life" found in 4.1 billion-year-old rocks in Western Australia.

——参考译文：早期的生命物证包括在格陵兰西南部发现的 37 亿年前变质沉积岩中的石墨非生物物质，以及在西澳大利亚发现的 41 亿年前岩石中的"生物生命遗迹"。

（2）In soil science, biotic material is often referred to as organic matter.

——be referred to ...　被称作为……。
——参考译文：在土壤科学中，生物材料通常被称为有机物质。

Exercises

1. Question for discussion

（1）Please list some examples of biotic materials.

（2）Why the biotic materials, and processed biotic materials are so popular?

（3）What are the biological substances in the soil?

2. Translate the following into Chinese

（1）Biotic material or biological derived material is any material that originates from living organisms.

（2）Examples of biotic materials are wood, straw, humus, manure, bark, crude oil, cotton, spider silk, chitin, fibrin, and bone.

（3）The use of biotic materials, and processed biotic materials (bio-based material) as alternative natural materials, over synthetics is popular with those who are environmentally conscious because such materials are usually biodegradable, renewable, and the processing is commonly understood and has minimal environmental impact.

3. Translate the following into English

（1）证据　　　　　　　（2）纤维蛋白
（3）合成　　　　　　　（4）可生物降解
（5）最小的

(6) 地球的生物多样性一直在不断扩大，除非出现大规模的生物灭绝。

(7) 许多燃料都具有生物来源，并且可以大致分为化石燃料和生物燃料。

Unit 1.3

Text

The structure of the atom

这篇课文介绍了原子结构，包括构成物质的原子种类及其排列方式，并阐述了几个基本且重要的概念：原子结构、电子排布、元素周期表，以及各种类型的初级和次级键。材料的重要特性取决于原子的几何排列及原子或分子之间的相互作用。熟练掌握原子结构以及原子和分子的键合方式，对于正确选择工程材料和开发新型先进材料至关重要。

Some important properties of materials depend on the geometric arrangement of atoms and the interactions that exist between the constituent atoms or molecules. The atomic structure includes all the atoms and their arrangements, which make up the building blocks of matter. All structures at the nanoscopic, microscopic, and macroscopic levels are created from these building blocks. The insight gained by understanding the atomic structure and the bonding configurations of atoms and molecules is crucial for the proper selection of engineering materials and the development of new advanced materials. For the purpose of the subsequent discussion, this chapter covers several basic and important concepts: atomic structure, electronic configuration, and the periodic table, as well as the various types of primary and secondary interatomic bonds that bind together the atoms that make up a material.

Each atom consists of a nucleus surrounded by electrons. The nucleus contains neutrons and protons with a net positive charge. The masses of these subatomic particles are infinitely small; protons and neutrons have approximately the same mass, 1.67×10^{-27} kg, which is significantly larger than that of an electron, 9.11×10^{-31} kg. Each chemical element is characterized by the number of protons in the nucleus, or the atomic number (Z). Each element has its own characteristic atomic number that defines the element. For example, by definition, any atom with six protons is a carbon atom. In a neutral atom, the atomic number or proton number is also equal to the number of electrons in its charge cloud.

The atomic mass (A) of a specific atom can be expressed as the sum of the masses of protons and neutrons in the nucleus. Although the number of protons is the same for all atoms of a particular element, the number of neutrons (N) may be variable. Therefore, atoms of some elements have two or more different atomic masses, which are called isotopes. Atomic mass units (amu) can be used to calculate atomic weights. An amu is defined as exactly 1/12 of the mass of a carbon atom with 6 protons and 6 neutrons. This also indicates that the mass of a neutron or a proton is close to 1 amu. Thus, a carbon-12 atom by itself has an atomic mass of 12 amu. The atomic mass is also the mass in grams of the Avogadro number NA of atoms. The quantity $NA = 6.02 \times 10^{23}$ atoms/mol is the number of atoms or molecules in a mole.

Therefore, the atomic mass has the unit of g/mol. As an example, one mole of iron contains 6.02×10^{23} atoms and has a mass of 55.847 g, or 55.847 amu.

Electrons in atoms

In the latter part of the nineteenth century, people realized that many phenomena involving electrons in solids could not be explained by classical mechanics. A set of principles and laws governing systems of atomic and subatomic solids was subsequently established and is known as quantum mechanics. An understanding of the behavior of electrons in atoms and crystalline solids necessarily involves a discussion of the concepts of quantum mechanics.

Atomic models

An early outgrowth of quantum mechanics was the simplified Bohr model of the atom, in which electrons are assumed to revolve around the nucleus in discrete orbits, and the position of any given electron is more or less well defined in its orbit. This atomic model is represented in **Figure 1.3.1**. Another important quantum mechanical principle states that the energy of an electron is quantized; that is, the electron is only allowed to have a specific value of energy. An electron can change energy, but in doing so, it must make a quantum jump either to an allowed higher energy (absorption energy) or to lower energy (emission energy). It is often convenient to think of these allowed electron energies as being related to energy levels or states. These states do not vary continuously with energy; that is, adjacent states are separated by a finite amount of energy. For example, the allowed states of a Bohr hydrogen atom are represented in **Figure 1.3.2(a)**. These energies are considered negative, and the zero reference is the unbound or free electron. Of course, a single electron associated with a hydrogen atom will fill only one of these states. Thus, the Bohr model represents an early attempt to figure electrons in atoms, in terms of both position (electron orbitals) and energy (quantized energy levels).

This Bohr model was eventually found to have some obvious limitations, as it could not explain some phenomena involving electrons. One solution was to adopt a wave-mechanical model, in which the electron is considered to exhibit both wave-like and particle-like characteristics. Thus, the Bohr model was further refined by wave mechanics, in which three new quantum numbers were introduced, giving rise to electron subshells within each shell. **Figures 1.3.2(a)** and (**b**) illustrate a comparison of these two models on this basis, using the hydrogen atom as an example. In this model, the electron is no longer considered as a particle

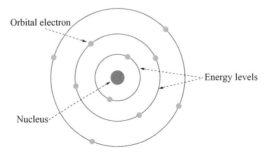

Figure 1.3.1　Schematic representation of the Bohr atom.

moving in discrete orbits. Instead, the position is considered as the probability of the electron in different positions around the nucleus. In other words, the positions are described by a probability distribution or electron cloud. **Figure 1.3.3** compares the Bohr model and the wave-mechanical model of the hydrogen atom. Both models are used throughout the book, and the choice depends on which model allows for a simpler explanation.

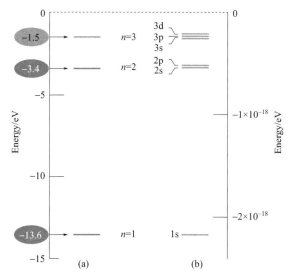

Figure 1.3.2 (a) The first three electron energy states for the Bohr hydrogen atom. (b) Electron energy states of the first three shells of the wave-mechanical hydrogen atom.

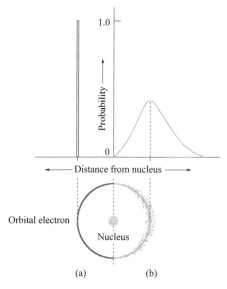

Figure 1.3.3 Comparison of the (a) Bohr and (b) wave-mechanical atom models in terms of electron distribution.

Quantum numbers

Using wave mechanics, each electron in an atom is characterized by four parameters called quantum numbers. The first quantum numbers are the principal quantum number n, the

Azimuthal quantum number l, the magnetic quantum number m_1, and the spin quantum number m_s.

The principal quantum number, n, is the most important in determining the energy level of the electron under consideration. It has only integer values of l or greater than l, i. e. $n=1,2,3,\ldots$. Each principal energy level, also called shell, represents a collection of subshells and orbitals with the same principal number n. As n increases, the energy of the electron under consideration also increases, indicating that the electron is less tightly bound to the nucleus (more easily ionized). Finally, as n increases, the probability of finding an electron further away from the nucleus also increases.

The second quantum number, l, represents the subshells, denoted by a lowercase letter, i. e., s, p, d, or f; it is related to the shape of the electron subshells. Moreover, the number of these subshells is restricted by the magnitude of n. When $n=1$, only one subshell is possible; however, when $n=2$, two different subshells are possible; when $n=3$ three different subshells are possible, and so on. The shape of the electron cloud or the boundary space of the orbit is determined by this number. The quantum number l can be expressed as an integer from 0 to $n-1$.

The magnetic quantum number, m_1, determines the number of energy states for each subshell. For the s subshell, there is only one energy state, while for the p, d, and f subshells, three, five, and seven states exist, respectively. In the absence of an external magnetic field, the states within each subshell are the same. However, when a magnetic field is applied, the states within these subshells split, with each state having a slightly different energy.

The spin quantum number, m_s, can be taken as $+1/2$ or $-1/2$. That is, the spin moment must be in the up or down orientation. Furthermore, according to Pauli's exclusion principle, no more than two electrons in an atom can simultaneously occupy the same orbital, and these two electrons must have opposite spins. In other words, no two electrons can have the same group of four quantum numbers.

The periodic table

In the periodic table, elements are classified according to their ground-state electron configuration (**Figure 1.3.4**). Thus, the atom of a particular element (e. g. Li with three electrons) contains one more electron than the element in front of it (He with two electrons). The arrangement allows all elements arranged in each column or group have similar valence electron structures, as well as chemical and physical properties. These properties change gradually, moving horizontally in each period and vertically in each column.

The elements located in group 0, the rightmost group, are noble gases, which have filled electron shells and stable electron configurations. Group VIIA and VIA elements lack one and two electrons, respectively, and do not have stable structures. Group VIIA elements (F, Cl, Br, I, and At) are sometimes referred to as halogens. The alkali and alkaline earth metals (Li, Na, K, Be, Mg, Ca, etc.) are labeled as group IA and IIA and have one and two electrons over stable structures, respectively. The elements in the three long cycles, groups IIIB to IIB, are known as transition metals, which have partially filled d-electron states and, in

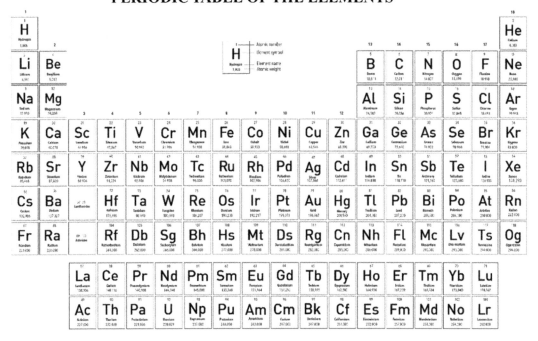

Figure 1.3.4　The periodic table of the elements. (The numbers in parentheses are the atomic weights of the most stable or common isotopes.)

some cases, one or two electrons in the next higher energy shell. Groups IIIA, IVA, and VA (B, Si, Ge, As, etc.) show characteristics intermediate between metals and nonmetals by the valence electron structures.

As seen in the periodic table, some elements belong to the metal classification. These elements are sometimes referred to as electropositive, indicating that they can give up a few valence electrons to become positively charged ions. In addition, elements located on the right side of the table are electronegative; that is, they readily accept electrons to form negatively charged ions, or sometimes they share electrons with other atoms. **Figure 1.3.5** shows the electronegativity values for the various elements arranged in the periodic table. In general, the electronegativity increases from left to right and from bottom to top. An atom is more likely to accept electrons if its outer shell is less "shielded" from the nucleus.

Primary bonds

The driving force behind the formation of bonds between atoms is that each atom seeks to be in the most stable state. By bonding with other atoms, the potential energy of each bonded atom is reduced, resulting in a more stable state.

The ionic bond

Ionic bonding is perhaps the easiest to describe and visualize. It is always found in compounds composed of metallic and nonmetallic elements, which are located at the horizontal

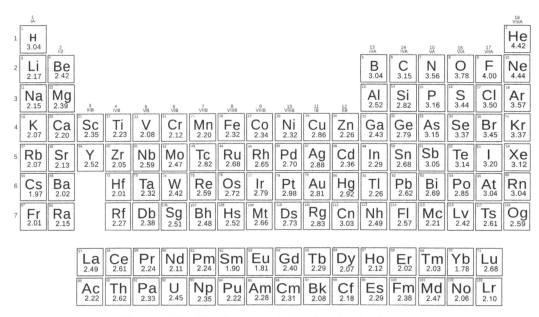

Figure 1.3.5 The electronegativity values for the elements.

ends of the periodic table. Metals and nonmetals are bonded by electron transfer and ionic bonding. Ionic bonding is usually observed between atoms with large differences in electronegativity. The atoms of metallic elements readily give up their valence electrons to non-metallic atoms. In this process, all atoms acquire a stable or inert gas configuration and, in addition, an electrical charge; that is, they become ions. As an example, let us consider the ionic bond between a metal Li with electronegativity 1.0 and a non-metal F with electronegativity 4.0. Briefly, the lithium atom loses an electron to form a cation, Li^+. In this process, the radius is reduced from $r=0.157$ nm for the Li^+ atom to $r=0.060$ nm for the Li cation. This decrease in size is due to the fact that (1) after ionization, the leading electron is no longer in the $n=2$ state, but in the $n=1$ state, and (2) the equilibrium between the positive and negative electron clouds is lost and the nucleus can exert stronger forces on the electrons, thus pulling them closer. On the contrary, the F atom gains the electrons lost by Li and forms the anion F^-. In this case, the radius increases from $r=0.071$ nm for the F atom to $r=0.136$ nm for F^-. It can be generalized that when a metal forms a cation, its radius decreases, while a non-metal forms an anion, its radius increases.

The covalent bond

In a covalent bond, the stable electron configuration is assumed by the sharing of electrons between neighboring atoms. Two covalently bonded atoms will each contribute at least one electron, and the shared electron can be considered to belong to both atoms. Covalent bonding is schematically illustrated in **Figure 1.3.6** for a molecule of methane (CH_4). The carbon atom has four valence electrons, whereas each of the four hydrogen atoms has a single valence electron. Each hydrogen atom can acquire a helium electron configuration (two 1-valence electrons) when the carbon atom shares an electron with it. Carbon now has four additional shared electrons, one for each hydrogen, for a total of eight valence electrons, and the electronic structure of neon. Covalent bonding is directional, which means it is between specific atoms and may exist only in the direction between one atom and another atom involved in electron sharing.

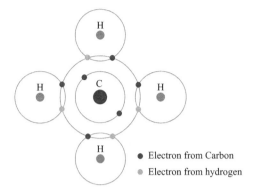

Figure 1.3.6 Diagram of covalent bonding in a methane molecule (CH_4).

The number of covalent bonds that is possible for a particular atom is determined by the number of valence electrons. For N valence electrons, an atom can covalently bond with at most $8-N$ other atoms. For example, $N=7$ for chlorine, and $8-N=1$, which means that one Cl atom can bond to only one other atom, as in Cl_2. Similarly, for carbon, $N=4$, and each carbon atom has $8-4$, or four, electrons to share. Diamond is simply the three-dimensional interconnecting structure wherein each carbon atom covalently bonds with four other carbon atoms.

It is possible to have partially ionic and partially covalent interatomic bonds; in fact, very few compounds exhibit pure ionic or covalent bonds. For a compound, the extent of any kind of bonding depends on the relative position of the constituent atoms in the periodic table (**Figure 1.3.4**) or the difference in their electronegativity (**Figure 1.3.5**). The wider the separation (both horizontally—relative to Group IVA—and vertically) from the lower-left to the upper-right corner (i. e., the greater the difference in electronegativity), the more ionic the bond is. Conversely, the closer the distance between the atoms (i. e., the smaller the electronegativity difference), the higher the degree of covalency. The ionicity percentage (%IC) of the bond between elements A and B can be approximated as

$$\text{\% ionic character} = \{1-\exp[-0.25(X_A-X_B)^2]\} \times 100 \quad (1.3.1)$$

where X_A and X_B are the electronegativities for the respective elements.

The metallic bond

It has been observed that during solidification, starting from a molten state, atoms of metals are packed tightly together in an organized and repetitive manner to reduce their energy and reach a more stable state in the form of a solid, thus forming metallic bonds. Metallic bonds are found in metals and their alloys. A relatively simple model has been proposed that is very close to the bonding scheme. Usually the metallic material has one, two, or at most three valence electrons. In this model, these valence electrons are not bound to any particular atom in the solid, but are more or less free to drift throughout the metal. They can be regarded to belong to the whole metal or to form an "ocean of electrons" or an "electron cloud". The remaining non-valent electrons and nuclei form the so-called ionic nuclei, which possess a net positive charge equal to the total valence electron charge of each atom.

Figure 1.3.7 shows a schematic diagram of the metallic bond. The free electrons shield the positively charged ion nuclei from mutually repulsive electrostatic forces that they would otherwise exert on each other; thus, the metallic bond is non-directional. In addition, these free electrons act as a kind of "glue" that holds the ion nuclei together. Metallic bonds are three-dimensional and non-directional, similar to ionic bonds. However, since no anion is involved, there is no restriction of electroneutrality. In addition, metal cations are not rigidly held in place as they are in ionic solids. In contrast to oriented covalent bonds, there are no local electron pairs shared between atoms. Therefore, metallic bonds are weaker than covalent bonds.

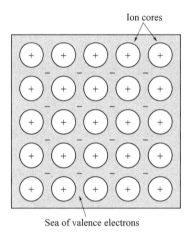

Figure 1.3.7 Schematic diagram of the metallic bonding.

The secondary bonds or Van der Waals bonds

So far, we have considered only primary bonding between atoms and showed that it depends on the interaction of their valence electrons. The driving force for atomic bonding is the reduction of the energy of the bonding electrons. Compared to primary bonding, secondary bonding is relatively weak. Almost all atoms or molecules have secondary bonds between them, but their existence is masked if any of the three primary bonds are present. Noble gases have a stable electronic structure and, in addition, the fact that molecules in the molecular structure are covalently bonded to each other proves the existence of secondary bonds.

The driving force for secondary bonding is the attraction of the electric dipole contained in the atoms or molecules. Essentially, an electric dipole exists whenever there is some separation of the positive and negative parts of an atom or molecule. As shown in **Figure 1.3.8**, the Coulombic attraction between the positive end of one dipole and the negative region of the neighboring dipole leads to bonding. Dipole interactions occur between induced dipoles, between induced dipoles and polar molecules (with permanent dipoles), and between polar molecules. Although the binding energies of secondary bonds are weak, they become important when they are the only bonds available to bind atoms or molecules together. In general, there are two main types of secondary bonds between atoms or molecules that involve electric dipoles: fluctuating dipoles and permanent dipoles.

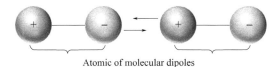

Figure 1.3.8　Schematic illustration of van der Waals bonding between two dipoles.

Fluctuating induced dipole bonds

Very weak secondary bonding can occur between the atoms of noble gas elements with complete outer valence electron shells (s^2 for helium and s^2p^6 for Ne, Ar, Kr, Xe, and Rn). These binding forces arise because of the asymmetric distribution of the electronic charges in these atoms producing an electric dipole. At any instant, there is a high probability that one side of the atom will have more electronic charge than the other (**Figure 1.3.9**). Thus, in a given atom, the electron charge cloud will change over time, creating a "fluctuating dipole". The fluctuating dipoles of nearby atoms can attract each other and form weak interatomic non-directional bonds. The liquefaction, and in some cases solidification, of noble gases and other electrically neutral and symmetric molecules such as H_2 and Cl_2, is achieved because of this type of bonding. In materials where induced dipole bonds dominate, the melting and boiling temperatures are extremely low. These are the weakest of all possible intermolecular bonds.

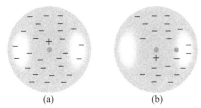

Figure 1.3.9　Electron charge distribution in atoms of noble gases:
(a) Idealized symmetric charge distribution in which negative and positive charge centers are superimposed at the center, (b) The actual asymmetric distribution of electrons leads to a temporary dipole.

Permanent dipole bonds

If the molecule contains a permanent dipole, then weak bonding forces arise between covalently bonded molecules. For example, the methane molecule CH_4, whose four C—H bonds are arranged in a tetrahedral structure, has a zero dipole moment due to the symmetrical arrangement. That is, the vectors of its four dipole moments add up to zero. In contrast, the chloromethane molecule CH_3Cl has an asymmetric tetrahedral arrangement consisting of three C—H bonds and one C—Cl bond, resulting in a net dipole moment of 2.0 dbar. Replacing one hydrogen atom in methane with a chlorine atom raises the boiling point of methane from $-128℃$ to $-14℃$ for chloromethane. The much higher boiling point of chloromethane is due to the permanent dipole bonding between chloromethane molecules.

Hydrogen bond

Hydrogen bonding is a special case of permanent dipole-dipole interactions between polar molecules. Hydrogen bonding occurs when a polar bond containing a hydrogen atom, O—H or N—H, interacts with electronegative atoms O, N, F, or Cl. For example, the water molecule H_2O has a permanent dipole moment of 1.84 due to its asymmetric structure with two hydrogen atoms at an angle of 105° to the oxygen atom [**Figure 1.3.10(a)**]. The hydrogen atom region of the water molecule has a positive charge center, while the oxygen atom has a negative charge center in the other end region [**Figure 1.3.10(a)**]. In hydrogen bonding between water molecules, the negatively charged region of one molecule is attracted by Coulomb forces to the positively charged part of the other molecule [**Figure 1.3.10 (b)**].

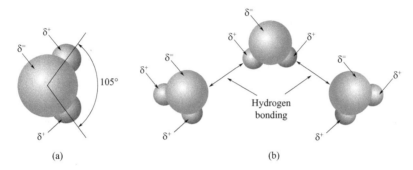

Figure 1.3.10 (a) The permanent dipole of the water molecule, (b) The formation of hydrogen bonds between water molecules due to permanent dipole attraction.

Many common molecules are composed of groups of atoms bound together by strong covalent bonds; these include elemental diatomic molecules (F_2, O_2, H_2, etc.) as well as a large number of compounds (H_2O, CO_2, HNO_3, C_6H_6, CH_4, etc.). In the condensed liquid and solid states, the bonds between molecules are weak secondary bonds. Therefore, the melting and boiling temperatures of molecular materials are relatively low. Those small molecules consisting of a few atoms are mostly gases at ordinary or ambient temperatures and pressures. On the other hand, many modern polymers are molecular materials consisting of very large molecules in solid form. Some of their properties depend to a large extent on the presence of van der Waals and hydrogen bonds.

Summary

• The properties of a material depend on its properties, which in turn are a function of its structure. In addition, the structure is determined by the way the material is processed.

• Most engineering materials fall into three broad categories: metals, polymers, and ceramic materials. Metals and alloys have good strength and ductility. Ceramics serve as excellent electrical and thermal insulators and are usually resistant to damage from high temperatures and corrosive environments, but are mechanically brittle. Polymers have relatively low strength; however, the ratio of strength to weight is very favorable. In addition, there are composite materials, represented by different combinations of metals, ceramics and polymers. Nanomaterials are distinguished by their size, i. e., less than 100 nm. They can be metals, polymers, ceramics or composites.

• Important properties of materials depend on the geometric arrangement of atoms and the interactions that exist between the constituent atoms or molecules. The insight gained by understanding the atomic structure and the bonding conformations of atoms and molecules is essential for the proper selection of engineering materials and the development of new advanced materials.

(**Selected from:** Callister Jr W D, Rethwisch D G. Fundamentals of Materials Science and Engineering: An Integrated Approach [M]. John Wiley & Sons, 2020.

Hofmann S. On Beyond Uranium: Journey to the End of the Periodic Table [M]. CRC Press, 2018.

Callister W D, Rethwisch D G. Materials Science and Engineering: An Introduction [M]. John Wiley &Sons, 2007.)

New Words and Expressions

microscopic $adj.$ 极小的，微小的
isotope $n.$ 同位素
Avogadro number   阿伏伽德罗常数
quantum mechanics   量子力学
crystalline $adj.$ 结晶的，晶状的
Bohr model   玻尔模型
wave mechanics   波动力学
probability distribution   概率分布
subshells $n.$ 亚层，子外壳，子壳，电子亚层
magnetic $adj.$ 磁性的，磁的
electronegativity $n.$ 电负性
exclusion principle   不相容原理
halogen $n.$ 卤素
alkaline $adj.$ 碱性的，含碱的
shielded $adj.$ 隔离的，屏蔽的
potential energy   势能
diamond $n.$ 金刚石
schematic diagram   原理图，示意图
dipole $n.$ 偶极子
methane $n.$ 甲烷
asymmetric $adj.$ 不对称的，非对称的

Notes

(1) The insight gained by understanding the atomic structure and the bonding configurations of atoms and molecules is crucial for the proper selection of engineering materials and the development of new advanced materials.

— gained by … 为过去分词 gained 引导的定语从句修饰 insight。
— bonding configurations 理解为成键方式，如离子键、共价键等。
— 参考译文：深入了解原子结构以及原子分子的成键方式，对于正确选择工程材料和开发新的先进材料至关重要。

（2）An early outgrowth of quantum mechanics was the simplified Bohr model of the atom, in which electrons are assumed to revolve around the nucleus in discrete orbits, and the position of any given electron is more or less well defined in its orbit.

— discrete orbits 指离散轨道，此时轨道的能级是确定的。
— in which 引导的非限制性定语从句，代表前面的玻尔模型。
— 参考译文：量子力学的一个早期产物是简化的玻尔原子模型，该模型假设电子在离散的轨道上围绕原子核旋转，并且任何给定的电子在其轨道上的位置一定程度上都是确定的。

（3）It has only integer values of 1 or greater than 1, i.e. $n=1,2,3,\ldots$ Each principal energy level is also called a shell and represents a collection of subshells and orbitals with the same principal number n.

— it 表示前一句话提到的 principal quantum number 即主量子数。
— 参考译文：它只能取值 1 或大于 1 的整数值，即 $n=1,2,3,\cdots\cdots$ 每个主要能级也被称为一个电子层，代表了具有相同主量子数 n 的电子亚层和轨道的集合。

（4）The arrangement gives all elements arranged in each column or group have similar valence electron structures, as well as chemical and physical properties.

— valence electron structures 价电子结构。
— 参考译文：这种排列方式使每一列或每组中元素都具有相似的价电子结构，以及化学和物理性质。

（5）In addition, the elements located on the right side of the table are electronegative; that is, they readily accept electrons to form negatively charged ions, or sometimes they share electrons with other atoms.

— 参考译文：此外，位于元素周期表右侧的元素是电负性的；也就是说，它们很容易得到电子形成带负电的离子，或者有时它们与其他原子共用电子。

（6）Covalent bonding is directional, which means it is between specific atoms and may exist only in the direction between one atom and another atom involved in electron sharing.

— covalent bonding 共价键，含有此键的典型分子有：H_2O、N_2 等。
— 参考译文：共价键是有方向性的，这意味着它是在特定原子之间，并且可能只存在于参与电子共用的一个原子和另一个原子之间的方向上。

（7）Dipole interactions occur between induced dipoles, between induced dipoles and polar molecules (with permanent dipoles), and between polar molecules.

— induced dipoles 诱导偶极子。

—polar molecules 极性分子，如 H_2O。

—参考译文：偶极相互作用发生在诱导偶极之间、诱导偶极和极性分子（具有永久偶极子）之间以及极性分子之间。

(8) Important properties of materials depend on the geometric arrangement of atoms and the interactions that exist between the constituent atoms or molecules.

—参考译文：材料的重要特性取决于原子的几何排列以及组成原子或分子之间存在的相互作用。

Exercises

1. Question for discussion

(1) What are the masses of protons and neutrons in the atom?

(2) Please define the carbon atom.

(3) What is the significance of the discovery of Bohr model?

(4) Please define the hydrogen bond.

(5) What are the two main types of secondary bonds between atoms or molecules?

(6) Please list the differences between metallic bonds and covalent bonds.

(7) What is the factor which will affect the radius of a metal?

2. Translate the following into Chinese

(1) Some important properties of materials depend on the geometric arrangement of atoms and the interactions that exist between the constituent atoms or molecules.

(2) Each atom consists of a nucleus surrounded by electrons.

(3) Each element has its own characteristic atomic number that defines the element.

(4) A set of principles and laws governing systems of atomic and subatomic solids was subsequently established and is known as quantum mechanics.

(5) Thus, the Bohr model represents an early attempt to describe electrons in atoms, in terms of both position (electron orbitals) and energy (quantized energy levels).

3. Translate the following into English

(1) 纳米级   (2) 洞察力

(3) 键合构型   (4) 随后的

(5) 亚原子   (6) 光电子

(7) 吸收能量   (8) 释放能量

(9) 范围   (10) 水平的

(11) 阳离子

(12) 因此，分子材料的熔化和沸腾温度相对较低。

(13) 如果分子包含永久偶极子，则在共价键合的分子之间会产生弱键合力。

(14) 附近原子的振荡偶极子可以相互吸引，形成弱的原子间非定向键。

(15) 金属阳离子不像它们在离子固体中那样被固定在特定的位置。

Reading Material

"Why didn't we think to do this earlier?" chemists thrilled by speedy atomic structures

这篇课文主要介绍了近年来有机材料原子结构研究技术的发展与变革。X 射线晶体学技术可用于原子结构的表征,但只能研究大晶体,并且工作周期长精度低。因此,研究人员使用电子束代替 X 射线,研究无机及蛋白质等生物大分子的结构。课文还简述了这种方法的局限性并对其应用进行了展望。

Organic chemists, make sure you're sitting comfortably. The structure of small organic molecules, such as those used in drugs, can be deduced in minutes rather than weeks, thanks to a technique that uses beams of electrons to quickly reveal how atoms are arranged.

The technique, called 3D electron diffraction, has been used by some inorganic chemists and material scientists since the mid-2000s to deduce the structure of molecules. But organic chemists, for whom the implications could be transformative, did not adopt it widely. In mid-October, two papers appeared online describing a way to use the same technique for drugs, making it much faster and easier to work out the structure of these small organic molecules than current techniques allow. I think there are a lot of people smacking their heads, saying, "Why didn't we think to do this earlier?" says John Rubinstein, a structural biologist at the University of Toronto who uses related techniques to study large molecules such as proteins.

Existing methods for finding the structure of small molecules require scientists to grow crystals for analysis, a laborious process that can take weeks or months. "Something that was a real barrier to their research is now basically removed," adds Rubinstein.

Knowing how atoms are arranged in a molecule is necessary for understanding that substance's function. Chemists working to develop new drugs, for example, depend on this structure to understand how the drug acts in the body—and how it could be tweaked to bind more strongly to its therapeutic target or to reduce side effects.

Long history

A technique called X-ray crystallography has been used for decades to deduce this arrangement. But it can take weeks of work—and is not always successful.

First, scientists need to coax the molecules to crystallize. They then blast the crystal with an X-ray beam. The crystal's lattice structure causes the X-rays to diffract, and a special detector records the resulting pattern. Scientists then use software to analyse the pattern and work out the structure of the molecule.

The challenges arise because X-ray diffraction works only with large crystals, and these can take weeks or even months to form. And some molecules are so hard to crystallize in the first place that it might not be possible to analyse them this way at all.

One alternative is to replace X-rays with electron beams, which can produce diffraction patterns using much smaller crystals. In 2007 and 2008, respectively, crystallographers at the

Johannes Gutenberg University in Mainz, Germany, and Stockholm University developed the first methods for detecting the 3D structures of molecules automatically using electron diffraction. Previously, scientists had to laboriously merge multiple 2D diffraction patterns together to get this 3D structure (**Figure 1.3.11**).

Initially, the technique was used mainly with inorganic structures, which aren't affected by radiation as much as organic molecules are. Then, in 2013, Tamir Gonen, a structural biologist at the University of California, Los Angeles, developed a version of electron diffraction called MicroED, which could be used on large biomolecules such as proteins.

Now, in the two latest papers, Gonen's team and another group from Switzerland have demonstrated that electron diffraction can also be used to work out the structure of smaller organic molecules. It's not quite the first time this has been done, says Xiaodong Zou, a structural chemist at Stockholm University. But it is an important demonstration of just how fast and easy this kind of analysis can be.

Eureka papers

In the first paper, published on October 16th in Angewandte Chemie International Edition, a team led by crystallographer Tim Grüne at the Paul Scherrer Institute in Switzerland reports the creation of a prototype device for finding the structure of small molecules, using the beam from an electron microscope and a compatible detector. The diffraction patterns are analysed by software that is already used in X-ray crystallography. "Everything is composed of parts which have existed before," says Grüne. "It's just really the smooth integration of the system."

The team used its set-up to find the structure of the painkiller paracetamol from tiny crystals formed from the powder inside capsules. These crystals were just a few micrometres long—much smaller than those can be analysed using X-ray diffraction.

In the second paper, a preprint uploaded to the ChemRxiv server on October 17th, Gonen's group adapted the MicroED technique to solve the structure of small molecules instead of proteins. Gonen says that making this shift was "trivial". The main tweaks concerned the preparation of the samples, he says: whereas fragile proteins need to be treated with care, in this case, all he had to do was grind down pharmaceutical powders. The team used this adapted version of MicroED to find the structure of powders of pharmaceuticals including ibuprofen (**Figure 1.3.11**) and the anti-epileptic drug carbamazepine. These crystals were some 100 nanometres wide—a billion times smaller in volume than those used in X-ray crystallography—and their structure could be resolved in under 30 minutes.

Rubinstein says it's surprising that a technique already used in other fields hasn't yet been widely adopted by organic chemists. "It's this great solution that's been sitting almost in plain sight," he says.

Gonen puts the oversight down to a lack of communication between disciplines. It was only when he began speaking with chemists, he says, that he became aware that they struggled with growing large crystals to analyse small molecules, leading him to realize he had a solution for them. "As a protein crystallographer, I never really thought very carefully about small molecules," he says. "For us, small molecules are the things we try to get rid of."

Figure 1.3.11 3D structure for ibuprofen.

Excitement and limitations

The technique has caused lots of excitement, but it does have some limitations. The 3D structure of some molecules, for example, results in mirror-image molecules that can have different chemical effects — but it is challenging to distinguish between these structures using electron diffraction. Differentiation of these mirror-image structures would require further development of the analysis software, says Grüne.

One obvious application for electron crystallography is elucidating potential candidates for drug development. But Gonen says the technique could have other applications, such as forensics, when quick identification of a substance can be vital.

Grüne is optimistic that his work will encourage hardware manufacturers to create new devices built specifically for electron crystallography. At the moment, researchers tend to rely on electron microscopes to generate the electron beams, but these are expensive and include components, such as lenses, that are not necessary for electron diffraction. They are also not optimized to work with the other pieces of equipment used in the analysis. With a purpose-built device, he says, "one could just solve structures with the push of the button".

(**Selected from:** Warren M. "Why didn't We Think to do This Earlier?" Chemists Thrilled by Speedy Atomic Structures [J]. Nature, 2018, 563(7729): 16-18.)

New Words and Expressions

diffraction   n. 衍射
therapeutic target   治疗靶点
ibuprofen   n. 布洛芬，异丁苯丙酸（镇痛消炎药）
X-ray crystallography   X 射线晶体学
lattice structure   晶格结构，点阵结构
crystallographer   n. 晶体学家
prototype   n. 原型，最初形态
detector   n. 探测器，检测器

painkiller   *n*. 止痛药
capsule   *n*. 胶囊，太空舱
pharmaceutical   *adj*. 制药的；*n*. 药物
anti-epileptic   *n*. 抗癫痫，抗癫痫药
mirror-image   *n*. 镜像

Notes

(1) The structure of small organic molecules, such as those used in drugs, can be deduced in minutes rather than weeks, thanks to a technique that uses beams of electrons to quickly reveal how atoms are arranged.

——参考译文：得益于电子束能快速揭示原子排列方式，药物中有机小分子的结构就能在数分钟（而非数周）内被推导出来。

(2) Chemists working to develop new drugs, for example, depend on this structure to understand how the drug acts in the body—and how it could be tweaked to bind more strongly to its therapeutic target or to reduce side effects.

——act 在这里指药物起作用。it 代指前面的 drug。
——参考译文：例如，开发新药的化学家们依靠这种结构来了解药物在体内如何起作用，以及如何对其进行调整，使其与治疗靶点更紧密地结合或减少副作用。

(3) These crystals were some 100 nanometres wide—a billion times smaller in volume than those used in X-ray crystallography—and their structure could be resolved in under 30 minutes.

——a billion times smaller 表示比……小十亿倍。
——参考译文：这些晶体尺寸约 100 纳米——体积比 X 射线晶体学中使用的晶体小十亿倍——它们的结构可以在 30 分钟内解析出来。

(4) It was only when he began speaking with chemists, he says, that he became aware that they struggled with growing large crystals to analyse small molecules, leading him to realize he had a solution for them.

——struggled with 可意译为很难。
——参考译文：他（Gonen）说，当他开始与化学家交谈时，他才认识到他们很难通过培养大晶体来分析小分子，这让他意识到他有解决这些问题的办法。

Chapter 2
Metallic Materials and Alloys

【本章导读】
　　本章介绍金属材料及合金，包括五个单元，共有五篇课文和五篇阅读材料。五篇课文分别介绍金属及合金的概念、金属材料的热平衡相图、金属的典型物理性质、超合金与非铁合金、金属基复合材料。五篇阅读材料分别讲解储氢合金的电化学应用、海水热交换器中钛管泄漏的失效分析、不锈钢中金属钼元素的抗腐蚀作用、高熵合金、铝基复合材料的力学行为。

Unit 2.1

Text

Introduction to metals and alloys

　　这篇课文介绍金属与合金相关的基础知识，阐述了金属的概念、化学键和晶体结构，介绍了合金的定义、制备方法及其基本物理性质。课文描述了金属材料力学性质的主要参数和典型特征。

Metals

　　A metal is a material that, when freshly prepared, polished, or fractured, shows a lustrous appearance, and conducts electricity and heat relatively well. Metals are typically malleable (they can be hammered into thin sheets) or ductile (can be drawn into wires). The atomic bonding in this group of materials is metallic and thus nondirectional in nature. Consequently, there are minimal restrictions as to the number and position of nearest-neighbor atoms; this leads to relatively large numbers of nearest neighbors and dense atomic packings for most metallic crystal structures. Also, for metals, when we use the hard-sphere model for the crystal structure, each sphere represents an ion core. Three relatively simple crystal structures are found for most of the common metals: face-centered cubic, body-centered cubic, and hexagonal close-packed.

Alloys

　　Examples of alloys include materials such as brass, pewter, phosphor bronze, amalgam, and steel. Complete solid solution alloys give single solid phase microstructure. Partial solutions give two or more phases that may or may not be homogeneous in distribution, depending on thermal history. An alloy's properties are usually different from those of its component elements. Alloy constituents are usually measured by mass. An alloy is usually classified as either substitutional or interstitial, depending on its atomic arrangement. In a substitutional alloy, the atoms from each element can occupy the same sites as their counterpart. In interstitial alloys, the atoms do not occupy the same sites. Alloys can be further classified as homogeneous (consisting of a single phase), heterogeneous (consisting of two or more phases), or intermetallic (where there is no distinct boundary between phases).

Alloying a metal involves combining it with one or more other metals or non-metals, which often enhances its properties. For example, steel is stronger than iron, its primary element. Physical properties (density, reactivity, conductivity) of an alloy may not differ greatly from those of its constituent elements, but its engineering properties (tensile strength and shear strength) may be substantially different. Unlike pure metals, most alloys do not have a single melting point; rather, they have a melting range in which the substance is a mixture of solid and liquid. However, for most alloys, there is one particular proportion of constituents, known as the "eutectic mixture", at which the alloy has a unique melting point.

Modulus of elasticity

The modulus of elasticity is a measure of a metal's stiffness or rigidity, which is a ratio of stress to strain of a material in the elastic region. It is a good indication of the atom bond strength in crystalline materials. The uniaxial modulus of elasticity is often referred to as Young's modulus. Since the atom bond strength decreases with increasing temperature the moduli also decrease as temperature increases.

Tensile strength

Tensile strength, also referred to as ultimate tensile strength, is the maximum resistance of a material to deformation in a tensile test carried to rupture. As stress is continuously applied to a body, a point will be reached where stress and strain are no longer related in a linear manner. In addition, if the force is released, the bar will not return to its original length or shape since some permanent deformation has taken place. The elastic limit extends from the point of origin to the proportional limit, where departure from a linear relationship between stress and strain occurs.

Yield strength

Yield strength is the stress at which plastic deformation is fully developed in some portion of the material. Strain-aging types of metallic materials, such as annealed or normalized low-carbon steels, show a sudden transition from elastic to plastic behavior as the applied stress reaches a critical value. This gives a true yield point, an observable physical phenomenon from which the stress can be determined quite accurately.

Elongation

Elongation is a measure of ductility, as measured by the percentage of elongation.

Increasing the gauge length of a specimen will decrease the percent of elongation to fracture. This is because after the neck forms all subsequent deformation takes place in the vicinity of the neck. The behavior around the neck is the same regardless of the length of the specimen.

Hardness

The hardness test is the most utilized mechanical property test of all methods available. These tests do not require much time and are very informative since hardness is related to strength. Hardness tests do not have the precision of other tests. Of the various hardness tests and hardness scales available they all have one thing in common—their hardness numbers are relative. There is no such thing as an absolute hardness number as in yield strength, for example. The two most common tests used for metals are Rockwell and Brinell, with the former being the most popular.

Thermal Conductivity

Thermal conductivity is the quantity of heat flow under steady state conditions through unit area per unit temperature gradient in the direction perpendicular to the area. The heat flow through a wall per unit of area is called the thermal flux, J, which is proportional to the thermal gradient, the proportionality constant being the thermal conductivity. The equation is

$$J = K \frac{\Delta T}{\Delta X} \qquad (2.1.1)$$

where $\Delta T/\Delta X$ is the thermal gradient (i.e., temperature change) per unit thickness; X and K is the thermal conductivity. The thermal conductivity is expressed as Btu ft/hr. ft^2 °F in English units and as kcal m/sm^2 K in the metric system. These values are used when calculating the heat transfer through a metal wall.

Thermal expansion coefficient

The thermal expansion coefficient represents a change in dimension per unit temperature change. Values are usually given as in./in. °F or cm/cm℃. Thermal expansion can be expressed as change in either volume, area, or length, with the last being the most frequently used. Thermal expansion data are usually reported to three significant figures and can be measured accurately by a number of means.

(**Selected from:** Wikipedia, Metal, Wikimedia Foundation [DB/OL].(2023-02-06) [2023-02-22]. https://en.wikipedia.org/wiki/Metal.

Wikipedia, Alloy, Wikimedia Foundation [DB/OL].(2023-02-18)[2023-02-22]. https://en.wikipedia.org/wiki/Alloy.

Schweitzer P A, P E. Metallic Materials Physical, Mechanical, and Corrosion Properties vol 19 [M]. CRC press, 2003.)

New Words and Expressions

homogeneous   *adj.* 同种类的，同性质的，同形态（固态、液态或气态）
heterogeneous   *adj.* 由很多种类组成的，（化学）不均一的，多相的

tensile   *adj*. 拉力的，可伸长的
eutectic   *adj*. 共熔的；*n*. 共熔合金
modulus   *n*. 系数，模数
elasticity   *n*. 弹性，弹力
deformation   *n*. 变形
proportional   *adj*. 成比例的，（数）成常比的
ductility   *n*. 延展性，柔软性
vicinity   *n*. 周围地区，附近
polish   *v*. 抛光
fracture   *n*. 断裂
restriction   *n*. 可限制的
hexagonal   *adj*. 六角形的
brass   *n*. 黄铜
pewter   *n*. 锡
phosphor bronze   *n*. 磷青铜
amalgam   *n*. 汞合金
substitutional   *adj*. 替代的
interstitial   *adj*. 间质的，组织间隙的；*n*. （物）填隙
elongation   *n*. 伸长

Notes

(1) Unlike pure metals, most alloys do not have a single melting point; rather, they have a melting range in which the substance is a mixture of solid and liquid.

—rather 在这里表达转折的意思，意为相反。in which 引导从句修饰 melting range。
—参考译文：与纯金属不同，大多数合金没有固定的熔点。相反，它们的熔点有一定范围，在这个范围内物质是由固体和液体组成的混合物。

(2) Since the atom bond strength decreases with increasing temperature, the moduli also decrease as temperature increases.

—参考译文：因为原子键强随温度的升高而降低，所以模量也随温度的升高而减小。

(3) Of the various hardness tests and hardness scales available they all have one thing in common—their hardness numbers are relative.

—这里破折号替代了冒号，用于强调后面的信息，增强句子的可读性。
—参考译文：现有的各种硬度测试和硬度标尺都有一个共同点——它们的硬度值是相对的。

(4) Thermal conductivity is the quantity of heat flow under steady state conditions through unit area per unit temperature gradient in the direction perpendicular to the area.

—temperature gradient   温度梯度。
—参考译文：导热率是在稳态条件下单位面积内沿垂直于面积方向的单位温度梯度的热流量。

Exercises

1. Questions for discussion

(1) What are the differences between metals and alloys?

(2) What are the most common crystal structures in metallic structures?

(3) List several common alloys according to the text.

(4) How are alloys classified?

2. Translate the following into Chinese

(1) Partial solutions give two or more phases that may or may not be homogeneous in distribution, depending on thermal history.

(2) Tensile strength, also referred to as ultimate tensile strength, is the maximum resistance of a material to deformation in a tensile test carried to rupture.

(3) Strain-aging types of metallic materials, such as annealed or normalizedlow-carbon steels, show a sudden transition from elastic to plastic behavior as the applied stress reaches a critical value.

3. Translate the following into English

(1) 面心立方　　　　　(2) 晶体材料

(3) 杨氏模量　　　　　(4) 极限抗拉强度

(5) 热通量　　　　　　(6) 热膨胀系数

(7) 合金化是指将一种金属与另一种或多种其他金属或非金属相结合,(其结果)通常会增强其性能。

(8) 此外,在力被撤去后,由于物体已经发生了部分永久变形,将不会恢复到原来的长度或形状。

(9) 热膨胀可以用体积、面积或长度的变化来表示,后者是最常用的。

4. Reading comprehension

(1) About crystal structure of metallic materials, all the following statements are true EXCEPT ____.

(A) Relatively large numbers of nearest neighbors

(B) Dense atomic packings

(C) Bonds are directional in nature

(D) Bonds are nondirectional in nature

(2) Which of the following is TRUE according to the text? ____

(A) Atom bond strength in metallic materials decreases with increasing temperature.

(B) Moduli decrease as temperature increases.

(C) Increasing the gauge length of a specimen will decrease the percent of elongation.

(D) A, B, and C.

(3) According to the text, metal's stiffness or rigidity can be measured by ____.

(A) yield strength　　　　(B) modulus of elasticity

(C) tensile Strength　　　(D) hardness

Reading Material
The hydrogen storage alloys for electrochemical applications

这篇课文主要介绍储氢合金相关的知识。课文讲解了储氢合金的基本结构和电化学储氢基本原理，并重点介绍了储氢合金的一个重要应用，即镍氢电池的负极材料。课文展望了该领域的发展并做了总结。

Introduction

Hydrogen storage alloys are important for a few electrochemical applications, especially in the energy storage area. The basic of electrochemical use of the hydrogen storage alloy can be described as follows: when hydrogen enters the lattice of most transition metals, interstitial metal hydride (MH) is formed. The electrons accompanying the hydrogen atoms form a metal-hydrogen band right below the Fermi level, which indicates that the interstitial MH is metallic in nature. While protons in the interstitial MH hop between neighboring occupation sites by quantum mechanical tunneling, the electrons remain within a short distance (3—10 angstroms) of the protons to maintain local charge neutrality. Under the influence of an electric field, electrons and protons will move in opposite directions. In an electrochemical environment, a voltage is applied to cause electrons to flow, and the charges are balanced out by moving conductive ions through a highly alkaline aqueous electrolyte with good ionic conductivity. During charge, a negative voltage (with respect to the counter electrode) is applied to the metal/metal hydride electrode current collector, and electrons enter the metal through the current collector to neutralize the protons from the splitting of water that occurs at the metal/electrolyte interface [**Figure 2.1.1(a)**]. This electrochemical charging process is characterized by the half reaction:

$$M + H_2O + e^- \longrightarrow MH + OH^-$$

During discharge, protons in the MH leave the surface and recombine with OH^- in the alkaline electrolyte to form H_2O, and charge neutrality pushes the electrons out of the MH through the current collector, performing electrical work in the attached circuitry [**Figure 2.1.1(b)**]. The electrochemical discharge process is given by the half reaction:

$$MH + OH^- \longrightarrow M + H_2O + e^-$$

The standard potential of this redox half reaction depends on the chosen MH and is usually as low as possible to maximize the amount of stored energy without exceeding the hydrogen evolution potential (−0.83 V *versus* standard hydrogen electrode). Zn is an exception. With a complete 3d shell, Zn is a natural prohibitor for hydrogen evolution and thus a more negative voltage is possible, which increases the operation voltage of Ni-Zn battery.

The most important electrochemical application for MH is the negative electrode material for nickel metal hydride (NiMH) batteries. Together with a counter electrode from the $Ni(OH)_2/NiOOH$ system, which has been used in NiCd and NiFe batteries as early as 1901 by Thomas Edison, the NiMH battery was first demonstrated by researchers in Battelle in 1967

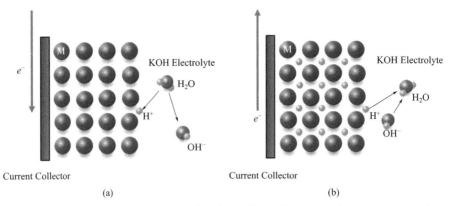

Figure 2.1.1 Schematics showing the electrochemical reactions between water and metal hydride during (a) charge and (b) discharge.
(Due to the alkaline nature of the electrolyte, protons cannot desorb or absorb from the surface of metal without the incorporation of water and OH^-)

with a mixed TiNi + Ti_2Ni alloy as the negative electrode. Commercialization of the NiMH battery was independently realized by Ovonic Battery Company, Sanyo, and Matsushita in 1989 with AB_2 and AB_5 MH alloys. NiMH battery development started from small cylindrical cells (0.7 to 5 Ah) for portable electronic devices and progressed to 100 Ah prismatic cells for electric vehicle applications. The first commercially available electric vehicle in the modern era was the EV1 produced by General Motors in 1999. It was powered by a 26.4 kW · h NiMH battery pack. Since then, NiMH batteries have powered more than 5 million hybrid electric vehicles made by Toyota, Honda, Ford, and other automakers, demonstrating the robustness and longevity of the NiMH battery. Recently, the NiMH battery has ventured into the stationary application market with advantages in long service life, a wide temperature range, low costs averaged over the service life, abuse immunity, and environmental friendliness. Several reviews on the topic of MH used in NiMH batteries are available. In this report, we present the recent progress since the last review made in 2010.

Besides NiMH batteries, MH (most commonly the misch metal-based AB5 MH alloy) can also be used in other electrochemical applications such as lithium-ion based batteries and metal-air batteries. Metal hydride electrodes have a potential window of 0.1 to 0.5 V versus Li^+/Li and the lowest polarization among conversion electrodes. These MH electrodes have shown the capability for greater capacity and can be used as anode electrodes in lithium-ion battery. An air-MH battery that utilizes a misch metal-based AB_5 alloy in conjunction with a perovskite oxide-based cathode has been demonstrated by several research groups. New types of V-flow/NiMH and lead-acid/NiMH hybrid batteries have been developed at the University of Hong Kong. Pd-treated $LaNi_{4.7}Al_{0.3}$ has been used in a Ni-hydrogen battery. Another application of $LaNi_5$ is the use as a cathode in a photo-electrochemical cell for water decomposition.

Conclusions

Hydrogen storage alloys for electrochemical application have been extensively studied for many years. We have presented a review of recent research activities on metal hydride alloys for nickel metal hydride battery and also provided an overview of the use of metal hydrides in other electrochemical applications. AB_5 and AB_2 alloys are very well established systems. In order to potentially dominate the future electric vehicle and stationary applications, self-discharge, low-temperature performance, and cycle stability become more important to study, and the trend of recent research reflects the efforts on improving the aforementioned properties. Superlattice A_2B_7-type alloy, which possesses the advantages of both AB_5 and AB_2 and low self-discharge capability, is likely to be the next generation of metal hydride alloy used as the negative electrode material in nickel metal hydride batteries and has attracted much attention. Although the Ti-Ni alloy system is difficult to process and has poorer high-rate performance, its much lower raw material cost makes the system one that merits further studies for improvement. The Mg-Ni alloy system holds great promise in achieving very high capacity, and recent research efforts have concentrated on improving its kinetics and cycle capability for the purpose of practical implementation. Laves phase-related BCC solid solution has high capacity; enhancing its stability is currently the most essential topic. The incorporation of quasicrystals by various fabrication methods remains an interesting subject. Zr-Ni alloy systems were systematically investigated in the last few years. Although their performance might not be satisfactory for electrochemical applications at the present time, further elemental modifications or use as a composite modifier can assist in realizing their potential.

(**Selected from:** Young K-h, Nei J. The Current Status of Hydrogen Storage Alloy Development for Electrochemical Applications [J]. Materials, 2013, 6(10):4574-4608.)

New Words and Expressions

proton   *n*. 质子
quantum   *n*. 量子
hydride   *n*. 氢化物
alkaline   *adj*. 碱性的
charge neutrality   电中性
prismatic cell   方形蓄电池
in conjunction with...   连同，共同，与……协力
polarization   *n*. 极化，偏振
superlattice   *n*. 超晶格
negative   *adj*. 消极的，负面的
desorb   *v*. 使……释放出吸收之物
incorporation   *n*. 结合，合并
prohibitor   *n*. 抑制剂
immunity   *n*. 免疫，免疫力
satisfactory   *adj*. 令人满意的，够好的，可以的
electrochemical   *adj*. 电化学的

Notes

(1) During charge, a negative voltage (with respect to the counter electrode) is applied to

the metal/metal hydride electrode current collector, and electrons enter the metal through the current collector to neutralize the protons from the splitting of water that occurs at the metal/electrolyte interface.

—参考译文：在充电过程中，在金属/金属氢化物的电极集流体上施加负电压（相对于对电极），电子通过集流体进入金属，中和在金属/电解质（液）界面上由水分解而产生的质子。

（2）Recently, the NiMH battery has ventured into the stationary application market with advantages in long service life, a wide temperature range, low costs averaged over the service life, abuse immunity, and environmental friendliness.

—参考译文：近年来，镍氢电池凭借服役寿命长、温度范围宽、服役期内平均成本低、抗滥用、环境友好等优点，已经进入了固定设施应用市场。

（3）Superlattice A_2B_7-type alloy, which possesses the advantages of both AB_5 and AB_2 and low self-discharge capability, is likely to be the next generation of metal hydride alloy used as the negative electrode material in nickel metal hydride batteries and has attracted much attention.

—参考译文：A_2B_7型超晶格合金兼具AB_5和AB_2的优点并具备低自放电特性，（因其）有望作为用于镍氢电池负极材料的下一代金属氢化物合金而备受关注。

（4）Although the Ti-Ni alloy system is difficult to process and has poorer high-rate performance, its much lower raw material cost makes the system one that merits further studies for improvement.

—参考译文：虽然 Ti-Ni 合金体系加工难度大、高倍率下性能差，但其原材料成本较低，（因此）值得进一步研究改进。

Unit 2.2

Text

Thermal equilibrium diagram

这篇课文介绍热平衡图/相图相关的基础知识。课文阐述了相图的定义和意义，说明了相图对应材料的热平衡状态，并分别介绍了单组分、双组分、三组分系统，给出了相图的实例。

Phase diagrams and the systems they describe are often classified based on the number of components (typically elements) in the system. A unary phase diagram plots the phase changes of one element as a function of temperature and pressure. A binary diagram plots the phase changes as a function of temperature for a system with varying composition of two components. While other extensive and intensive properties influence the phase structure, materials scientists typically hold these properties constant for practical ease of use and interpretation. Phase diagrams are usually constructed with a constant pressure of one atmosphere. Phase diagrams are useful graphical representations that show the phases in equilibrium present in the system at various specified compositions, temperatures, and pressures. It should be recognized that phase diagrams represent equilibrium conditions for an alloy, which means that very slow heating and cooling rates are used to generate data for their construction. The equilibrium states that are represented on phase diagrams are known as heterogeneous equilibria, because they refer to the coexistence of different states of matter (gas, liquid, and/or solid phases with different crystal structures). When two or more phases are in mutual equilibrium, each phase must be in the lowest free-energy state possible under the restrictions imposed by its environment. This equilibrium condition means that each phase is in an internally homogeneous state with a chemical composition that is identical everywhere within each phase, and that the molecular and atomic species of which the phase is composed (if more than one) must be present in equilibrium proportions.

Unary systems

A system containing only one pure metal is referred to as a unary system, which can exist as a solid, liquid, and/or gas, depending on the specific combination of temperature and pressure. Assuming a constant atmospheric pressure, a metal melts when heated to a specific temperature and boils with further heating to a specific boiling temperature. Through evaporation, metal atoms leave the container as a vapor. In conditions where matter can enter or leave the system, these systems are known as open systems.

Binary systems

A binary phase diagram plots the different states of matter as a function of temperature for a system at constant pressure with varying composition of two components (or elements). The addition of an alloying element represents another degree of freedom (or variable), which thus allows two distinct forms (or phases) to coexist under equilibrium conditions. The mixed (i. e., heterogeneous) equilibria phases in binary alloys also can occur in either solids or when liquid and solid phases change during melting or solidification.

Binary systems have two elements of varying concentration. Unlike pure metals, alloys do not necessarily have a unique melting point. Instead, most alloys have a melting range. This behavior can be seen in the phase diagram for copper-nickel (**Figure 2.2.1**). The solidus line is where the solid phase begins to melt, and the new phase consists of a mixture of a solid-solution phase and some liquid phase. This solid-liquid mix in a two-phase region is referred to as the mushy zone. Like the two-phase system of a liquid and vapor phase in a closed system, the liquid phase is in equilibrium with the solid when the average rate of atoms leaving the liquid equals the rate joining the liquid from the solid phase. Temperature and alloy concentration changes the relative amount of liquid and solid phase in the mushy zone. The reason for the mushy zone can be understood in qualitative terms by examining the nickel-copper phase diagram (**Figure 2.2.1**). Pure copper has a lower melting point than nickel. Therefore, in a nickel-copper alloy, the copper atoms melt before the regions rich in nickel. The mushy zone becomes 100% liquid when the temperature is raised above the liquidus line. As one might expect, the liquidus line converges to the melting points of the pure metals in an alloy phase diagram. The liquidus denotes for each possible alloy composition the temperature at which freezing begins during cooling or, equivalently, at which melting is completed on heating. The lower curve, called the solidus, indicates the temperatures at which melting begins on heating or at which freezing is completed on cooling. Above the liquidus every alloy is molten, and this region of the diagram is, accordingly, labeled L for the liquid phase or liquid solution. Below the solidus all alloys are solid, and this region is labeled α because it is customary to use Greek letters to designate different solid phases. At temperatures between the two curves, the liquid and solid phases both are present in equilibrium, as is indicated by the designation L+α.

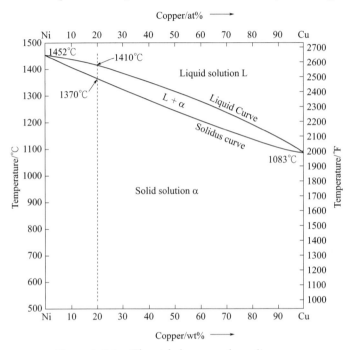

Figure 2.2.1　The nickel-copper phase diagram.

However, it also is important to understand that some specific alloy compositions do not have a mushy zone. That is, some alloys have a unique composition with a specific melting point that is lower than the melting points of the two pure metals in the alloy. This type of alloy is referred to as a eutectic alloy (where the term eutectic is taken from the Greek word for "easily melted"). Eutectic alloys are important because complete melting occurs at a low temperature. For example, cast irons are based on compositions around the iron-carbon eutectic composition of iron with 4.30wt% C. This improves castability by lowering the temperature required for melting and also promotes better solidification during casting. Another industrially significant eutectic is in the lead-tin system for solders.

Ternary systems

When more than two components exist in a system, illustrating equilibrium conditions graphically in two dimensions becomes more complicated. One option is to add a third composition dimension to the base, forming a solid diagram having binary diagrams as its vertical sides. This can be represented as a modified isometric projection. Here, boundaries of single-phase fields (liquidus, solidus, and solvus lines in the binary diagrams) become surfaces; single-and two-phase areas become volumes; three-phase lines become volumes; and four-phase points, can exist as an invariant plane. The composition of a binary eutectic liquid, which is a point in a two-component system, becomes a line in a ternary diagram.

The Gibbs phase rule

The construction of phase diagrams is greatly facilitated by certain rules that come from thermodynamics. Foremost among these is Gibbs' phase rule, which applies to all states of matter (solid, liquid, and gaseous) under equilibrium conditions. The maximum number of phases (P) that can coexist in a chemical system, or alloy, is:

$$P = C - F + 2$$

where F is the number of degrees of freedom, and C is the number of chemical components (usually elements for alloys). The phases are the homogeneous parts of a system that, having definite bounding surfaces, are conceivably separable by mechanical means alone, for example, a gas, liquid, and solid. The degrees of freedom, F, are those externally controllable conditions of temperature, pressure, and composition, which are independently variable and which must be specified in order to completely define the equilibrium state of the system.

(**Selected from:** Okamoto H, Schlesinger M E, Mueller E M. Introduction to Phase Diagrams vol 3 [M/CD]. ASM International, 2016.)

New Words and Expressions

equilibrium   *n*. 均衡，平静
coexistence   *n*. 共存
imposed   *v*. 把……强加于
mushy zone   糊状区
solder   *n*. 焊料，接合物；*v*. 焊接，使连接在一起
vertical   *adj*. 垂直的，纵向的
intensive   *adj*. 密集的；*n*. 加强器
extensive   *adj*. 广阔的
proportion   *n*. 部分，份额，比例，匀称
atmospheric   *adj*. 大气的，大气层的
concentration   *n*. 浓度
mushy   *adj*. 糊状的
converge   *v*. 汇集，聚集，相同
thermodynamics   *n*. 热力学

Notes

(1) It should be recognized that phase diagrams represent equilibrium conditions for an

alloy, which means that very slow heating and cooling rates are used to generate data for their construction.

——参考译文：应该认识到，相图代表合金的平衡条件，这代表在绘制相图时采用的数据是在非常缓慢的加热和冷却速率下得到的。

(2) This equilibrium condition means that each phase is in an internally homogeneous state with a chemical composition that is identical everywhere within each phase, and that the molecular and atomic species of which the phase is composed (if more than one) must be present in equilibrium proportions.

——参考译文：这种平衡条件意味着每一相都处于内部均质状态，各相的化学成分都相同，而且组成该相的分子和原子种类（如果不止一种）必须比例均衡。

(3) Like the two-phase system of a liquid and vapor phase in a closed system, the liquid phase is in equilibrium with the solid when the average rate of atoms leaving the liquid equals the rate joining the liquid from the solid phase.

——参考译文：与一个由液相和气相组成的两相封闭系统类似，当原子离开液相的平均速率与从固相转变为液相的速率相等时，液相与固相处于平衡状态。

(4) The liquidus denotes for each possible alloy composition the temperature at which freezing begins during cooling or, equivalently, at which melting is completed on heating.

——这里连续用到两个 at which，都是用来修饰 temperature。
——参考译文：液相线表示每一种可能的合金成分在冷却过程中开始凝固的温度，或者表示在加热过程中完全熔化的温度。

(5) One option is to add a third composition dimension to the base, forming a solid diagram having binary diagrams as its vertical sides.

——在英文写作中，经常用动词现在时形式直接引导一个状语从句，省去连接词，表示原因、结果、让步、条件等。
——参考译文：（其中）一种选择是在（二元相图的）基础上添加第三元组分，形成一个以二元相图为投影的立体相图。

(6) The composition of a binary eutectic liquid, which is a point in a two-component system, becomes a line in a ternary diagram.

——参考译文：二元共晶液体的组分在二元体系中用一个点表示，而在三元相图中变成了一条线。

Exercises

1. Questions for discussion

(1) What is the main difference between an open and a close system?
(2) Briefly describe liquidus, solidus line and mushy zone.
(3) What are the most important characteristics of eutectic alloys?

2. Translate the following into Chinese

(1) While other extensive and intensive properties influence the phase structure, materials scientists typically hold these properties constant for practical ease of use and interpretation.

(2) A binary phase diagram plots the different states of matter as a function of temperature for a system at constant pressure with varying composition of two components (or elements).

(3) The mushy zone becomes 100% liquid when the temperature is raised above the liquidus line.

3. Translate the following into English

(1) 二元的   (2) 三元的
(3) 习惯性地   (4) 焊料
(5) 自由度   (6) 平衡
(7) 相互的   (8) 固化
(9) 最重要的   (10) 等轴测
(11) 预测   (12) 会聚

(13) 当两个或两个以上的相处于相互平衡状态时，在其环境的约束下，每个相必须处于自由能最低的状态。

(14) 假设（在）一个恒定的大气压（下），金属会在加热到一个特定的温度时熔化，并在进一步加热到一个特定的温度时沸腾。

(15) 较低的曲线称为固相线，表示加热时熔化开始或冷却时凝固结束的温度。

4. Reading comprehension

(1) Abouteutectic alloys, all the following statements are true EXCEPT ____.
(A) It complete melting occurs at a low temperature
(B) It improves castability by increasing the temperature required for melting
(C) It promotes better solidification during casting
(D) One industrial eutectic is lead-tin system for solders

(2) A copper-nickel alloy of composition 70 wt% Ni-30 wt% Cu is slowly heated from a temperature of 1300℃ (2370 ℉).
(A) At what temperature does the first liquid phase form?
(B) What is the composition of this liquid phase?
(C) At what temperature does complete melting of the alloy occur?
(D) What is the composition of the last solid remaining prior to complete melting?

(3) According to Gibbs phase rule, for binary systems, when three phases are present, there is (are) ____ degree(s) of freedom.
(A) 0   (B) 1
(C) 2   (D) 3

Reading Material
Failure analysis on leaked titanium tubes of seawater heat exchangers

这篇课文介绍了一个失效分析的案例。分析对象是核电站热交换器中在工作期间发生泄漏的钛管。研究人员采用多种微结构和形貌测试手段研究了钛管发生泄漏的原因,并针对失效提出了应对措施。

Introduction

The vigorous development of nuclear power industry in China is not independent of her great endeavor in ensuring safety of the structures, systems and components (SSCs), for which besides optimization of design and improvement of construction for the upcoming Generation III reactors, focuses are also laid on in-service inspection, online monitoring, periodic testing, and failure analysis etc. for the sake of effective and efficient ageing management of the operational reactors. As one of the cooling systems in the BOP (balance of plant), the recirculating cooling water (RCW) system that utilizes seawater as the ultimate heat sink for the moderator water seems not as significant as those systems in the nuclear island. However, safety of the equipment and components in this system, of course also including other systems in the BOP, still deserves special attention. Once failures occur on them, not only economic losses due to shutdown for overhaul, but also potential social panics about seawater pollution by the radioactive wastes might be aroused.

In our previous works, comprehensive failure analysis was conducted on a large number of leaked titanium tubes of the seawater heat exchangers in the RCW system of one coastal nuclear power plant in China, and the root causes were finally determined as the synergetic effect from both the electrochemical corrosions as galvanic corrosion, crevice corrosion, hydrogen blistering and hydrogen embrittlement, and the mechanical degradations as erosion, clogging, fretting and surface damage. The countermeasures proposed were then promptly adopted by the operator, and have been proved wholly effective from then on. However, a few individual tubes were found leaked again after several years. Although normal operation of the heat exchangers would not be affected on account that the number of tubes in each heat exchanger is 4932 and the acceptable limit of leakage rate is commonly 5%, analysis to identify the failure causes was still necessary for securing the concept of "absolute safety" for this nuclear power plant.

To this end, this paper addresses failure analysis on one of these repeatedly leaked tubes of the heat exchangers in the RCW system of this nuclear power plant. By referring to our past experiences, investigation from four aspects were carried out, including examination of the base materials, inspection of the environmental media, observation of the defect morphologies, and detection of the micro-area compositions. Based on the analysis results, the root causes were then identified, meanwhile the relevant mechanisms were discussed and the pertinent countermeasures were proposed. Considering that failure analysis always ends up at the proposal of suggestions while the effectiveness is rarely reported, achievement of this paper that is concerned with a recurring failure case will provide not only the further improvement of the countermeasures, but also a demonstration of the sense of responsibility in engineering failure analysis.

Failure analysis

Based on our previous works, causes from two aspects including electrochemical corrosions and mechanical degradations were generally blamed for leakage of the titanium tubes in these RCW heat exchangers. Since the element of titanium was not detected in the sediment through X-ray fluorescence (XRF) and X-ray diffraction (XRD), in other words, no evidence of corrosions was found. Hence, it can be concluded that on one hand, the countermeasures we proposed before for prevention of electrochemical corrosions were really effective, and on the other hand, focuses should be mainly laid on the mechanical degradations this time.

According to the SEM micrographs of the three locations on the rupture of this leaked tube, at least five types of mechanical degradation morphologies could be observed, including fluid erosion [**Figure 2.2.2(a)**], particle impact [**Figure 2.2.2(b)**], micro cracking, and abrasive erosion in both oriented manner and disorder manner. It was easy to conclude that these morphologies were primarily induced by the flowing natural seawater that contained particles of sea sands in the tube side of the RCW heat exchangers. However, different to the morphologies from pure erosion, the morphologies in this case were really diverse. Furthermore, limited to the geographical environments of this nuclear power plant, such an operational condition (i.e. seawater contains sands) was common, but why failure only occurred on a few individual tubes instead of a large number of tubes?

In our previous works, it was found that mechanical damages were occasionally left on the inner wall of some tubes due to improper operation of the drawing heads during installation. After consulting the staff on the field, we learnt that this issue was basically impossible to be avoided and remedied since there are currently no feasible and effective means to find out the in-service tubes that are embedded with mechanical damages until they are perforated. Let's look back to the diverse morphologies of the rupture presented above. It was obvious that they were resulted from the turbulence rather than the pure erosion from the straight flow, which must be induced by the localized surface geometry change of the inner wall of the tube, e.g. existence of pre-existing damages from the drawing heads. When turbulence was formed, vortex erosion would be initiated at such locations, whose extent is several times that of the pure erosion, and generally causes diverse erosion traces. Now it can be concluded that the mechanical damages on the inner wall of some titanium tubes from drawing heads were actually the initial cause for the turbulence, which then led to multiple mechanical degradation mechanisms from vortex erosion and resulted in diverse morphologies. Once the resultant microscopic defects like pits, scratches, cracks, drop off of base material etc. were formed, their extent would be self-aggravated due to the increasing complexity of the localized geometry. As a result, with aggregation of these defects, the wall thicknesses of the affected titanium tubes would be thinned and even perforated, and eventually led to the macroscopic degradation morphologies like cutting edge [**Figure 2.2.2(b)**], curled strip [**Figure 2.2.2(c)**] and antler-shaped fractures

[**Figure 2.2.2(d)**]. So far, the failure causes and relevant mechanisms of these individual leaked tubes have been identified. With regard to the corrosion of the plug, it was clear that unqualification of the screw for its lower chromium content than the specification requirement should be mainly ascribed to. Corrosion of the plug, as well as other carbon steel equipment in the RCW system, would release iron-based corrosion products like FeO(OH) into the tube side, but of course it had nearly no effect on the initiation and progression of erosion on the titanium tubes.

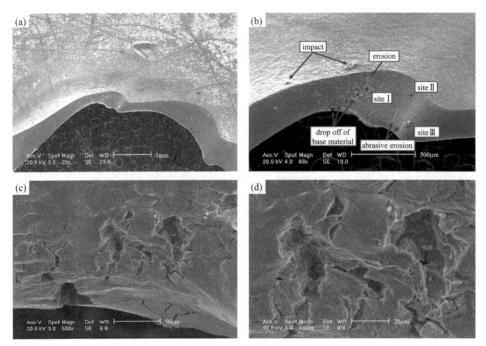

Figure 2.2.2　SEM micrographs of the location A on the leaked tube: (a) Total morphology; (b) Diverse degradation traces on the thinned edge; (c) and (d) Micro cracks.

To conclude, the recurrence of leakage on several titanium tubes in the RCW heat exchangers was an individual case, whose procedure can be briefly classified into three consecutive stages: (1) mechanical damages from drawing heads, (2) vortex erosion from turbulence, and (3) diverse microscopic and macroscopic degradation morphologies and perforation. That is to say, the countermeasures we proposed before were really effective, but online monitoring should be further strengthened for leakage detection of the titanium tubes of these RCW heat exchangers in the future.

Conclusions and recommendations

(1) No evidence of electrochemical corrosions was detected on the leaked titanium tubes, that is to say, the causes from mechanical degradations should be mainly ascribed to.

(2) The pre-existing mechanical damages on the inner wall of some titanium tubes induced by improper operation of the drawing heads during installation were the root cause for leakage.

(3) The synergetic effect of impact, erosion, cracking, and abrasive erosion in both oriented and disorder manners from the turbulence of seawater that contained natural sea sands

finally resulted in the diverse defect morphologies.

Based on the failure analysis results presented above, several countermeasures were then proposed as follows.

(1) The level of emphasis on online monitoring of the titanium tubes should be improved, once leakage was detected, the leaked tubes must be immediately plugged in the next routine downtime.

(2) Close attention should also be paid to quality of the desalinated water, once the contents of detrimental impurities exceed the upper limits, pertinent measures must be implemented at once.

(**Selected from:** Mortensen A, Llorca J. Metal Matrix Composites [J]. Annual Review of Materials Science, 2010, 40: 243-270.)

New Words and Expressions

  degradation *n*. 恶化，衰退，降解
  fluorescence *n*. 荧光
  diffraction *n*.（光，声等的）衍射
  erosion *n*. 侵蚀，腐蚀
  vortex *n*. 涡流，漩涡
  titanium *n*. 钛（金属元素，符号 Ti，原子序号 22）
  aggregation *n*. 聚合，聚集
  turbulence *n*. 湍流，紊流
  perforation *n*. 穿孔，贯穿
  synergetic effect *n*. 协同效应
  morphology *n*. 形态，形貌
  leakage *n*. 泄漏，渗漏物

Notes

(1) Since the element of titanium was not detected in the sediment through X-ray fluorescence (XRF) and X-ray diffraction (XRD), in other words, no evidence of corrosions was found.

  —XRF 和 XRD 都是材料表征的重要技术。
  —参考译文：通过 X 射线荧光和 X 射线衍射技术，未在沉积物中检测到钛元素，换句话说，没有发现腐蚀的证据。

(2) According to the SEM micrographs of the three locations on the rupture of this leaked tube, at least five types of mechanical degradation morphologies could be observed, including fluid erosion, particle impact, micro cracking, and abrasive erosion in both oriented manner and disorder manner.

  —SEM 扫描电子显微镜，是一种对材料进行微观形貌表征的重要设备。
  —参考译文：从该泄漏管破裂处三个部位的 SEM 显微照片上可以观察到至少五类机械降解形貌，包括流体冲刷、颗粒冲击、微裂纹以及有序和无序的磨粒磨损。

(3) Furthermore, limited to the geographical environments of this nuclear power plant, such an operational condition (i. e. seawater contains sands) was common, but why failure only occurred on a few individual tubes instead of a large number of tubes?

——参考译文：此外，受限于核电站的地理环境，这种海水含砂工况很常见。但为什么只有个别钛管失效而不是大量钛管（失效）呢？

(4) Now it can be concluded that the mechanical damages on the inner wall of some titanium tubes from drawing heads were actually the initial cause for the turbulence, which then led to multiple mechanical degradation mechanisms from vortex erosion and resulted in diverse morphologies.

——参考译文：现在可以得出结论，牵引头对部分钛管内壁造成的力学损伤实际上是引起湍流（腐蚀）的起因，这诱发了因涡流侵蚀引起的多种机械降解机制，进而造成了多种（腐蚀）形貌。

(5) As a result, with aggregation of these defects, the wall thicknesses of the affected titanium tubes would be thinned and even perforated, and eventually led to the macroscopic degradation morphologies like cutting edge, curled strip and antler-shaped fractures.

——参考译文：因此，这些缺陷的聚集会使受影响的钛管壁变薄甚至穿孔，最终形成肉眼可见的降解形貌，如刃口、卷边和鹿角形破口。

Unit 2.3

Text

Characters of metallic materials—ductility, malleability, and corrosion

这篇课文主要说明金属材料的一些典型性质，包括金属的延展性和金属的腐蚀。针对金属腐蚀，课文简要介绍了阴极保护等腐蚀防护方案。

Ductility

Ductility is more commonly defined as the ability of a material to deform easily upon the application of a tensile force, or as the ability of a material to withstand plastic deformation without rupture. Ductility may also be thought of in terms of bendability and crushability. Ductile materials show large deformation before fracture. The lack of ductility is often termed brittleness. Usually, if two materials have the same strength and hardness, the one that has the higher ductility is more desirable. The ductility of many metals can change if conditions are altered. An increase in temperature will increase ductility. A decrease in temperature will cause a decrease in ductility and a change from ductile to brittle behavior. Cold-working also tends to make metals less ductile. Cold-working is performed in a temperature region and over a time interval to obtain plastic deformation, but not relieving the strain hardening. Minor additions of impurities to metals, either deliberate or unintentional, can have a marked effect on the change from ductile to brittle behavior. The heating of a cold-worked metal to or above the temperature at which metal atoms return to their equilibrium positions will increase the ductility of that metal. This process is called annealing. Ductility is desirable in the high temperature and high pressure applications in reactor plants because of the added stresses on the metals. High ductility in these applications helps prevent brittle fracture.

Malleability

Where ductility is the ability of a material to deform easily upon the application of a tensile force, malleability is the ability of a metal to exhibit large deformation or plastic response when being subjected to compressive force. Uniform compressive force causes deformation in the manner shown in **Figure 2.3.1**. The material contracts axially with the force and expands laterally. Restraint due to friction at the contact faces induces axial tension on the outside. Tensile forces operate around the circumference with the lateral expansion or increasing girth. Plastic flow at the center of the material also induces tension. Therefore, the criterion of fracture (that is, the limit of plastic deformation) for a plastic material is likely to depend on tensile rather than compressive stress. Temperature change may modify both the plastic flow mode and the fracture mode.

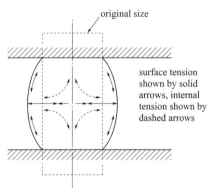

Figure 2.3.1 Malleable deformation of a cylinder under uniform axial compression.

Corrosion

Corrosion is defined as the destructive and unintentional attack on a metal; it is electrochemical and ordinarily begins at the surface. The problem of metallic corrosion is significant; in economic terms, it has been estimated that approximately 5% of an industrialized nation's income is spent on corrosion prevention and the maintenance or replacement of products lost or contaminated as a result of corrosion reactions. The consequences of corrosion are all too common. Familiar examples include the rusting of automotive body panels and radiator and exhaust components. For metallic materials, the corrosion process is normally electrochemical, that is, a chemical reaction in which there is transfer of electrons from one chemical species to another. Metal atoms characteristically lose or give up electrons in what is called an oxidation reaction. For example, a hypothetical metal M that has a valence of n (or n valence electrons) may experience oxidation according to the reaction

$$M \longrightarrow M^{n+} + ne^-$$

in which M becomes an $n+$ positively charged ion and in the process loses its n valence electrons; e^- is used to symbolize an electron. Examples in which metals oxidize are

$$Fe \longrightarrow Fe^{2+} + 2e^-$$
$$Al \longrightarrow Al^{3+} + 3e^-$$

The site at which oxidation takes place is called the anode; oxidation is sometimes called an anodic reaction. The electrons generated from each metal atom that is oxidized must be transferred to and become a part of another chemical species in what is termed a reduction reaction. For example, some metals undergo corrosion in acid solutions, which have a high concentration of hydrogen (H^+) ions; the H^+ ions are reduced as follows:

$$2H^+ + 2e^- \longrightarrow H_2$$

and hydrogen gas (H_2) is evolved. The electromotive force (emf) series (**Table 2.3.1**) is generated by coupling the standard hydrogen electrode to standard half-cells for various metals and ranking them according to measured voltage. **Table 2.3.1** shows the corrosion tendencies for the several metals; those at the top (i.e., gold and platinum) are noble, or chemically inert. As one moves down the table, the metals become increasingly more active—that is, more susceptible to oxidation. Sodium and potassium have the highest reactivities.

Table 2.3.1 The standard emf series.

	Electrode Reaction	Standard Electrode Potential/V^0
	$Au^{3+} + 3e^- \longrightarrow Au$	+1.420
	$O_2 + 4H^+ + 4e^- \longrightarrow 2H_2O$	+1.229
	$Pt^{2+} + 2e^- \longrightarrow Pt$	~+1.2
	$Ag^+ + e^- \longrightarrow Ag$	+0.800
Increasingly inert (cathodic)	$Fe^{3+} + e^- \longrightarrow Fe^{2+}$	+0.771
	$O_2 + 2H_2O + 4e^- \longrightarrow 4OH^-$	+0.401
	$Cu^{2+} + 2e^- \longrightarrow Cu$	+0.340
	$2H^+ + 2e^- \longrightarrow H_2$	0.000
	$Pb^{2+} + 2e^- \longrightarrow Pb$	−0.126
	$Sn^{2+} + 2e^- \longrightarrow Sn$	−0.136
	$Ni^{2+} + 2e^- \longrightarrow Ni$	−0.250
	$Co^{2+} + 2e^- \longrightarrow Co$	−0.277
	$Cd^{2+} + 2e^- \longrightarrow Cd$	−0.403
	$Fe^{2+} + 2e^- \longrightarrow Fe$	−0.440
Increasingly active (anodic)	$Cr^{3+} + 3e^- \longrightarrow Cr$	−0.744
	$Zn^{2+} + 2e^- \longrightarrow Zn$	−0.763
	$Al^{3+} + 3e^- \longrightarrow Al$	−1.662
	$Mg^{2+} + 2e^- \longrightarrow Mg$	−2.363
	$Na^+ + e^- \longrightarrow Na$	−2.714
	$K^+ + e^- \longrightarrow K$	−2.924

Corrosion prevention

Inhibitors are substances that, when added in relatively low concentrations to the environment, decrease its corrosiveness. The specific inhibitor depends on both the alloy and the corrosive environment. Several mechanisms may account for the effectiveness of inhibitors. Some react with and virtually eliminate a chemically active species in the solution (such as dissolved oxygen). Other inhibitor molecules attach themselves to the corroding surface and interfere with either the oxidation or the reduction reaction or form a very thin protective coating. Inhibitors are normally used in closed systems such as automobile radiators and steam boilers. In addition, the design should allow for complete drainage in the case of a shutdown, and easy washing. Because dissolved oxygen may enhance the corrosivity of many solutions, the design should, if possible, include provision for the exclusion of air. Physical barriers to corrosion are applied on surfaces in the form of films and coatings. A large diversity of metallic and nonmetallic coating materials is available. It is essential that the coating maintain a high degree of surface adhesion, which undoubtedly requires some preapplication surface treatment. In most cases, the coating must be virtually nonreactive in the corrosive environment and resistant to mechanical damage that exposes the bare metal to the corrosive environment. All three material types—metals, ceramics, and polymers—are used as coatings for metals.

Cathodic protection

One of the most effective means of corrosion prevention iscathodic protection; it can be used for all eight different forms of corrosion discussed earlier and in some situations may completely stop corrosion. Again, oxidation or corrosion of a metal M occurs by the generalized reaction:

$$M \longrightarrow M^{n+} + ne^-$$

Cathodic protection simply involves supplying, from an external source, electrons to the metal to be protected, making it a cathode; the preceding reaction is thus forced in the reverse (or reduction) direction. One cathodic protection technique employs a galvanic couple: the metal to be protected is electrically connected to another metal that is more reactive in the particular environment. The latter experiences oxidation and, upon giving up electrons, protects the first metal from corrosion. The oxidized metal is often called asacrificial anode, and magnesium and zinc are commonly used because they lie at the anodic end of the galvanic series. The process of galvanizingis simply one in which a layer of zinc is applied to the surface of steel by hot dipping. In the atmosphere and most aqueous environments, zinc is anodic to and will thus cathodically protect the steel if there is any surface damage. Any corrosion of the zinc coating will proceed at an extremely slow rate because the ratio of the anode-to-cathode surface area is quite large. For another method of cathodic protection, the source of electrons is an impressed current from an external DC power source. The negative terminal of the power source is connected to the structure to be protected. The other terminal is joined to an inert anode (often graphite), which, in this case, is buried in the soil; high-conductivity backfill material provides good electrical contact between the anode and the surrounding soil. A current path exists between the cathode and the anode through the intervening soil, completing the electrical circuit. Cathodic protection is especially useful in preventing corrosion of water heaters, underground tanks and pipes, and marine equipment.

(**Selected from:** Callister W D, Rethwisch D G. Materials Science and Engineering - An Introduction: vol 1[M]. 10th ed. Wiley, 2018.)

New Words and Expressions

malleability   *n*. 可锻性，展延性
corrosion   *n*. 腐蚀，侵蚀，腐蚀产生的物质
exhaust   *v*. 耗尽，排放；*n*. 废气，排气装置
inhibitor   *n*. 抑制剂，抗化剂，抑制者
drainage   *n*. 排水系统，排水，污水
adhesion   *n*. 黏附，黏附（力），固定
cathodic   *adj*. 阴极的
anodic   *adj*. 阳极的
galvanic   *adj*. 电流的，触电似的
electrode   *n*. 电极
intervening   *adj*. 介于中间的

Notes

(1) Minor additions of impurities to metals, either deliberate or unintentional, can have a marked effect on the change from ductile to brittle behavior.

一参考译文：有意或无意地加入少量杂质都会对金属从韧性到脆性的转变产生显著影响。

(2) The problem of metallic corrosion is significant; in economic terms, it has been estimated that approximately 5% of an industrialized nation's income is spent on corrosion prevention and the maintenance or replacement of products lost or contaminated as a result of corrosion reactions.

一参考译文：金属腐蚀是一个重要的问题。从经济角度来看，据估计一个工业化国家把大约5%的收入用于防腐蚀、维护或更换由于腐蚀反应而损失或污染的产品上。

(3) Cathodic protection simply involves supplying, from an external source, electrons to the metal to be protected, making it a cathode; the preceding reaction is thus forced in the reverse (or reduction) direction.

一参考译文：阴极保护指利用外接电源向被保护的金属提供电子，使其成为阴极，因此迫使反应向相反（或还原）方向进行。

(4) The other terminal is joined to an inert anode (often graphite), which, in this case, is

buried in the soil; high-conductivity backfill material provides good electrical contact between the anode and the surrounding soil.

——参考译文：另一端连接到惰性阳极（通常是石墨）上，在这种情况下，惰性阳极被埋在土壤中，高导电性的回填材料在阳极和周围土壤之间提供良好的电接触。

Exercises

1. Questions for discussion

(1) What are the differences between malleability and ductility of metallic materials?

(2) One-half of an electrochemical cell consists of a pure nickel electrode in a solution of Ni^{2+} ions; the other half is a cadmium electrode immersed in a Cd^{2+} solution. If the cell is a standard one, write the spontaneous overall reaction and calculate the voltage that is generated.

(3) Briefly describe mechanism of cathodic protection.

2. Translate the following into Chinese

(1) The heating of a cold-worked metal to or above the temperature at which metal atoms return to their equilibrium positions will increase the ductility of that metal.

(2) Where ductility is the ability of a material to deform easily upon the application of a tensile force, malleability is the ability of a metal to exhibit large deformation or plastic response when being subjected to compressive force.

(3) Any corrosion of the zinc coating will proceed at an extremely slow rate because the ratio of the anode-to-cathode surface area is quite large.

(4) It is essential that the coating maintain a high degree of surface adhesion, which undoubtedly requires some preapplication surface treatment.

3. Translate the following into English

(1) 电动势 (2) 电偶
(3) 电路 (4) 保护涂层
(5) 牺牲阳极 (6) 压缩力
(7) 破坏性的 (8) 干涉
(9) 脆性断裂 (10) 抑制
(11) 散热器 (12) 蒸汽锅炉

(13) 通常，如果两种材料具有相同的强度和硬度，其中具有较高延展性的材料更为理想。

(14) 温度的变化可能同时改变塑性流动模式和断裂模式。

(15) 抑制剂是以较低浓度添加到环境中即可降低其腐蚀性的物质。

4. Reading comprehension

(1) According to the text, the following statements about ductility are true EXCEPT ____.

(A) An increase in temperature will increase ductility.

(B) An increase in temperature will cause a change from ductile to brittle behavior.

(C) Ductility is desirable in the high temperature and high pressure applications.

(D) High ductility helps prevent brittle fracture.

(2) For an electrochemical cell consists of a lead electrode in a solution of Pb^{2+} ions; to generate the largest overall cell voltage, the other half should be a ____ electrode immersed in a ____ solution.

(A) Ni; Ni^{2+}

(B) Cd; Cd^{2+}

(C) Zn; Zn^{2+}

(D) Al; Al^{3+}

(3) According to the text, the following statements related to inhibitors are true EXCEPT ____.

(A) Inhibitors react with and virtually eliminate a chemically active species in the solution.

(B) The design of inhibitors should allow for complete drainage in the case of a shutdown, and easy washing.

(C) Inhibitors are rarely used in closed systems such as automobile radiators and steam boilers.

(D) Inhibitor molecules attach themselves to the corroding surface and interfere to form a very thin protective coating.

Reading Material

Revisiting the effect of molybdenum on pitting resistance of stainless steels

这篇课文主要介绍金属 Mo 在防止不锈钢局部腐蚀方面的作用。除了传统的 Cr 等元素，研究表明 Mo 对阻止不锈钢局部腐蚀有重要的作用。这一现象主要基于两种可能原因：钝化层的形成和对溶解动力学的影响。课文还展望了该领域的未来研究方向。

Stainless steels, which refer to steels containing a certain amount of chromium, are widely used in all walks of life because of their excellent corrosion resistance. Generally, the corrosion resistance of stainless steels mainly comes from the effect of chromium, which helps to form a continuous and dense oxide film. The higher the chromium content in stainless steel, the better the corrosion resistance. Besides Cr, adding other appropriate alloyed elements is a very important method to improve the localized corrosion resistance of stainless steels. Alloyed elements can be beneficial to corrosion resistance only if there is a certain chromium content in stainless steels. Well-known helpful alloyed element includes Ni, Mo, N, and Cu. Chromium is the basis of the corrosion resistance of stainless steels; however, some researches have shown that in some extreme environments such as acid environment, chromium may lose its protective effect on stainless steels. In the later stage of localized corrosion development, chromium may even accelerate the corrosion rate, but at this time, the presence of molybdenum can make stainless steels still have localized corrosion resistance. When the content of chromium in stainless steel is sufficient, molybdenum has an obvious effect on improving the corrosion resistance of stainless steel. And the higher the chromium content in stainless steel, the more obvious the effect of molybdenum.

For typical localized corrosion of stainless steels, such as pitting corrosion, crevice corrosion and stress corrosion cracking, the presence of molybdenum has a significant inhibition effect. There are many types of elements in stainless steel, and the interaction between the elements is complex; localized corrosion of stainless steel involves the rupture of the passive film, corrosion growth kinetics and other processes, so the mechanism itself is complicated; environmental factors such as pH and halogen ion concentration have a various influence on localized corrosion. In such a situation, although many experiments have been conducted to investigate the role of molybdenum in the resistance of stainless steel to localized corrosion, there is still no unanimous opinion among researchers so far on the mechanism of molybdenum's effect on stainless steel. This paper attempts to state some important experimental facts obtained so far and the possible mechanism of the effect of molybdenum based on these experiments, and then carry out some necessary discussion.

Two schools of thought

In general, existing investigations tend to focus on two aspects: passive film and dissolution kinetics. Specifically, some investigators believe that the main role of molybdenum is to change the composition of the passive film, thus enhancing the stability of the passive film and improving the localized corrosion resistance; the other is that molybdenum affects the dissolution kinetics of localized corrosion, which reduces the active dissolution rate and makes continuous propagation of localized corrosion difficult to occur.

Mo enhancing the passive film

The resistance of stainless steel to localized corrosion mainly comes from the uniform and dense passive film on its surface (mainly composed of chromium oxide), and the breakdown of the passive film is a prerequisite for localized corrosion to occur. Therefore, when considering the role of molybdenum in stainless steel, many researchers naturally propose that molybdenum may enhance the property of passive film. Some experiments proved that molybdenum did exist in the passive film and changed the structure of the passive film, and thus the localized corrosion resistance of the passive film was enhanced. Furthermore, some researchers found that the thickness of the passive film increased when molybdenum was present in the passive film, or molybdenum could reduce the active sites on the surface of the passive film. Some other researchers have found that the presence of molybdenum made it difficult for chloride ions to pass through the passive film, thus preventing the damage caused by aggressive chloride ions attacking the local passive film.

Scientists measured the polarization curve of Fe19Cr and Fe24Cr2Mo alloy at $0.5 \text{ mol} \cdot \text{L}^{-1}$ H_2SO_4 at 25℃ and the results showed that the molybdenum-containing specimens had higher corrosion potential, smaller maximum current and lower passivation potential. Through analyzing the passive films of the specimens polarized to different stages by electron spectroscopy for chemical analysis (ESCA) technique (**Figure 2.3.2**), the following facts were found: the inner layer of passive film of Fe-Cr-Mo alloy was Cr_2O_3, and the outer layer was hydroxides of Fe, Cr and Mo. During anodic dissolution, Mo and Cr were enriched on the surface, thus increasing the corrosion resistance of the passivated film (manifesting as a smaller dissolution current density). Scientists studied the effect of molybdenum on the pitting resistance of stainless steel under alkaline conditions, and detected the presence of molybdenum in the passive films of molybdenum-containing specimens.

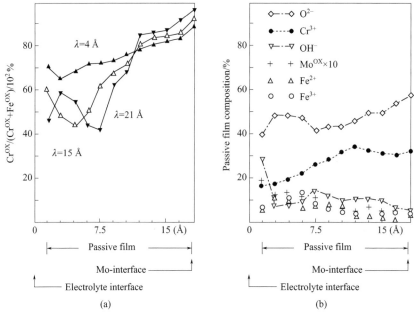

Figure 2.3.2 (a) Calculated $Cr^{3+}/(Cr^{3+}+Fe^{2+}+Fe^{3+})$ ratios.
(b) Composition of the passive film in depth.

Mo hindering the dissolution kinetics

The dissolution process after the rupture of the passive film, i.e., after the localized corrosion nucleation, is also the focus of many experimental studies. With regard to the effect of molybdenum on dissolution kinetics, it can simply be argued that adding molybdenum to stainless steels is similar to the adding of a molybdate inhibitor to the solution. At the beginning of the localized corrosion initiation, molybdenum dissolves into the solution as molybdate ions, which can form insoluble molybdate with cations such as iron ions, thus reducing the concentration of metal cations in the corrosive system and also reducing the concentration of aggressive chloride ions. As a result, the corrosion current density is reduced and the expansion of localized corrosion is hindered, making the nucleation site of the localized corrosion easier to repassivate (more difficult to propagate). With the propagation of localized corrosion, the concentration of hydrogen ion increases gradually. At a certain time, molybdate ion may react with hydrogen ion to form MoO_2, which consumes hydrogen ion and reduces the system's aggressiveness, preventing the further expansion of localized corrosion.

In the solution containing bromide ions, some researchers found the role of molybdate inhibitor not as effective. Corresponding to this, it was also found that in the similar solution, the role of the alloyed molybdenum in stainless steel was also not as effective, which confirmed from the other side that the alloyed molybdenum and the molybdate inhibitor may have the same mechanism to resist localized corrosion. When the molybdenum content is relatively low, stainless steels in chloride solution have smaller pitting potential (E_{pit}) and larger dissolution current than those in bromide solution; however, when the molybdenum content is relatively high, the results become the opposite. The experiment provided a very direct evidence for the effect of molybdenum on dissolution kinetics. In the same condition (same temperature, potential, solution pH, chloride ion concentration, etc.), compared with stainless steel without molybdenum (SS304GF), the transient current density of metastable pitting of stainless steel with molybdenum (SS316GF) was small, indicating that the metastable pitting grew slowly and the size of metastable pits were small. Thus the stable pitting was more difficult to form. Some latest research showed that the resistance to localized corrosion of stainless steels containing Mo and N is improved due to the chemical interaction between nitrogen and molybdenum which hindered active dissolution.

Outlook

Under the conditions that the breakdown of the passive film is the critical step, if there are experiments proving that the presence or content of molybdenum in stainless steel does not change the results such as pitting nucleation frequency and passivation current density, these experiments may prove that Mo cannot enhance the performance of passive film or the enhancement is not significant. But there is no such evidence at present. Under the conditions that the pitting growth kinetics is the critical step, if there are experiments proving that the presence or content of molybdenum in stainless steel does not change the results such as E_{pit} and active dissolution current density, then it may be proved that the effect of Mo on activation dissolution kinetics can be ignored. But there is no such evidence at present, either. It should be noted that the design of these experiments must be based on the premise of figuring out under what circumstances the breakdown of passive film or pitting growth kinetics becomes the critical step, which is obviously not easy.

To further reveal the detailed effect of molybdenum, the following aspects may be worthy of attention in future research. First, since all the commercial stainless steels have rather complex microstructure, they are not proper specimens for us to explore the fundamental and microscopic mechanisms. Adopting a series of carefully designed Fe-Cr-Ni-Mo alloys may be helpful. Second, pitting is a stochastic and abrupt process and it is not easy to follow a single pit in a bulk specimen. Thus, usage of one dimensional artificial pit and controlling the pit geometry may better reveal the role of Mo in the dissolution process. In addition, the adoption of advanced *in situ* electrochemical techniques and characterization methods will be a strong tool to monitor the real-time changes of passive film and the local pitting environment.

(**Selected from:** Sun Y, Tan X, Lei LL, et al. Revisiting the Effect of Molybdenum on Pitting Resistance of Stainless Steels [J]. Tungsten, 2021, 3 (3): 329-337.)

New Words and Expressions

molybdenum　*n*. 钼（金属元素，符号 Mo，原子序号 42）
localized　*adj*. 局部的，小范围的
pitting corrosion　点状腐蚀，斑蚀
crevice corrosion　缝隙腐蚀，裂隙腐蚀，接触腐蚀
prerequisite　*n*. 先决条件，前提；*adj*. 先决的，必备的
rupture　*n*. 破裂，决裂；*vi*. 破裂；*vt*. 使破裂
dissolution　*n*. 分解，溶解
premise　*n*. 前提，假设；*v*. 以……为基础，
microscopic　*adj*. 微小的，微观的
artificial　*adj*. 人造的，人工的
metastable　*n*. 亚稳的，介稳的，亚稳（态）的；*adj*.（化）亚稳的

Notes

（1）The higher the chromium content in stainless steel, the better the corrosion resistance.

—这是一个复合句（表示一方的程度随着另一方的变化而变化）。其中前面的句子是状语从句，后面的句子是主句。

—参考译文：不锈钢中铬含量越高，耐腐蚀性越好。

（2）In such a situation, although many experiments have been conducted to investigate the role of molybdenum in the resistance of stainless steel to localized corrosion, there is still no unanimous opinion among researchers so far on the mechanism of molybdenum's effect on stainless steel.

—参考译文：在这种情况下，虽然有很多实验研究了钼在不锈钢抗局部腐蚀中的作用，但到目前为止研究人员还没有就钼对不锈钢的作用机制达成一致意见。

（3）Specifically, some investigators believe that the main role of molybdenum is to change the composition of the passive film, thus enhancing the stability of the passive film and improving the localized corrosion resistance; the other is that molybdenum affects the dissolution kinetics of localized corrosion, which reduces the active dissolution rate and makes continuous propagation of localized corrosion difficult to occur.

—参考译文：具体而言，有研究者认为钼的主要作用是改变钝化膜的成分，从而提高钝化膜的稳定性及局部耐蚀性；另一方面钼影响了局部腐蚀的溶解动力学，降低了活性溶解速率，限制了局部腐蚀的持续蔓延。

（4）Through analyzing the passive films of the specimens polarized to different stages by electron spectroscopy for chemical analysis (ESCA) technique, the following facts were found: the inner layer of passive film of Fe-Cr-Mo alloy was Cr_2O_3, and the outer layer was hydroxides of Fe, Cr and Mo.

—参考译文：通过电子能谱化学分析（ESCA）技术对不同极化阶段样品的钝化膜进行

分析，发现 Fe-Cr-Mo 合金的钝化膜内层为 Cr_2O_3，外层为 Fe、Cr、Mo 的氢氧化物。

(5) Corresponding to this, it was also found that in the similar solution, the role of the alloyed molybdenum in stainless steel was also not as effective, which confirmed from the other side that the alloyed molybdenum and the molybdate inhibitor may have the same mechanism to resist localized corrosion.

—参考译文：与此相对应，人们还发现在相似的溶液中，不锈钢中的合金钼没有起到有效作用，这从另一方面证实了合金钼和钼酸盐抑制剂可能具有相同的抗局部腐蚀机制。

Unit 2.4

Text

Superalloys and nonferrous alloys

钢和其他一些铁基合金的应用非常广泛，但是也有密度大、导电率低、在一些环境中容易腐蚀等缺点，因此非铁基的合金也值得详细研究。这篇课文主要介绍非铁基合金和超合金的主要特点，列举了一些典型超合金的成分，并说明了部分非铁合金的应用场景。

Nonferrous alloys

Steel and other ferrous alloys are consumed in exceedingly large quantities because they have such a wide range of mechanical properties, may be fabricated with relative ease, and are economical to produce. However, they have some distinct limitations, chiefly (1) a relatively high density, (2) a comparatively low electrical conductivity, and (3) an inherent susceptibility to corrosion in some common environments. Thus, for many applications, it is advantageous or even necessary to use other alloys that have more suitable property combinations. Alloy systems are classified either according to the base metal or according to some specific characteristic that a group of alloys share. On occasion, a distinction is made between cast and wrought alloys. Alloys that are so brittle that forming or shaping by appreciable deformation is not possible typically are cast; these are classified ascast alloys. However, those that are amenable to mechanical deformation are termed wrought alloys. In addition, the heat-treatability of an alloy system is mentioned frequently. "Heat-treatable" designates an alloy whose mechanical strength is improved by precipitation hardening or a martensitic transformation (normally the former), both of which involve specific heat-treating procedures.

Superalloys

The superalloys have superlative combinations of properties. Most are used in aircraft turbine components, which must withstand exposure to severely oxidizing environments and high temperatures for reasonable time periods. Mechanical integrity under these conditions is critical; in this regard, density is an important consideration because centrifugal stresses are diminished in rotating members when the density is reduced. These materials are classified according to the predominant metal(s) in the alloy, of which there are three groups: iron-nickel, nickel, and cobalt. Other alloying elements include the refractory metals (Nb, Mo, W, Ta), chromium, and titanium. Furthermore, these alloys are also categorized as wrought or cast. Compositions of several of them are presented in **Table 2.4.1**. In addition to turbine applications, superalloys are used in nuclear reactors and petrochemical equipment.

Table 2.4.1 Compositions for Several Superalloys.

Alloy name	Composition/wt%									
	Ni	Fe	Co	Cr	Mo	W	Ti	Al	C	other
Iron-Nickel (wrought)										
A-286	26	55.2	—	15	1.25	—	2.0	0.2	0.04	0.005 B, 0.3 V
Incoloy 925	44	29	—	20.5	2.8	—	2.1	0.2	0.01	1.8 Cu
Nickel (wrought)										
Incoloy 925	52.5	18.5	—	19	3.0	—	0.9	0.5	0.05	5.1 Nb, 0.15 max Cu
Waspaloy	57	2.0 max	13.5	19.5	4.3	—	3.0	1.4	0.07	0.006 B, 0.09 Zr
Nickel (cast)										
Rene 80	60	—	9.5	14	4	4	5	3	0.17	0.015 B, 0.03 Zr
Mar-M-247	59	0.5	10	8.25	0.7	10	1	5.5	0.15	0.015 B, 3 Ta, 0.05 Zr, 1.5 Hf
Cobalt (wrought)										
Haynes 25 (L-605)	10	1	54	20	—	15	—	—	0.1	
Cobalt (cast)										
X-40	10	1.5	57.5	22	—	7.5	—	—	0.50	0.5 Mn, 0.5 Si

Miscellaneous nonferrous alloys

The preceding discussion covers the vast majority of nonferrous alloys; however, a number of others are found in a variety of engineering applications, and a brief mention of these is worthwhile. Nickel and its alloys are highly resistant to corrosion in many environments, especially those that are basic (alkaline). Nickel is often coated or plated on some metals that are susceptible to corrosion as a protective measure. Monel, a nickel-based alloy containing approximately 65 wt% Ni and 28 wt% Cu (the balance is iron), has very high strength and is extremely corrosion resistant; it is used in pumps, valves, and other components that are in contact with acid and petroleum solutions. As already mentioned, nickel is one of the principal alloying elements in stainless steels and one of the major constituents in the superalloys. Lead, tin, and their alloys find some use as engineering materials. Both lead and tin are mechanically soft and weak, have low melting temperatures, are quite resistant to many corrosion environments, and have recrystallization temperatures below room temperature. Some common solders are lead-tin alloys, which have low melting temperatures. Applications for lead and its alloys include X-ray shields and storage batteries. The primary use of tin is as a very thin coating on the inside of plain carbon steel cans (tin cans) that are used for food containers; this coating inhibits chemical reactions between the steel and the food products.

Unalloyed zinc also is a relatively soft metal having a low melting temperature and a subambient recrystallization temperature. Chemically, it is reactive in a number of common environments and, therefore, susceptible to corrosion. Galvanized steel is just plain carbon steel that has been coated with a thin zinc layer; the zinc preferentially corrodes and protects the steel. Typical applications of galvanized steel are familiar (sheet metal, fences, screen, screws, etc.). Common applications of zinc alloys include padlocks, plumbing fixtures, automotive parts (door handles and grilles), and office equipment. Although zirconium is relatively abundant in the Earth's crust, not until quite recent times were commercial refining techniques developed. Zirconium and its alloys are ductile and have other mechanical characteristics that are comparable to those of titanium alloys and the austenitic stainless steels. However, the primary asset of these alloys is their resistance to corrosion in a host of corrosive media, including superheated water. Furthermore, zirconium is transparent to thermal neutrons, so that its alloys have been used as cladding for uranium fuel in water-cooled nuclear reactors. In terms of cost, these alloys are also often the materials of choice for heat exchangers, reactor vessels, and piping systems for the chemical-processing and nuclear industries. They are also used in incendiary ordnance and in sealing devices for vacuum tubes.

(**Selected from:** Callister W D, Rethwisch D G. Materials Science and Engineering - An Introduction: vol 1[M]. 10th ed. Wiley, 2018.)

New Words and Expressions

nonferrous *adj.* 非铁的，不含铁的
susceptibility *n.* 敏感性
miscellaneous *adj.* 混杂的，各种各样的，多方面的
cast alloys 铸造合金
wrought alloys 锻制合金
martensitic *adj.* 马氏体的
integrity *n.* 完整性，完全
centrifugal *adj.* 离心的
petrochemical *n.* 石油化学产品；*adj.* 石油化工的
zirconium *n.* 锆（金属元素，符号 Zr，原子序号 40）
neutron *n.* 中子

Notes

(1) "Heat-treatable" designates an alloy whose mechanical strength is improved by precipitation hardening or a martensitic transformation (normally the former), both of which involve specific heat-treating procedures.

——martensitic 马氏体的。马氏体（martensite）是黑色金属材料的一种组织名称，是碳在 α-Fe 中的过饱和固溶体。

——参考译文："可热处理"指的是合金可以通过沉淀硬化或马氏体转变（通常是前者）来提高机械强度，这两种方法都涉及特定的热处理步骤。

(2) Most are used in aircraft turbine components, which must withstand exposure to severely oxidizing environments and high temperatures for reasonable time periods.

——参考译文：大多数（超合金）被用于飞机涡轮部件，这些部件必须能承受在严重氧化和高温环境下暴露一定时间。

(3) These materials are classified according to the predominant metal (s) in the alloy, of which there are three groups: iron-nickel, nickel, and cobalt.

——参考译文：这些材料根据合金中的主要金属成分可分为三类：铁-镍、镍和钴。

(4) Nickel is often coated or plated on some metals that are susceptible to corrosion as a protective measure. Monel, a nickel-based alloy containing approximately 65 wt% Ni and 28 wt% Cu (the balance is iron), has very high strength and is extremely corrosion resistant; it is used in pumps, valves, and other components that are in contact with acid and petroleum solutions.

——based... 基于……的。

——参考译文：镍通常被作为一种保护剂涂或镀在一些易受腐蚀的金属上。蒙乃尔（Monel）是一种镍基合金，含有大约 65% 的镍和 28% 的铜（其余为铁），具有很高的强度和极强的耐腐蚀性，可用于泵、阀门和其他与酸和石油溶液接触的部件。

(5) Chemically, it is reactive in a number of common environments and, therefore, susceptible to corrosion.

——a number of 许多。

——参考译文：在化学层面，它（锌）在许多常见环境下都活泼，因此易被腐蚀。

Exercises

1. Questions for discussion

(1) Briefly describe heat-treatable alloys.

(2) What is the principal difference between wrought and cast alloys?

(3) What are the main limitations of ferrous alloys?

2. Translate the following into Chinese

(1) Furthermore, zirconium is transparent to thermal neutrons, so that its alloys have

been used as cladding for uranium fuel in water-cooled nuclear reactors.

(2) Zirconium and its alloys are ductile and have other mechanical characteristics that are comparable to those of titanium alloys and the austenitic stainless steels.

(3) Alloy systems are classified either according to the base metal or according to some specific characteristic that a group of alloys share.

3. Translate the following into English

(1) 有利的   (2) 高温合金

(3) 栅栏   (4) 过热的

(5) 军械   (6) 涡轮机

(7) 奥氏体的   (8) 重结晶

(9) 如上所述,镍是不锈钢的主要合金元素之一,也是高温合金的主要成分之一。

(10) 然而,这些合金的主要优点是它们能抵抗包括过热水在内的许多腐蚀性介质的腐蚀。

(11) 非合金锌也是一种相对较软的金属,具有较低的熔点和低温再结晶温度。

Reading Material

High-entropy alloys

这篇课文主要介绍一种前沿新材料——高熵合金。高熵合金通常包含五种或以上的成分,且每种成分含量都较高。早期研究大多关注多相合金,但近年来单相合金也引起了极大的关注。课文介绍了高熵合金相关研究进展,还展望了未来的研究方向。

Scientists hypothesized that the presence of multiple (five or more) elements in near-equiatomic proportions would increase the configurational entropy of mixing by an amount sufficient to overcome the enthalpies of compound formation, thereby deterring the formation of potentially harmful intermetallics. This was a counterintuitive notion because the conventional view—likely based on binary phase diagrams in which solid solutions are typically found at the ends and compounds near the centers — was that the greater the number of elements in concentrated alloys, the higher the probability that some of the elements would react to form compounds. But Yeh and colleagues reasoned that, as the number of elements in an alloy increased, the entropic contribution to the total free energy would overcome the enthalpic contribution and, thereby, stabilize solid solutions. They coined a catchy new name, high-entropy alloys (HEAs), for this class of materials containing five or more elements in relatively high concentrations (5～35 at%).

Interestingly, most of the HEAs investigated in the early years were multi-phase alloys rather than single-phase, solid solutions. In recent years though, increased attention has been paid to a few model, single-phase alloys to probe their behavior in greater detail and uncover fundamental mechanisms without the confounding effects of secondary phases. Investigation of single-phase alloys allows variables such as the number, types and concentrations of alloying elements to be systematically varied and directly correlated with physical and mechanical properties, such as elastic constants, stacking-fault energies, diffusion coefficients, strength and ductility. Such studies, although important in their own right, have broader applicability to the understanding of certain multi-phase alloys, such as the advanced γ-γ' Ni-based superalloys used in jet engines and power generation, in which the matrix phase (γ) is a complex, solid solution consisting of multiple elements in relatively high concentrations, with none of them being the majority element. At least some of the rate-controlling mechanisms that govern the overall properties of the γ-γ' composite (such as dislocation climb) occur within the γ phase or at the γ-γ' phase boundaries. Thus, a fundamental understanding of the energetics and kinetics of point defects, dislocations and diffusion in complex solid solutions is relevant also to understanding the behavior of multi-phase alloys. Finally, complex solid solutions are important from a basic scientific viewpoint because of the conceptual advances needed to bridge the gap between the relatively well understood dilute solid solutions and the poorly understood concentrated solid solutions.

The appeal of the entropy hypothesis lies in its straightforward, testable prediction: alloys comprising five or more elements would form single-phase solid solutions. However, when put to the test, it turned out that, in several nominally similar, five-element alloys, all of them except for one contained multiple intermetallic phases, showing that configurational entropy is not a useful predictor of which alloys are simple solid solutions. In addition to configurational entropy, one needs to take into account other entropic contributions, as well as mixing enthalpies. A further complication was that Yeh and co-workers' analysis was performed at the melting point, whereas most alloy microstructures tend to be examined at room temperature. Depending on diffusion rates, phase-transformation kinetics and cooling rates, room-temperature and high-temperature microstructures may be quite different. Indeed, HEAs often contain multiple phases when examined at room temperature, and many of the second phases are intermetallics that can be brittle and hinder mechanical performance. If the goal is to avoid the formation of such phases, simply mixing together five or more elements in near-equiatomic concentrations is unlikely to be a useful approach. Even multi-component alloys that are initially single phase after solidification tend to separate into multiple metallic and intermetallic phases when annealed at intermediate temperatures. This phase instability has only recently been recognized as a common feature of many (perhaps most) HEAs and needs to be accounted for when considering alloys for high-temperature applications.

Because phase relationships are governed by the total free energy, the original high-entropy criterion could conceivably be supplemented with a "low-enthalpy" criterion for solid-solution stability. In fact, there are many alloys containing fewer than five elements (medium entropy alloys) that are single-phase solid solutions; these alloys may well be enthalpy-stabilized rather than entropy-stabilized. This relaxes the original high entropy criterion that alloys should contain five or more elements in near-equiatomic ratios. It also provides a wider composition space for investigation that includes alloys with fewer elements, as well alloys deviating from equiatomic compositions. However, determining the enthalpy is not as easy as estimating the configurational entropy of an ideal solid solution. Therefore, although one could christen these alloys "low-enthalpy alloys", it is less useful as a classification scheme. HEAs have taken the field of materials science, or, more precisely, metallurgy, by storm; indeed, they are likely to be central to the field of structural, and perhaps functional, materials for another decade or longer. As we discussed, the high-entropy phenomenon has been principally driven by three main factors: the exceptional mechanical properties of a few specific alloys, most notably the FCC CrCoNi-based alloys, especially for cryogenic temperature applications; the search for new refractory HEAs that can operate with sustained strength at ever-increasing temperatures, a difficult pursuit because of the limited ductility and oxidation resistance of these materials; and the prospect of discovering new materials with unprecedented properties, or combinations of properties, from the seemingly unbounded list of possible multiple-principal-element materials, although this is a non-trivial exercise in light of the methods currently available to venture into uncharted materials, composition and microstructural space. At present, the field has been dominated by the first factor, as the Cantor alloy and its derivatives have displayed truly compelling mechanical properties, coupled with now several successful attempts at microstructural optimization and new mechanistic interpretations to underpin their impressive mechanical performance. However, we suspect that the latter two factors will ultimately come to the fore as the quest for higher-temperature structural materials remains a focus in materials science, which is driven by advances in aerospace and now hypersonic activities; the possible discovery of new, hitherto unknown supermaterials will forever be a worthy cause.

However, in the short term, several specific issues remain to be resolved pertaining to what may make HEAs distinct from conventional alloys. One is the role of local chemical order on the macroscopic properties of HEAs. Density functional theory (DFT) -based Monte Carlo simulations on equimolar CrCoMo with several hundred atoms and molecular dynamics simulations with a few million atoms using a newly developed classical potential for CrCoMo both predict that the degree of local chemical order in CrCoNi has a marked influence on its stacking-fault energy, twinning energy, energy difference between the FCC and HCP phases and formation energy for point defects, all of which are parameters that are known to markedly affect the strength and deformation of CrCoNi-based HEAs. However, the role of local chemical order in HEAs remains an open issue because it has yet to be convincingly confirmed experimentally. Only one study, using extended X-ray adsorption fine structure on CrCoNi, has suggested that Cr atoms display a preference to bonding to Ni and Co atoms, rather than other Cr atoms, which is consistent with the DFT-based Monte Carlo predictions. However, the fascinating notion of tuning atomic order to achieve enhanced macroscale mechanical properties awaits more convincing experimental confirmation that such local chemical order actually exists in apparently random HEAs solid solutions.

Further opportunities lie in the exploration of thermodynamics, microstructure evolution and properties of HEAs with near-zero stacking-fault energies. Metastable HEAs with small stacking-fault energy and deformation-driven twinning and martensite formation have been studied; yet, even more complex substructures have been recently reported for dual-phase, FCC-HCP HEAs with near-zero or slightly negative stacking-fault energy. In these materials, the matrix can assume both the HCP and the FCC structure, owing to the energetic equivalence of the co-existing phases. The deformation-driven nanoscale refinement observed in such alloys seems to be further promoted by the very small coherent interface energies among the different phases and microstructure ingredients, that is, the FCC phase, the HCP phase, twins and stacking faults. The similarity in phase energy can also lead to a bidirectional TRIP effect, in which FCC-structured matrix portions transform under load into HCP regions and vice versa, depending on the local, micromechanical stresses. This effect leads to an extreme microstructure refinement down to the nanometer regime. Interesting effects in HEAs may be expected owing to the multi-element co-decoration of lattice defects. Whereas traditional segregation phenomena such as Cottrell atmospheres, grain-boundary segregation and Suzuki decoration are well known, they may work differently when involving many interacting species that co-segregate. The thermodynamic and kinetic competition among the decorating species, as well as their interactions once segregated, may lead to novel structures and phenomena.

Another avenue of interesting research pertains to the role of interstitial elements in HEAs. Several groups studied the influence of C on phase stability, carbide formation and mechanical properties, but other elements such as N, O, H and B are also interesting doping candidates in HEAs. For example, it was recently shown that O, which is normally a rather harmful interstitial element in metals, aids in forming ordered O-Ti-Zr-rich clusters in an O-doped TiZrHfNb HEA, enhancing its strength and ductility. Even the most harmful interstitial element in metallic alloys, H, has been found to lend a certain resistance to hydrogen embrittlement in HEAs. This effect was attributed to the fact that H reduces the material's stacking-fault energy, thus providing higher local strain hardening. To aid future research efforts, several theoretical and experimental tools need to be improved.

In general, we believe that more attention should be placed on potential innovative applications for HEAs. Most efforts in the field are focused on mechanical properties but recent works also revealed unexpected features of HEAs that might be interesting for magnetic, invar or catalytic applications. Particularly, HEAs with multi-functional properties might lead to new processes or products. Thus, it will be worth exploring where specific properties may be expected that are better than those of established material classes, such as steels or aluminum alloys.

In this regard, it is pertinent to note that nature effectively designs structural architectures to create multi-functional materials with unusual combinations of properties that are often difficult to achieve in a single, synthetic material; for example, seashells possess both strength and ductility, which are often mutually exclusive properties in man-made materials. HEAs offer the possibility of multi-functionality but without the need for bottom-up (atomic-scale) processing. For example, strain hardening to attain strength and ductility in CrCoNi-based HEAs is promoted not simply through structure but also via composition and the associated mechanism tuning. HEAs with excellent mechanical properties remain to be engineered to achieve other properties, such as corrosion and/or oxidation resistance or resistance to stress-corrosion cracking, for example by alloy additions to change the composition of the protective self-passivating oxides that form on their surface. Understanding and quantifying diffusion in multi-component systems is another area of research that needs to be revived to develop creep-resistant HEAs.

These challenges will undoubtedly provide innumerable promising pursuits for materials scientists and engineers in the never-ending mission to design and develop superior materials to meet the needs of next-generation engineering applications, particularly for energy and transportation.

(**Selected from:** George E P, Raabe D, Ritchie R O. High-entropy alloys[J]. Nature Reviews Materials, 2019, 4 (8): 515-534.)

New Words and Expressions

configurational entropy　构型熵
enthalpy　*n.* 焓
point defects　点缺陷
dislocations　*n.* 位错
microstructure　*n.* 显微结构，微观结构
solidification　*n.* 凝固，团结，浓缩
criterion　*n.* 尺度，标准，准则
metallurgy　*n.* 冶金，冶金学
hypersonic　*adj.* 极超音速的
thermodynamic　*adj.* 热力学的
kinetic　*adj.* 动力学的
seemingly　*adv.* 貌似，看似
in light of　根据，鉴于，从……观点
innumerable　*adj.* 无数的，数不清的
uncharted　*adj.* 未知的
venture into　冒险进入，大胆尝试某事

Notes

(1) This was a counterintuitive notion because the conventional view — likely based on binary phase diagrams in which solid solutions are typically found at the ends and compounds near the centers — was that the greater the number of elements in concentrated alloys, the higher the probability that some of the elements would react to form compounds.

——这里破折号的作用是解释和强调。可以先理解破折号之外的主句，再理解破折号内的解释或强调。

——参考译文：这是一个违反直觉的观点，因为根据传统观点——就像在二元相图中，固溶体通常在（二元相图的）最后，化合物靠近（二元相图的）中心——合金元素的数量越多，（其中）一些元素发生反应形成化合物的概率越高。

(2) In these materials, the matrix can assume both the HCP and the FCC structure, owing to the energetic equivalence of the co-existing phases.

——参考译文：在这些材料中，由于共存相的能量相等，基体可以同时呈现密排六方和

面心立方结构。

(3) HEAs have taken the field of materials science, or, more precisely, metallurgy, by storm; indeed, they are likely to be central to the field of structural, and perhaps functional materials for another decade or longer.

—taken by storm　在这里是席卷……的意思。
—参考译文：高熵合金已经席卷了材料科学领域，或者更准确地说是冶金领域。事实上，在未来十年或更长的时间内它们很可能成为结构材料和功能材料领域的中心。

(4) As we discussed, the high-entropy phenomenon has been principally driven by three main factors: the exceptional mechanical properties of a few specific alloys, most notably the FCC CrCoNi-based alloys, especially for cryogenic temperature applications; the search for new refractory HEAs that can operate with sustained strength at ever-increasing temperatures, a difficult pursuit because of the limited ductility and oxidation resistance of these materials; and the prospect of discovering new materials with unprecedented properties, or combinations of properties, from the seemingly unbounded list of possible multiple-principal-element materials.

—FCC face-centered cubic　面心立方。
—参考译文：正如我们所讨论的，高熵现象（的发展）主要由三个因素驱动：(1) 一些具有卓越力学性能的合金，尤其是可以在低温下应用的面心立方 CrCoNi 基合金；(2) 由于材料的延展性和抗氧化性有限，很难找到在持续升温过程中仍保持强度的新型耐火高熵合金；(3) 从看似无数种多主元素材料中，发现具有前所未有的性能或性能组合的新材料具有远大前景。

(5) However, we suspect that the latter two factors will ultimately come to the fore as the quest for higher-temperature structural materials remains a focus in materials science, which is driven by advances in aerospace and now hypersonic activities; the possible discovery of new, hitherto unknown supermaterials will forever be a worthy cause.

—come to the fore　涌现。
—参考译文：然而，我们猜想后两个因素最终会出现。由于航空航天的进步和现代高超音速活动的推动，追求更高温度的结构材料仍然是材料科学的焦点。发现新的、迄今为止未知的超材料将永远有意义。

Unit 2.5

Text

Metal-matrix composites

这篇课文介绍金属基复合材料的基础知识。课文介绍了金属基复合材料的概念，从多个角度解释了金属基复合材料的特色、制备方法和形变机制。课文还说明了金属基复合材料的两种常见的增强方式：颗粒增强以及纤维增强。

Introduction to metal matrix composite

Metals and alloys are generally produced and shaped in bulk form but can also be intimately combined with another material that serves to improve their performance. The resulting material is a metal matrix composite (MMC). This class of composites encompasses many different materials that can be distinguished according to their base metal (e. g. , aluminum, copper, titanium); according to the other, reinforcement, phase (e. g. , fibers, particles, whiskers); or according to their manufacturing process (e. g. , powder metallurgy, diffusion bonding, infiltration, stir casting). Bulk metals and alloys are economical high-performance materials. Why, then, combine these with another phase, as doing so will always add cost and cause complications (in recycling, notably)? Among the stronger drivers of MMC technology are the following basic facts.

The first is that the composite approach to materials design makes it possible to go beyond boundaries drawn in property space by basic attributes of the main materials classes. A classical example is the specific modulus of metals, defined as elastic modulus E divided by density ρ: (E/ρ). This parameter is a measure of the performance, in weight-critical deformation-limited structural applications, of linear elastic components subjected to uniaxial stress. Now, the main engineering metals and alloys have roughly the same $(E/\rho) \approx 26$ MJ · kg^{-1}. So essentially the only way to exceed this limit in a metallic material is to replace a significant fraction of the metal atoms with a phase made of atoms that are (a) situated in the top rows of Mendeleev's periodic table and (b) strongly bonded to one another. Examples are ceramics such as Al_2O_3, B_4C, or SiC and certain variants of carbon (e. g. , high-modulus carbon fibers or diamond).

The second reason why MMCs are interesting has to do with processing. Making a composite is the only approach by which a significant volume fraction of oxide or carbide can be introduced into some important metals. Iron is very frequently- and easily—reinforced with a wide array of carbides, nitrides, or (more rarely) oxides because carbon, nitrogen, and oxygen are soluble in this molten metal. Liquid aluminum, magnesium, and copper, in contrast, have essentially no solubility for carbon. Therefore, the only way to introduce a carbide into these metals is by making a composite; the same holds for aluminum with oxides or nitrides. In other words, MMC technology holds the key, for aluminum, magnesium, or copper, to much of the wide spectrum of microstructures and properties available with iron.

A third reason for interest in MMCs has little to do with metals. Some phases, ceramics in particular, have far better properties in finely divided form. Notably, micrometer-sized ceramic fibers or ceramic particles can be much stronger than bulk ceramics. Additionally, small, single crystalline ceramic particles can be excellent conductors of heat. Carbon, too, can in finely divided form be very strong, stiff, and a good conductor of heat (as diamond, for example). Taking advantage of this in macroscopic materials therefore calls for the incorporation of such finely divided nonmetallic phases (as fibers, platelets, films, or particles) within a composite material, the matrix of which can advantageously be made of metal. Indeed, metals conduct better, are stronger, tougher, and more environmentally resistant than polymers — and of course are much tougher than ceramics.

Processing and materials

The past ten years have seen a significant enlargement in the palette of reinforcements that are combined with metal. New particles have appeared, starting with fine-scale submicrometer (or nano-) particles. These tend to be intrinsically strong while also bringing Orowan hardening or grain refinement to the metallic matrix. The difficulty with nanosized reinforcements is in processing: Nanoparticles are much harder to incorporate or to distribute uniformly within metal. Capillary forces, often important in MMC processing, are much higher with nanoparticles (because these forces scale with the inverse of the reinforcement size).

Despite the challenges posed in processing, some laboratories and companies have managed to successfully incorporate nanoparticles or nanofibers into metal and to produce engineering-quality (i.e., nonporous) MMCs with attractive properties by pressure infiltration, by stir casting, by electrodeposition, or by powder metallurgy. Another, relatively well-established, pathway to the production of nanoscale MMC structures is by attrition, generally involving ball milling followed by consolidation. Other methods that have been recently developed to this end are more chemical in nature; these are *in situ* methods, exemplified by internal oxidation. Here, nanoscale reinforcements are created by internal reaction between precursor materials that are finely combined within the composite.

Mechanisms of deformation and fracture

From the structural viewpoint, a main limitation to the industrial application of MMCs has been the embrittlement associated with the addition, to the metal, of brittle ceramic reinforcements. The physical phenomena that control this behavior occur at the micrometer scale and above in MMCs. As mentioned above, these phenomena can now be observed and quantified accurately using novel three-dimensional microstructural characterization techniques. In addition, deformation and failure processes within MMCs can now be simulated using several novel tools and methods, including discrete dislocation (DD) dynamics, large-scale finite element (FE) simulation, and nonlinear homogenization theory. Driven by this combination of strong engineering motivation and novel methods, research on the nonlinear deformation and fracture of MMCs has led to significant advances in this general area during the past decade.

Particle-reinforced composites

The dispersion of stiff ceramic particles or fibers within a ductile metallic matrix leads to an increase in flow stress of the metal by load transfer across a strong interface from the matrix to the reinforcement. Constraint imposed by the ceramic reinforcements on matrix plastic deformation induces large tensile hydrostatic stresses in the matrix. This enhances the load carried by the reinforcements and hence the composite flow stress but also triggers the early development of internal damage in the form of particle fracture, interface decohesion, and/or matrix void growth. The challenge to find how the microstructure of these materials can be tailored to improve their ductility and toughness has been addressed through a combination of simulation and experiment.

Fiber-reinforced composites

In continuous parallel fiber-reinforced composites, progressive fiber fracture controls the longitudinal tensile strength. Two extreme situations can be found. Under global load sharing (GLS) conditions, the effect of the local stress concentration around the broken fibers is negligible: Load shed by fractured fibers is redistributed equally among the intact fibers in the remaining cross-sectional area of the composite at that location. The stress-strain curve then exhibits a smooth maximum, which arises from the competition between (a) hardening induced by matrix and fiber deformation and (b) softening caused by fiber fragmentation. Under these assumptions, elegant analytical solutions and numerical models have been developed to predict the composite strength, accounting for matrix/fiber load transfer and fiber fracture statistics. GLS, however, represents an idealization because it ignores damage localization. Evidence of damage localization in MMCs is provided by *in situ* monitoring of damage using X-ray synchrotron radiation in a Ti/SiC composite tested in tension: Fiber fracture is localized near the fracture surface, whereas most of the sample remains free of damage. In addition, GLS predicts failure when the tangent modulus reaches zero and does not account for size effects; both predictions are opposed to available experimental results in most fiber-reinforced MMCs.

(**Selected from:** Mortensen A, Llorca J. Metal Matrix Composites [J]. *Annual Review of Materials Science*, 2010, 40: 243-270.)

New Words and Expressions

encompass　*v.* 包括，包含
reinforcement　*n.* 加强，加固
Mendeleev　门捷列夫（俄国化学家）
consolidation　*n.* 合并，固结
tangent　*n.* 切线，切面，正切；*adj.* 切线的，相切的
infiltration　*n.* 渗透，渗透物
attrition　*n.* 摩擦，磨损

Notes

(1) Despite the challenges posed in processing, some laboratories and companies have managed to successfully incorporate nanoparticles or nanofibers into metal and to produce engineering-quality (i.e., nonporous) MMCs with attractive properties by pressure infiltration, by stir casting, by electrodeposition, or by powder metallurgy.

一参考译文：尽管在加工过程中存在挑战，但部分实验室和公司已经成功地将纳米颗粒或纳米纤维加入金属中，并通过压力渗透、搅拌铸造、电沉积或粉末冶金等方法生产出（符合）工程质量（即无孔）的、性能优异的金属基复合材料。

(2) In addition, deformation and failure processes within MMCs can now be simulated using several novel tools and methods, including discrete dislocation (DD) dynamics, large-scale finite element (FE) simulation, and nonlinear homogenization theory.

——参考译文：此外，金属基复合材料的形变和失效过程现在可以使用一些新工具和新手段来模拟，包括离散位错动力学、大规模有限元模拟和非线性均匀化理论。

(3) Evidence of damage localization in MMCs is provided by *in situ* monitoring of damage using X-ray synchrotron radiation in a Ti/SiC composite tested in tension: Fiber fracture is localized near the fracture surface, whereas most of the sample remains free of damage.

——*in situ* 源于拉丁语，指在原地，原位置，在科研文献中一般指原位。

——参考译文：采用 X 射线同步辐射对拉伸后的 Ti/SiC 复合材料进行原位损伤监测，为金属基复合材料的局部损伤提供了证据：纤维断裂集中在断口附近，而大部分样品仍未发生损伤。

Exercises

1. Questions for discussion

(1) Briefly introduce a method to fabricate of nanoscale MMC structures.

(2) List three combinations that form MMC structures.

(3) Briefly explain the reason why MMCs attract attention in processing.

2. Translate the following into Chinese

(1) The first is that the composite approach to materials design makes it possible to go beyond boundaries drawn in property space by basic attributes of the main materials classes.

(2) The past ten years have seen a significant enlargement in the palette of reinforcements that are combined with metal.

(3) From the structural viewpoint, a main limitation to the industrial application of MMCs has been the embrittlement associated with the addition, to the metal, of brittle ceramic reinforcements.

3. Translate the following into English

(1) 晶须　　　　　(2) 毛细作用力
(3) 纳米尺度　　　(4) 前驱体
(5) 单轴向应力

(6) 过去的十年里，在强大的工程动力与新颖方法相结合的驱动下，对金属基复合材料非线性形变和断裂的研究已经取得了显著的进展。

(7) 在连续的平行纤维增强复合材料中，纤维的断裂影响着复合材料的纵向拉伸强度。

(8) 陶瓷增强体对基体塑性变形的约束引发了基体中较大的拉伸应力。

4. Reading comprehension

(1) According to the text, the following statements about particle-reinforced composites are true EXCEPT ＿＿＿.

(A) Dispersion of stiff ceramic particles can increase in flow stress of the metal.

(B) Ceramic reinforcements induce large compressive hydrostatic stresses in the matrix.

(C) Reinforcements may trigger the early development of internal damage in the form of particle fracture.

(D) Reinforcements can transfer load across a strong interface from the matrix to the reinforcement.

(2) Which of the following statement are TRUE according to the text? ____

(A) Nanoparticles can be incorporated within metal because of their size.

(B) Nanoparticles can be easily distributed uniformly within metal.

(C) Capillary forces increase with increasing the reinforcement size.

(D) Capillary forces are much higher with nanoparticles.

Reading Material
Mechanical behavior of particle reinforced metal matrix composites

这篇课文主要阐明了碳化硅（SiC）颗粒增强铝基复合材料力学性能与结构之间的关系，并基于实验数据建立了相应的模型。

Introduction

Particle reinforced metal matrix composites (MMCs) have the very large potential to provide ultrahigh mechanical behaviors, for example specific stiffness and specific strength, in the civil and defense applications as well as the automotive and aerospace industries. Considering the materials characteristics and producing process, the composite structures of particle reinforced MMCs largely depend on their reinforced particles, such as: the particle sizes, the particle shapes, the particle positions and the particle contents in the composites, in which the particle-matrix interfaces are also largely retained and they further affect the mechanical behaviors of particle reinforced MMCs. Therefore, the load-carrying of reinforced ceramic particles, e. g. silicon carbide and aluminum oxide, in metal matrix was very significant, and the effect of particle size were as well taken into account on the mechanical behaviors of the MMCs. However, it is very difficult to completely use experimental analysis to find the key parameters in the composite structures, which should be improved to optimize the overall tensile behavior of particle reinforced MMCs. This work is well attempted nowadays thorough composite structural modeling, from unit particle models to multiple particles models, from photograph analysis to direct geometric modeling, from 2D simple models to 3D complex models, which can take into account the composite structural characteristics of particle reinforced MMCs. Along with the composites structural modeling of particle reinforced MMCs, the morphologies of particle-matrix interfaces were established and the interfacial behaviors, for example widely applied cohesive failure model, were also introduced to perform mechanical behavior of particle reinforced MMCs. These structural models of particle reinforced MMCs are based upon experimental observations that can provide useful guidelines to a certain extent for optimum composite structures design. Therefore, a long way still exists to go before the potential of particle reinforced MMCs can be wholly achieved to develop new strong and lightweight materials in both material design and industrial applications. The present work aims to experimentally and numerically investigate the intrinsic relationship between the mechanical behavior and complex composite structure coupling with particle-matrix interfacial behaviors within the silicon carbide (SiC) particles reinforced aluminum (Al) matrix composites. Based on advanced particle size analysis technique, a large number of SiC particles are experimentally measured to present available SiC particular geometrical information, such as the particle size and particle aspect ratio. In the light of this statistical information of polyhedral SiC reinforced particles, a developed 3D structural modeling program can not only establish the structural models close to reality of SiC particles, but also reproduce the composite structures in line with the actual ones of particle reinforced MMCs. In these created structural models, the random dispersions of the sizes, the shapes, the positions and the contents of SiC particles can be realized according to the actual structural characteristics of SiC/Al composites. For numerically performing the mechanical behaviors of SiC/Al composites, elastoplastic mechanical properties with particle-matrix interfacial behaviors (adhesion, cohesive and frictions interfaces) are applied, and reasonable loads and boundary conditions are conducted. Both experimental and numerical results indicate that the volume fraction of SiC particles and particle-matrix interfacial behaviors play the significant role in the enhancement of mechanical behaviors of SiC/Al composites.

Production of SiC/Al composites and experimental testing

In this study, commercial α-SiC particles with an average particle size 13 μm were used as the reinforced materials [seen in **Figure 2.5.1(c)**], and 7A04 Al alloy with chemical components: Zn: 5.0—7.0, Mg: 1.8—2.8, Cu: 1.4—2.0, Mn: 0.2—0.6, Cr: 0.1—0.25 and Al: rest in weight percent was applied as the metal matrix. SiC/Al composites with different volume fractions of SiC particles were produced by a stir casting technique and then were extruded at high temperature to generate fine enough grains and to make the reinforced particles uniformly dispersed, shown in **Figure 2.5.1(a)**. **Figure 2.5.1(b)** provides the microscopic structural characteristics of produced SiC/Al composites, in which the dispersions of the particle shapes and particle positions of SiC particles are relatively uniform. **Figure 2.5.1(c)** presents a further magnified particle-matrix zone in SiC/Al composites, where two basic particle shapes: triangle shape and quadrangle shape can be extracted for the composite structural reproduction in the next part. Meanwhile, **Figure 2.5.1(d)** presents a reproduced composite structural model of 14 vol% SiC/Al composite with the model size of 100 μm $\times$ 100 μm $\times$ 20 μm.

Particle-matrix interfacial behavior

In order to study and assess the particle-matrix interfacial behaviors on mechanical

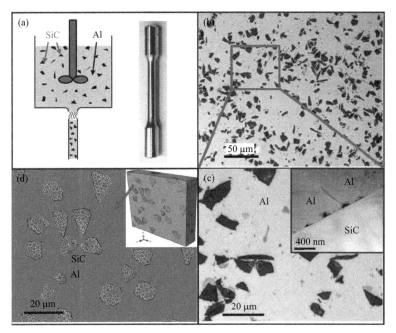

Figure 2.5.1 Production and structures of SiC/Al composites: (a) casting production, (b) microscopic characteristics, (c) particle-matrix interfaces, (d) structural modeling.

properties of SiC/Al composites, this work considers three basic interfacial behaviors: (i) adhesion interface; (ii) friction interface; (iii) cohesive interface. This handle is based on two aspects: one is the microstructural observations in TEM, seen in **Figure 2.5.1(c)**; and another is the SiC/Al phases compatible to form a coherent embedding of SiC particles in the matrix material. The complete adhesion interfaces are realized in the 3D structural models of SiC/Al composites and the friction interfaces are performed with the coefficient of friction of 0.3. For the cohesive interfaces, the cohesive zone model can characterize the crack initiation and propagation as an alternative criterion, in which the interfacial properties are specified by the cohesive interfacial strength t and fracture energy Γ, and the damage evolution is described by the damage factor D ranging from 0.0 to 1.0.

Towards the cohesive interfacial strength, Lloyd pointed out that larger particles give lower fracture strength, and the fracture strength is inversely proportional to the square root of the particle size. There are more defects around larger particles than smaller particles, and these defects lead to the degradation of the cohesive interface strength. Thus, we assume that the cohesive interfacial strength t follows the same particle size dependence, i.e., $\sigma_{max} \sim 1/d^{1/2}$, where d is the average particle size. For the average particle size of 13 μm, the cohesive interfacial strength can be determined as $t = 326$ MPa, while the fracture energy is set as $\Gamma = 91.9$ J/m^2. Based on the above different particle-matrix interfaces, the tensile processes of SiC/Al composites can be numerically performed.

Conclusion

In this work, the 3D composite structure modeling and both experimental and numerical tensile mechanical behaviors of SiC particles reinforced Al matrix composites are performed. In the numerical simulations, elastoplastic mechanical properties and three different particle-matrix interfacial behaviors: adhesion, cohesive and friction interfaces are considered. From the results, several main conclusions are provided.

(1) Based on microscopic structural characteristics of SiC particles, the particle size and aspect ratio distributions are experimentally measured from the particle size analysis of numerous SiC particles. Then, two typical 2D shapes (triangle and quadrangle shapes) with length ratio and changing factor are extruded and cut at all edges to form lots of different 3D polyhedral SiC particles. Considering the complexity of composite structures, a developed structural modeling program can establish the 3D structural models of SiC/Al composites with randomly dispersed SiC particles, in which the particle sizes, the particle shapes, the particle positions and the volume fraction of SiC particles can be reproduced in line with the actual composite structures of SiC/Al composite products.

(2) After the fabrications of SiC/Al composites with varying volume fractions of SiC particles, experimental uniaxial tensile tests of SiC/Al composites were applied. Using elastoplastic mechanical properties and particle-matrix interfacial behaviors, numerical uniaxial tensile simulations of SiC/Al composites are performed on the created 3D structural models. From the simulating work, numerical tensile stress-strain relations of SiC/Al composites present good agreement with the experimental results, in which the particle-matrix interfacial behavior and the volume fraction of SiC particles play the significant role in the enhancing mechanical behaviors of SiC/Al composites.

(3) The developed structural modeling program can effectively establish the structural models of reinforced particles, metal matrix and particle-matrix interfaces. Enhancing effect of SiC particles on mechanical behaviors of SiC/Al composites increases with the rising volume fraction of SiC particles coupling with the upwards interfacial strength and fracture energy. Moreover, the structural modeling and simulating works in this study can put forwards to establish the relationship between the actual composite structures and mechanical behaviors of particle reinforced MMCs.

(**Selected from:** Su Y, Ouyang Q, Zhang W, et al. Composite Structure Modeling and Mechanical Behavior of Particle Reinforced Metal Matrix Composites [J]. Materials Science and Engineering A, 2014, 597: 359-369.)

New Words and Expressions

optimum   *adj.* 最优的，最适宜的；*n.* 最佳条件
polyhedral   *adj.* 多面体的
elastoplastic   *adj.* 弹塑性的
aspect ratio   长宽比，长径比
quadrangle   *n.* 四边形
extrude   *v.* 挤出，压出
aluminum   *n.* 铝（金属元素，符号 Al，原子序号 13）
in line with   符合，与……一致

Notes

(1) Particle reinforced metal matrix composites (MMCs) have the very large potential to provide ultrahigh mechanical behaviors, for example, specific stiffness and specific strength, in the civil and defense applications as well as the automotive and aerospace industries.

——参考译文：颗粒增强金属基复合材料在民防、汽车和航空航天工业中具有非常大的应用潜力，（它）具备超高的力学特性，例如比刚度和比强度。

(2) From the simulating work, numerical tensile stress-strain relations of SiC/Al composites present good agreement with the experimental results, in which the particle-matrix interfacial behavior and the volume fraction of SiC particles play the significant role in the enhancing mechanical behaviors of SiC/Al composites.

——参考译文：模拟结果表明，SiC/Al 复合材料拉伸应力-应变的数学关系与实验结果吻合，其中颗粒-基体界面行为和 SiC 颗粒体积分数对提高 SiC/Al 复合材料的力学性能起着重要作用。

(3) Considering the complexity of composite structures, a developed structural modeling program can establish the 3D structural models of SiC/Al composites with randomly dispersed SiC particles, in which the particle sizes, the particle shapes, the particle positions and the volume fraction of SiC particles can be reproduced in line with the actual composite structures of SiC/Al composite products.

——参考译文：考虑到复合材料结构的复杂性，先进的结构建模程序可以构建 SiC 颗粒随机分布的 SiC/Al 复合材料的三维结构模型，其中 SiC 颗粒的尺寸、形状、位置和体积分数可以与 SiC/Al 复合材料的实际复合结构保持一致。

Chapter 3
Ceramics

【本章导读】
　　本章介绍陶瓷材料，包括四个单元，共有四篇课文和一篇阅读材料。四篇课文分别介绍陶瓷材料的基本概念、生物陶瓷、3D 打印技术制备陶瓷材料、固态陶瓷电解质材料。一篇阅读材料讲解了高熵陶瓷材料。

Unit 3.1

Text

Introduction to ceramics

　　这篇课文主要介绍陶瓷相关基础知识。课文讲解了陶瓷的基本定义和发展历史，介绍了陶瓷材料典型的物理化学性质，并分类说明了几类主要的陶瓷材料。作为示例，课文还给出了一些陶瓷材料的具体组分。

Definitions and histories

　　In traditional materials science, materials can be classified into five categories: metals, polymers, ceramics, semiconductors and composites. The first three are based on the types of atoms involved and the bonding between them, the fourth on the materials conductivity, and the last on the materials structure.

　　Ceramics are usually related to "mixed" binding— a combination of covalent, ionic, and sometimes metallic. They are composed of interconnected atomic arrays and no discrete molecules are involved. This property distinguishes ceramics from molecular solids such as iodine crystals (composed of discrete I_2 molecules) and paraffin wax (composed of long-chain alkane molecules). It also excludes ice, which consists of discrete H_2O molecules and usually behaves just like many ceramics. Most ceramics are metal or non-metal compounds. In this respect, ceramics can be defined as a nonmetallic, inorganic solid. The definition of ceramics here is not quite comprehensive. For example, glass — which behaves like a solid at room temperature and below but has the structure of a liquid—is actually a very important ceramic.

　　The word ceramic comes from the Greek Keramos, which means "potter's clay" or "pottery". Its origin is a Sanskrit term, meaning "to burn." The early Greeks used the word Keramos when describing products produced by heating clay-contained materials. For a long time, the term has included all products made of fired clay such as bricks, fireclay refractories, sanitary-ware and tableware.

In 1822, silica refractories were first manufactured. Although no clay was involved, the traditional shaping, drying, and firing processes were used to make silica refractories. Therefore, although the term "ceramic" retained the original meaning of clay products, it also began to include other products manufactured by the same process. The field of ceramics can be defined as the art and science of manufacturing and using solid objects that contain a ceramic as their basic component.

General properties

Generally, ceramics have the following properties.

Brittleness

The brittleness of ceramic can be recognized from everyday experience, e. g., dropping a glass cup or a dinner plate. At room temperature, as the mixed ionic-covalent bonding maintains the arranged atoms together, most ceramics are brittle. Nevertheless, when the temperature is over the glass transition temperature, the glass would exhibit a viscous liquid behavior but not brittle manner. That is why glass is easy to form intricate shapes.

Compressive strength

The compressive strength of ceramics is superior to the tensile strength. This characteristic is important when the ceramics components are used in load-bearing. The stress distributions should be carefully considered in the ceramics to guarantee that they are compressive. Ceramics generally have a weak toughness, although this property can be significantly enhanced by combining them with other components in composites.

Poor electrical and thermal conduction

Commonly, for ceramics, the valence electrons are not free as they are tied in bonds.

Chemical stability

Many ceramics are stable in harsh environments. For example, pyrex glass is widely used in chemistry laboratories because it not only can effectively withstand thermal shock owing to its low coefficient of thermal expansion (33×10^{-7} K^{-1}) but also is good resistance to many corrosive chemicals.

Transparence

Many ceramics such as sapphire watch covers, precious stones, and optical fibers are transparent due to their large band gap (E_g). In contrast, metals are transparent to visible light only when they are very thin, typically $< 0.1mm$.

Classification

Ceramics can be simply divided into traditional and advanced ceramics. Traditional ceramics are usually based on clay and silica, and mainly include bricks and tiles, whitewares, and pottery. Advanced ceramics mainly refer to newer materials with specific function, such as piezoelectric ceramics, high-entropy ceramics and bioceramics etc. Additionally, the ceramics produced by advanced manufacturing techniques (e. g., additive manufacturing or 3D/4D printing) also belong to advanced ceramics. They may exhibit superior mechanical properties and corrosion/oxidation resistance, or peculiarly electrical, optical and/or magnetic properties. The clay-based traditional ceramics have been used for over 25000 years, while advanced ceramics have generally been developed only within the last 100 years.

On the other hand, based on an application-classification scheme, ceramics can also be mainly categorized into following groups: glasses, structural clay products, refractories, abrasives, cements, and the newly developed advanced ceramics.

Typical ceramics

Glasses

The glasses are non-crystalline silicates containing other oxides, mainly CaO, Na_2O, K_2O, and Al_2O_3, which are regarded as a type of ceramics. The typical applications include containers, lenses, and fiberglass. The compositions of several common glass materials are provided in **Table 3.1.1**.

Table 3.1.1 The compositions of several common glass materials

Glass Type	Composition/wt%						Characteristics and Applications
	SiO_2	Na_2O	CaO	Al_2O_3	B_2O_3	Other	
Fused silica	>99.5						High melting temperature, very low coefficient of expansion (thermally shock resistant)
96% Silica (Vycor)	96				4		Thermal shock and chemically resistant-laboratory ware
Borosilicate (Pytex)	81	3.5		2.5	13		Thermal shock and chemically resistant-ovenware
Container (soda-lime)	74	16	5	1		4 MgO	Low melting temperature, easily worked, also durable
Fiberglass	55		16	15	10	4 MgO	Easily drawn into fibers-glass-resin composites
Optical flint	54	1				37 PbO, 8 K_2O	High density and high index of refraction-optical lenses
Glass-ceramic (Pyroceram)	43.5	14		30	5.5	6.5 TiO_2, 0.5 As_2O_3	Easily fabricated; 'strong'; resists thermal shock-ovenware

Most inorganic glasses can be transformed from amorphous to crystalline by appropriate high-temperature heating treatment. This process is termed as crystallization, and the obtained product is a fine-grained polycrystalline material, i. e., a glass-ceramic. In one sense, the formation of these small glass-ceramic grains experiences a phase transformation, involving nucleation and growth stages.

Glass-ceramic materials have the following characteristics: relatively high mechanical strength; low coefficients of thermal expansion (to avoid thermal shock); good resistance to high temperature; good dielectric properties; and satisfied biological compatibility. The most common uses for glass-ceramic materials are as ovenware, tableware, oven windows, and range tops—primarily due to their high strength and excellent resistance to thermal shock. They are also employed as electrical insulators and substrates for printed circuit boards and are utilized for architectural cladding and for heat exchangers and regenerators.

Clay products

Clay is one of the most widely used ceramic raw materials. This ingredient is found naturally in rich abundance. Another reason for its popularity is that clay products are conveniently produced. When clay and water are mixed in an appropriate proportion, the formed plastic mass is very amenable to shaping. The molded piece is dried to remove some moisture, and then fired at an elevated temperature to enhance its mechanical strength.

Clay-based products are mainly divided into two categories: structural clay products and whitewares. The former one includes building bricks, tiles, and sewer pipes while the latter one contains porcelain, pottery, tableware, china, and plumbing fixtures (sanitary ware).

Refractories

Refractories are materials capable of withstanding high temperatures without melting or decomposing and maintaining unreactive and inert when in contact with corrosive liquids and gases. Their typical applications include for furnaces, cement kilns, furnaces used for glass manufacturing and metallurgical heat treatments. The composition and processing methods determine the performance of a refractory ceramic. **Table 3.1.2** shows compositions for several commercial refractory materials.

Table 3.1.2 Compositions of five common ceramic refractory materials

Refractory Type	Composition/wt%							Apparent Porosity/%
	Al_2O_3	SiO_2	MgO	Cr_2O_3	Fe_2O_3	CaO	TiO_2	
Fireclay	25—45	70—50	0—1		0—1	0—1	1—2	10—25
High-alumina fireclay	90—50	10—45	0—1		0—1	0—1	1—4	18—25
Silica	0.2	96.3	0.6			2.2		25
Periclase	1.0	3.0	90.0	0.3	3.0	2.5		22
Periclase-chrome ore	9.0	5.0	73.0	8.2	2.0	2.2		21

Refractory is available in precast shapes, which is conveniently installed and used economically. Precast products mainly include bricks, crucibles, and furnace structural parts. About 70% of all refractories used in industry are in the form of preformed bricks with a variety of shapes. There are several different types of refractory brick. (1) Silica brick, which is made from naturally occurring sources of silica and bonded by adding 3.0%—3.5% CaO to promote liquid phase sintering. (2) Fireclay brick, which is made from kaolinite ($Al_2O_3 \cdot 2SiO_2 \cdot 2H_2O$) with 25%—45% Al_2O_3. (3) Dolomite brick, which is made from dolomite ($CaCO_3 \cdot MgCO_3$). (4) Chrome brick, which is made from naturally occurring chrome ore (containing 34% Al_2O_3 and 30% Cr_2O_3). (5) Zircon refractory brick ($ZrO_2 \cdot SiO_2$).

Abrasives

The abrasives are commonly used in grinding, polishing, lapping, drilling, cutting, sharpening, buffing, and sanding. The prime requisite for abrasive materials is hardness or wear resistance; usually with a Mohs hardness of at least 7. Additionally, a high degree of toughness and good resistance to high temperature are essential to ensure no occurrence of fracture and decomposition in abrasive particles. Abrasives are used in three types: i) bonded abrasives, ii) coated abrasives and iii) loose grains.

Bonded abrasives are typically bonded to a wheel (grinding, polishing, and cut-off wheels) and the abrasive action can be achieved by rotation of the wheel. The bonding materials include glassy ceramics, polymer resins, shellacs, and rubbers. The surface structure should be porous, as the porosity is benefit to a continual flow of air current or liquid coolants and thus prevents overheating. The bonded abrasives can be used in saws to cut concrete, asphalt and metals, and wheels for grinding, sharpening, and deburring.

Coated abrasives refer to those in which abrasive particles are firmly attached to some type of fabrics or paper backing material. The most well-known example is sandpaper. Typical backing materials include paper, rayon, cotton, polyester, and nylon. Polymers such as phenolics, epoxies, acrylates, and glues are commonly used for backing-particle adhesive. The coated abrasives are usually used in abrasive belts, hand-held abrasive tools, lapping of wood, ophthalmic equipment, glass, plastics, jewelry, and ceramic materials.

Loose abrasive grains are often required in grinding, lapping, and polishing wheels. Lapping and polishing have different objectives. The aim of lapping is to improve the accuracy of object's shape whereas the purpose of polishing is to reduce the surface roughness. Loose abrasive particles have sizes in the micron and submicron ranges and are not bonded to another surface but are free to roll or slide. Loose abrasive processes are typically employed in high-precision finishing operations and its typical applications include mechanical seals, jewel watch bearings, magnetic recording heads, electronic circuit substrates, surgical instruments, and optical fiber connectors.

Cements

Cements are powder ceramics that react chemically with a liquid (usually, though not necessarily water) to form a solid structure. A cement slurry is the suspension of this powder in the liquid phase. Some cement pastes need the presence of water to harden, while others can harden under the existence of air or carbon dioxide. The two fundamental reactions with CO_2 and H_2O are listed as following:

$$Ca(OH)_2 + CO_2 \longrightarrow CaCO_3 + H_2O$$
$$CaO \cdot Al_2O_3 + 10H_2O \longrightarrow CaO \cdot Al_2O_3 \cdot 10H_2O$$

Different water content can lead to different hydrate phases. If sulfur is involved in gypsum, the reaction becomes more complex and the mineral ettringite $[Ca_6Al_2(SO_4)_3(OH)_{12} \cdot 26H_2O]$ would be formed during the hydration process.

Portland cement is called ahydraulic cement because its hardness is generated by chemical reactions with water. It is utilized primarily to bind inert particles (sand and/or gravel) together into a cohesive mass. Other important cement materials, such as lime, are nonhydraulic—that is, carbon dioxide but not water is involved in the hardening reaction.

(**Selected from:** Callister W D, Rethwisch D G. Fundamentals of Materials Science and Engineering: An Integrated Approach [M]. 5th ed. Wiley, 2015.

Carter C B, Norton MG. Ceramic Materials: Science and Engineering [M]. 2nd ed. Springer, 2013.)

New Words and Expressions

covalent   *adj*. 共有原子价的，共价的
pottery   *n*. 陶瓷，陶器
discrete   *adj*. 分离的，不相关联的，分立式，非连续
clay   *n*. 黏土，陶土
brick   *n*. 砖，积木
refractory   *n*. 耐火材料，耐火物质
sanitary   *adj*. 卫生的，清洁的；*n*. 公共厕所
tableware   *n*. 餐具，食具
brittleness   *n*. 脆性，脆度
intricate   *adj*. 错综复杂的，难理解的
pyrex   *n*. 耐热玻璃，派热克斯玻璃（商标名称）
piezoelectric   *adj*. 压电的
porcelain   *n*. 瓷，瓷器
bearing   *n*. 举止，方位，轴承
sewer   *n*. 阴沟，污水管，下水道
tile   *n*. 瓦片，瓷砖，空心砖
gypsum   *n*. 石膏
kiln   *n*. 窑，炉
silicate   *n*. 硅酸盐，硅酸盐矿物
lime   *n*. 石灰
abrasive   *n*. （用来擦洗表面或使表面光滑的）磨料；*adj*. 研磨的
cement   *n*. 水泥
cladding   *n*. 镀层，保护层
whiteware   *n*. 洁具
lining   *n*. 衬层，内衬，衬里
fireclay   *n*. 耐火黏土
kaolinite   *n*. 高岭（石）
periclase   *n*. 方镁石
chrome ore   铬矿
precast   *adj*. 预制的，预先浇铸的
dolomite   *n*. 白云石
asphalt   *n*. 沥青，柏油
deburring   *n*. 修边，除去毛刺

ophthalmic   *adj.* 眼科的
phenolics   *n.* 酚醛树脂
hydraulic   *adj.* 水力的
submicron   *adj.* 亚微米的，亚微细粒的

Notes

(1) In this respect, ceramics can be defined as a nonmetallic, inorganic solid.

—in this respect 在这方面。

—参考译文：在这方面，陶瓷可以定义为一种非金属的无机固体。

(2) Although no clay was involved, the traditional shaping, drying, and firing process was used to make silica refractories.

—be used to 被用来。

—参考译文：尽管不涉及黏土，传统的成型、干燥和烧制工艺（可以）被用来制造二氧化硅耐火材料。

(3) Many ceramics such as sapphire watch covers, precious stones, and optical fibers are transparent due to their large band gap (E_g).

—参考译文：许多陶瓷，如蓝宝石表壳、宝石和光纤都是透明的，这归因于它们有大的带隙（E_g）。

(4) On the other hand, based on an application-classification scheme, ceramic can also be mainly categorized into following groups: glasses, structural clay products, refractories, abrasives, cements, and the newly developed advanced ceramics.

—on the other hand 另一方面。

—参考译文：另一方面，根据应用分类方案，陶瓷主要可分为玻璃、结构黏土制品、耐火材料、磨料、水泥和新开发的先进陶瓷。

(5) Additionally, a high degree of toughness and good resistance to high temperature are essential to ensure no occurrence of fracture and decomposition in abrasive particles.

—参考译文：此外，高韧性和良好的耐高温性是确保磨料颗粒不发生断裂和分解的必要条件。

(6) Portland cement is called a hydraulic cement because its hardness is generated by chemical reactions with water.

—参考译文：硅酸盐水泥被称为水硬性水泥，因为它的硬度是由与水的化学反应产生的。

Exercises

1. Questions for discussion

(1) Which of the following materials could be classified as a ceramic. Justify your answer.

(a) solid argon (Ar); (b) molybdenum disilicide ($MoSi_2$); (c) NaCl; (d) crystalline sulfur (S); (e) ice; (f) boron carbide (B_4C).

(2) Pearls and garnets are both examples of gems. We classify garnet as a ceramic. Would you classify pearl as a ceramic? Briefly justify your answer.

(3) What are traditional ceramics? What are advanced ceramics? Please give some examples.

(4) What is the difference between crystalline ceramics and non-crystalline ceramics?

2. Translate the following into Chinese

(1) The brittleness of ceramic can be recognized from everyday experience, e. g., dropping a glass cup or a dinner plate.

(2) Bonded abrasives are typically bonded to a wheel (grinding, polishing, and cut-off wheels) and the abrasive action can be achieved by rotation of the wheel.

(3) The most common uses for glass-ceramic materials are as ovenware, tableware, oven windows, and range tops—primarily due to their high strength and excellent resistance to thermal shock.

(4) Coated abrasives refer to those in which abrasive particles are firmly attached to some type of fabrics or paper backing material.

3. Translate the following into English

(1) 研磨　　　　　　(2) 织物
(3) 多孔的　　　　　(4) 冷却剂
(5) 冶金学的
(6) 它受欢迎的另一个原因是黏土产品生产方便。
(7) 然而当温度超过玻璃化转变温度时，玻璃将表现出黏性液体行为，而不是脆性。
(8) 水泥是粉末陶瓷与液体（通常是水但不一定必须是水）发生化学反应形成的固体结构。
(9) 工业中使用的所有耐火材料中约有 70% 是各种形状的预制砖。

Unit 3.2

Text

Bioceramics

这篇课文主要介绍生物陶瓷材料近年来的发展。课文讲解了几种生物陶瓷的概念、主要分类以及在生物医药应用上的优缺点，然后分类介绍了几种典型的生物陶瓷，并着重介绍了羟基磷灰石这种广泛应用的生物陶瓷。

Bioceramics

Ceramics that are used for the repair and reconstruction of human body parts are termed bioceramics. Currently, the most important application of bioceramics is in implants (e. g., alumina hip prosthesis). Owing to its very low reactivity in the body, alumina is regarded as an inert bioceramic. On the contrary, bioactive ceramics can bond directly with bone, although they are relatively weak in comparison with implanted high-strength alumina, zirconia and metal. Accordingly, they are usually utilized as coatings, depending on the toughness and mechanical strength of the substrate.

Bioceramics can be mainly categorized into nearly inert bioceramics, bioactive ceramics and resorbablebioceramics based on their chemical reactivity in the body. For a nearly inert bioceramics, when it is implanted into the body, a protective response would be initiated, leading to an encapsulation of nonadherent fibrous coating with a thickness of around 1mm. Over time, the implants would be completely isolated. In the case of bioactive ceramics such as HA coating, a bond would form across the implant/tissue interface, which imitates the natural repair process of human body. Resorba blebioceramics, such as tricalcium phosphate (TCP), can be dissolved and then replaced by the surrounding tissue in the body. Note that the dissolution products should not be toxic.

Compared to other implant materials, the bioceramics have the main advantage of biocompatibility but disadvantages of low toughness and high E (which may lead to stress shielding). The combination of bioceramics with other components to form a composite provides a good solution to increase its toughness. As a typical example, polyethylene (PE) - reinforced with HA particles composite exhibits a much higher toughness than pure HA particles. Moreover, the E of composite is more closely matched to that of a natural bone.

Ceramic Implants

There are two basic criteria for ceramic implants: i) compatible with the physiological environment, ii) matched mechanical properties of the replaced tissue.

Most bioceramic implants are in direct contact with bones. Bone is a living material composed of cells and a blood supply, which are encased in a solid composite structure. The composite consists of a flexible and tough of collagen and apatite of calcium and phosphate crystals, resembling calcium HA. The most concern in the use of bioceramics is cancellous (spongy bone) and cortical (compact bone).

The mechanical properties of the implant are significantly important. If the implant has a much larger E than the replaced bone, then a problem of stress shielding would occur. Stress shielding seriously weakens the surrounding bone, where the applied load is in compression. (Bone must be loaded in tension to stay healthy). Decreasing E to eliminate stress shielding is one of the primary driving forces for the development of bioceramic composites.

Alumina and Zirconia

Both Al_2O_3 and ZrO_2 are regarded as inert bioceramics as they have little chemical change when exposed to body fluids for a long time. Alumina with high-density and high-purity ($>$ 99.5%) is used in many implants, particularly as load-bearing hip prosthesis and dental implants.

Most alumina implants are in the form of very fine grained polycrystalline Al_2O_3, which can be made by pressing followed by sintering at temperatures ranging from 1600 to 1800℃. During this process, a small amount of MgO ($<0.5\%$) as a grain growth inhibitor is added, which allows a formation of high-density product by sintering without pressure.

Implant materials should last longer than patients. However, it is impossible to provide a definite lifetime prediction for each individual implant due to the probabilistic failure in ceramics. The investigations on aging and fatigue tests show that the production of Al_2O_3 implants should satisfy the highest standards of quality assurance, especially when used in orthopedicprostheses for young patients.

Although Al_2O_3 ceramics integrate good biocompatibility and excellent wear resistance, they have only limited flexural strength and weak toughness. Compared to Al_2O_3 ceramics, ZrO_2 ceramics have higher fracture toughness and flexural strength, and lower E. Nevertheless, there are some concerns with ZrO_2. i) The flexural strength and toughness of ZrO_2 ceramics will slightly deteriorate during long-term exposure to body fluids, which is related to the martensitic transformation from the tetragonal to the monoclinic phase. ii) The wear resistance of ZrO_2 is inferior to that of Al_2O_3. iii) ZrO_2 may contain low concentrations of long half-life radioactive elements, such as Th and U, which are hard to be separated out.

Bioactive glasses

Most bioactive glasses are silicates and based on four components of SiO_2, Na_2O, CaO, and P_2O_5. The first thoroughly studied bioactive glass is known as Bioglass® 45S5, consisting of 45 wt% SiO_2, 24.5 wt% Na_2O, 24.4 wt% CaO, and 6 wt% P_2O_5. However, Bioglass® 45S5 exhibits a random and two-dimensional sheet-like structure, which is different from that of the silicate glasses.

Bioactive glasses can be readily produced using the processes developed for other silicate glasses. The constituent oxides or compounds are first mixed in the appropriate ratios and then fused at high temperatures to generate a homogeneous melt. On cooling, a glass is made. Considering that bioactive glasses will be used inside the human body, it is very necessary to employ high-purity raw materials. Additionally, the fusing process is usually conducted in Pt or Pt alloy crucibles to avoid contamination.

The mechanical strength and the fracture toughness of bioactive glasses is relatively low. Consequently, bioactive glasses are usually used as coating on metals, in low-loaded or compressively loaded devices, but not in load-bearing applications. Cone-shaped plugs of bioactive glasses have been used in oral surgery to fill the jaw defects caused by tooth extraction. Bioactive glasses in the powder form have also been used for the treatments of patients with periodontal disease and paralysis of one of the vocal cords. The combination of bioactive glasses with natural bone can also be used in maxillofacial reconstruction.

Hydroxyapatite (HA)

The apatite family of minerals can be expressed as the formula $A_{10}(BO_4)_6X_2$. For HA, or more specific calcium hydroxyapatite, A, B and X represent Ca, P, and OH, respectively. The mineral parts of teeth and bones are composed of an apatite of calcium and phosphorus. Natural bone contains about 70% HA by weight.

HA is a hexagonal crystal structure (space group is P63/m) with $a=0.9432$ nm and $c=0.6881$ nm. The hydroxyl ions are located at the corners of the projected basal plane and occur at equal intervals along the column perpendicular to the basal plane and parallel to the c axis. A network of $(PO_4)^{3-}$ groups as the skeletal framework enables the stability of apatite. The Ca, PO_4, and OH groups in the HA structure can be substituted. Once substitution occurs, the lattice parameter as well as the properties of the HA crystal such as solubility would be changed. If the OH^- groups are replaced by F^-, the anions are closer to the surrounding Ca^{2+} ions. This substitution can stabilize the structure, which is proposed as one of the reasons that fluoridation is favorable to reducing tooth decay, as demonstrated by the investigation on the incorporation of F into HA and its impact on solubility.

HA is produced in one of two forms, i. e., dense or porous for use in biomedical applications. The simplest method used to prepare dense HA is to dry-press HA powder. A small amount of a binder may be added into powder to form mixture. After pressing, the green ceramic is sintered in air with the temperature up to 1300℃. Dense HA in both as block form and as particles can be used in many fields. One typical application is replacements for tooth roots following extraction. This implant helps to minimize alveolar ridge resorption and maintain ridge shape following tooth loss.

The advantage of porous HA is that it allows ingrowth of tissue into the pores and provides biological fixation of the implant. Porous HA should mimic the framework (or stromal component) of the bone when it is used as bone graft substitute. Three methods have been used to produce porous HA ceramics. The first one is to sinter HA powders, or a mixture of suitable reactant powders, with naphthalene particles that volatilize during heating to generate an interconnected porous network. The second one is to mix water-setting HA cements with sucrose granules. When the sucrose is dissolved prior to the cement setting, the porosity would be created. The last one is to employ the natural template. The calcium carbonate ($CaCO_3$) skeleton of reel-building corals as a suitable template can react with phosphate groups via the following chemical reaction:

$$10CaCO_3 + 6(NH_4)_2HPO_4 + 2H_2O \longrightarrow Ca_{10}(PO_4)_6(OH)_2 + 6(NH_2)CO_3 + 4H_2CO_3$$

The HA structure prepared by this way replicates the porous marine skeleton (i. e., interconnected porosity). HA grown on coral skeleton templates can be used to simulate the stroma of cortical bone and cancellous bone.

Bioceramic coatings

The combination of bioceramic coating onto the surface of a substrate allows us to achieve the bulk properties of the substrate and the surface properties of the coating. The three main purposes for applying a coating are: i) make the implant biocompatible, ii) turn a non-bioactive surface into a bioactive one, and iii) protect the substrate against corrosion. There are four different substrate-coating combinations: ceramic on ceramic, glass on ceramic, ceramic on metal and glass on metal.

Bioceramic layers can be coated on metallic substrates, which combines the fracture toughness of the metal with the ability of a bioactive surface to the surrounding tissue. Specifically, the metal implant coated by a bioceramic materials is favorable to stabilization of the implant in the surrounding bone, prolonging the functional life of the prosthesis.

Both HA and TCP are often used as ceramic coatings. The structure and properties of HA are introduced in the previous section. As a resorbable bioceramic, TCP is eventually dissolved and replaced by tissue when it is implanted into the body. The resorbable TCP bioceramics serves as scaffolding and allows tissue infiltration and eventual replacement. Essentially, this is the same function as bone grafts. TCP has been clinically utilized in dentistry and orthopedics. Specifically, bulk TCP material with dense or porous form, is employed for alveolar ridge augmentation, immediate tooth root replacement, and maxillofacial reconstruction. However, TCP cannot be used in load bearing application due to its poor mechanical properties. Therefore, TCP is usually employed as a coating on metal substrates.

(**Selected from:** Carter C B, Norton MG. Ceramic Materials: Science and Engineering [M]. 2nd ed. Springer, 2013.)

New Words and Expressions

hydroxyapatite   *n*. 羟磷灰石
collagen   *n*. 胶原蛋白，胶原
tricalcium phosphate   磷酸三钙
cortical   *adj*. 皮层的，皮质的
orthopedic   *adj*. （关节和脊柱）矫形的
prosthesis   *n*. 假体（如假肢、假眼或假牙）
periodontal   *adj*. 牙周的
paralysis   *n*. 麻痹，瘫痪，（活动、工作等）能力的完全丧失
maxillofacial   *adj*. 上颌面的
alveolar   *n*. 牙槽嵴
cancellous   *adj*. 罗眼状的，多孔的
naphthalene   *n*. 萘（用于制作卫生球等）
sucrose   *n*. 蔗糖
stroma   *n*. 叶绿体基质，基质

Notes

(1) Herein, new developments in ceramics such as bioceramics, high-entropy ceramics and 3D-printing technologies for ceramics will be introduced.

—参考译文：在此，（本文）将介绍生物陶瓷、高熵陶瓷和陶瓷3D打印技术等陶瓷领域的新发展。

(2) Owing to its very low reactivity in the body, alumina is regarded as an inert

bioceramic.

— owing to 由于。
— 参考译文：由于其在体内的反应活性非常低，氧化铝被认为是一种惰性生物陶瓷。

(3) On the contrary, bioactive ceramics can bond directly with bone, although they are relatively weak in comparison with implanted high-strength alumina, zirconia and metal.

— on the contrary 相反地。in comparison with... 与……相比。
— 参考译文：相反，生物活性陶瓷可以直接与骨骼结合，虽然与植入的高强度氧化铝、氧化锆和金属相比，它们（的机械强度）相对较弱。

(4) Over time, the implants would be completely isolated.

— 参考译文：随着时间的推移，植入物将被完全隔离。

(5) Compared to other implant materials, the bioceramics have the main advantage of biocompatibility but disadvantages of low toughness and high E (which may lead to stress shielding).

— compared to... 与……相比。stress shielding 应力屏蔽。
— 参考译文：与其他植入材料相比，生物陶瓷具有生物相容性的主要优点，但缺点是低韧性和高弹性（可能导致应力屏蔽）。

(6) Bone is a living material composed of cells and a blood supply, which are encased in a solid composite structure.

— 参考译文：骨骼是一种由细胞和血液供应组成的活体材料，细胞和血液供应被包裹在一个坚固的复合结构中。

(7) Alumina with high-density and high-purity (>99.5%) is used in many implants, particularly as load-bearing hip prosthesis and dental implants.

— 参考译文：高密度和高纯度（>99.5%）的氧化铝被用于许多植入物，特别是作为承重髋关节假体和牙科植入物。

Exercises

1. Questions for discussion

(1) Alumina (Al_2O_3) ceramic implants are required to have a small grain size (<4.5 mm).

(a) Why do you think a small grain size is important?

(b) How does the addition of MgO to the powder mixture help to keep the grain size small?

(c) Are there any other ways that could be used to limit the extent of grain growth?

(2) Explain why HA and TCP behave differently in the body.

(3) What is a bioactive ceramic?

(4) Write down the general formula for HA and state its percentage in bone by weight. How does its structure affect its properties?

2. Translate the following into Chinese

(1) Bioceramic layers can be coated on metallic substrates, which combines the fracture toughness of the metal with the ability of a bioactive surface to the surrounding tissue.

(2) When the sucrose is dissolved prior to the cement setting, the porosity would be created.

(3) The mineral parts of teeth and bones are composed of an apatite of calcium and phosphorus.

3. Translate the following into English

(1) 耐火材料　　　　　(2) 磨料

(3) 不依不饶的　　　　(4) 疲劳

(5) 有四种不同的基材-涂层组合：陶瓷上的陶瓷、陶瓷上的玻璃、金属上的陶瓷和金属上的玻璃。

(6) 具体而言，覆有生物陶瓷材料的金属植入物有利于植入物在周围骨骼中的稳定，延长假体的使用寿命。

(7) 生物陶瓷涂层与基材表面的结合使我们能够实现基材的整体性能和涂层的表面性能。

(8) 大多数氧化铝植入物是非常细晶粒的多晶 Al_2O_3 形式，可以在 1600～1800℃ 的温度范围内烧结制成。

Reading Material

High-entropy ceramics

这篇课文介绍了近些年来兴起的高熵陶瓷。课文从高熵陶瓷的晶格结构出发，说明由于其独特的结构和成分带来的不寻常性质，并介绍了高熵陶瓷一些可能的应用领域。

High-entropy ceramics (HECs) are defined as the solid solutions of inorganic compounds with five or more cations or anions in equal or near-equal atomic ratios. The concept of HECs is originated from that of high-entropy alloys, which are formed by five or more metal elements with equal or near-equal atomic fractions. HECs are long-range structurally ordered but compositionally disordered.

A variety of high-entropy ceramics with different crystal structures and compositions have been prepared and investigated. Based on the compositional anions, high-entropy ceramics can be classified into high-entropy oxides, carbides, borides, nitrides, sulfides, silicates and so on.

The crystal structures of high-entropy ceramics are generally determined by the individual constituents. For example, the crystal structure of high-entropy carbides (HfZrTaNbTiC) is a common face-centered cubic lattice with group space of Fd-3m, dubbed rock-salt phase. Within this crystal structure, Hf, Zr, Ta, Nb and Ti atoms are randomly located on a set of equal sites that originally occupied by metal atoms in individual carbides structure [**Figure 3.2.1(a)**]. Similarly, the high-entropy borides possess hexagonal lattice, which composed of layers of five types of metal atoms and layers of boron atoms alternately [**Figure 3.2.1(b)**]. Furthermore, the crystal lattice constants of high-entropy ceramics approximately equal to the average value of individual constituents.

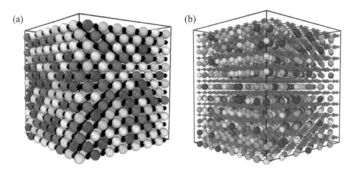

Figure 3.2.1 Crystal structure of (a) high-entropy carbide and (b) high-entropy diboride. Small balls are carbon or boron atoms, and colored big atoms are metal atoms.

In general, all atoms are precisely located in their own atomic sites in well-crystallized solids. However, affected by differences in atomic sizes, electronegativity, and constant of crystal lattice of individual constituents, a fluctuation from perfect site is usually observed, leading to a severe crystal lattice distortion. The existence and degree of lattice distortion can be precisely detected by X-ray absorption spectroscopy test based on synchrotron radiation system.

The unique structures and compositions endow HECs with unusual properties. There are four main effects exhibited in high-entropy ceramics, i.e., sluggish diffusion effect, cocktail effect, lattice distortion effect and high-entropy effect. One core property of HECs is superior phase stability originated from the high-entropy effect. Thermodynamically, the phase stability of HECs is determined by the Gibbs free energy, which can be defined as:

$$\Delta G_{mix} = \Delta H_{mix} - T\Delta S_{mix} \tag{3.2.1}$$

where H_{mix} is the mixing enthalpy of high-entropy ceramics, S_{mix} is the mixing entropy, which can be calculated as:

$$S_{mix} = -R \sum_{x_i} (x_i \sum_n c_j \ln c_j) \tag{3.2.2}$$

where R is the gas constant, x_i is the fraction of sublattice i in total atom sites, n is number of components in the sublattice, and c_j is atom fraction of j component. Larger configuration entropy favors a lower Gibbs free energy and hence leading to an excellent phase stability. Another distinct feature for HECs is hardness and strength. The severe lattice distortion caused by different atom radius and lattice constants of compositional individual ceramics results in pining of dislocations which provides excess strength.

HECs possess unusual thermal performances. Since lattice distortion and atoms arrange disorder widely exist in HECs, these structure features serve as efficient scattering sources for photons and electrons, and thus result in the reduction of thermal conductivity. In addition, some kinds of HECs, such as borides, nitrides and carbides, exhibit much better oxidation resistance than the individual ceramics.

HECs have some special electronic conductivity and dielectric property. For the electronic conductivity, the creation of lattice distortion has been widely adopted to reduce the electrical contribution to the thermal conductivity. Correspondingly, the electron scattering can be strengthened and thus the reduced electrical conductivity can be achieved in high-entropy borides and carbides ceramics. For dielectric property, there are more than one sublattices in which at least one sublattice site is occupied by multi-elements. On the one hand, the sublattice exhibits a long-range periodicity with the distortion that influences the behaviors of electrons, dipoles and band structure. On the other hand, randomly distributed multi-elemental atoms bring compositional disorder that can reduce short range order. The wide designable compositional space of high-entropy ceramics provides more flexible ways to tune the compositions, defects, degrees of disorder/order, band structures, in which dielectric properties can be easily modified.

The unique structures, compositions and superior physical and chemical features endow HECs with combined properties that cannot be achieved in traditional ceramics. This nascent family of materials shows promising applications in various fields including ultrahigh temperature, thermal protection and thermal insulation of hypersonic vehicles, thermal and environmental barrier coatings for engine components, irradiation resistant devices for nuclear energy, cutting tools, electromagnetic wave absorption and interference shielding, electrodes for energy storage and conversion devices.

(**Selected from:** Akrami S, Edalati P, Fuji M, et al. High-Entropy Ceramics: Review of Principles, Production and Applications [J]. Materials Science and Engineering: R: Reports, 2021, 146: 100644.)

New Words and Expressions

entropy   *n*. 熵（热力学函数）
carbide   *n*. 碳化物
boride   *n*. 硼化物
nitride   *n*. 氮化物
sulfide   *n*. 硫化物
electronegativity   *n*. 电负性
synchrotron   *n*. 同步加速器
thermodynamical   *adj*. 热力学的
enthalpy   *n*. 焓
hardness   *n*. 硬度
strength   *n*. 强度
dislocation   *n*. 位错
dielectric   *adj*. 介电的
dipole   *n*. 偶极，偶极子
hypersonic   *adj*. 超音速的

Notes

(1) High-entropy ceramics (HECs) are defined as the solid solutions of inorganic compounds with five or more cations or anions in equal or near-equal atomic rations.

—be defined as 被定义为；near-equal atomic ration 近等原子配比。

——参考译文：高熵陶瓷（HECs）被定义为具有五个或五个以上阳离子或阴离子亚晶格的无机化合物的固溶体，这些亚晶格的原子比例相等或接近相等。

（2）The crystal structures of high-entropy ceramics are generally determined by the individual constituents.

——参考译文：高熵陶瓷的晶体结构通常由单个组分决定。

（3）Similarly, the high-entropy borides possess hexagonal lattice, which composed of layers of five types of metal atoms and layers of boron atoms alternately.

——参考译文：同样，高熵硼化物具有六方晶格，由五种金属原子层和硼原子层交替组成。

（4）Furthermore, the crystal lattice constants of high-entropy ceramics approximately equal to the average value of individual constituents.

——参考译文：此外，高熵陶瓷的晶格常数近似等于各个组分的平均值。

（5）Larger configuration entropy favors a lower Gibbs free energy and hence leading to an excellent phase stability.

——参考译文：较大的构型熵有利于较低的吉布斯自由能，因此导致了优异的相稳定性。

（6）The severe lattice distortion caused by different atom radius and lattice constants of compositional individual ceramics results in pining of dislocations which provides excess strength.

——参考译文：不同组分陶瓷的原子半径和晶格常数的不同导致了严重的晶格畸变，（这）导致位错的钉扎，从而提供了额外的强度。

（7）The wide designable compositional space of high-entropy ceramics provides more flexible ways to tune the compositions, defects, degrees of disorder/order, band structures, in which dielectric properties can be easily modified.

——参考译文：高熵陶瓷广阔的成分可设计空间为调控成分、缺陷、无序/有序程度、能带结构提供了更为灵活的方式，其中介电特性可以很容易地被改变。

Unit 3.3

Text

3D-printing technologies for ceramics

这篇课文主要介绍用新型的增材制造技术——3D打印来制备陶瓷材料。3D打印制备陶瓷材料的技术途径主要有基于料浆的技术、基于粉末的技术、基于块体的技术。课文分别介绍了这三种技术并分析了它们的优缺点。

Three-dimensional (3D) printing is an additive manufacturing (AM) technology that fabricates a complex-shaped object from a digital model by depositing feedstock materials layer by layer. Compared with traditional fabrication methods of ceramics, the 3D printing technology owns advantages of high efficiency, high resolution, low cost, efficient utilization of material and most significant, capability of fabricating intricate 3D structures, which is extremely useful for fabricating precise components in various fields, such as construction, medicine and aerospace.

The 3D printing technique of ceramics was first reported by Marcus et al. and by Sachs et al. in the 1990s. To date, with the latest advances in material science and computer science, virtual 3D printing technologies have been specifically developed for ceramic manufacturing. According to the form of pre-processed feedstock, 3D printing technologies could be generally classified into slurry-based, powder-based and bulk solid-based methods. Note that slurry-based technologies use ceramic/polymer mixture with viscosities ranging from low-viscosity (mPa·s) ink with a low ceramic loading (up to 30 vol%) to high-viscosity (Pa·s) paste with a much greater ceramic loading (up to 60 vol%).

Slurry-based technologies

The stereolithography (SLA) technique is one of the earliest developed 3D printing technology and has been extensively used worldwide. During the SLA process, ceramic-containing slurry layers are radiated by a light source of a certain wavelength (usually in the ultraviolet range), causing the polymerization of the slurry layers with incorporation of the ceramic particles. After debinding the resulting polymer and sintering the structural material in a furnace, the final ceramic part is obtained. Because of the layer-by-layer polymerization process, the SLA technique is capable of fabricating parts of high surface quality at fine resolutions down to the micrometer scale. SLA has been extensively developed and applied to the fabrication of dense/cellular ceramic parts in various fields, ranging from parts with complex structures such as integrally cored casting molds, microelectronic components such as sensors and photonic crystals, to biomedical implants such as bone scaffolds and dental components.

The digital light processing (DLP) technique is in fact a mask-based SLA, in which an integral image is transferred to the photopolymerizable liquid surface by exposing the light source through a patterned mask once only. The ultra-fast light switching and integral projection allow the DLP 3D printing process time to be dramatically reduced as it is much faster than the conventional SLA point-line-layer scanning process. The higher efficiency of DLP over the conventional SLA process makes it a promising 3D printing technique for ceramic fabrication.

Inkjet printing (IJP) is a well-known method for creating two-dimensional digital text and images by means of ejecting liquid-phase materials (i.e. ink) in droplet form through printhead nozzles onto paper, plastic or other substrates. The application of IJP to printing ceramic components was first described by Blazdell and co-workers in 1995, using ZrO_2 and TiO_2 ceramic inks, both of volume fractions as small as 5 vol%. The performance and properties of IJP ceramic hinge on the ceramic powder and ink formulation, particularly the rheological characteristics such as dispersity, stability, viscosity and surface tension. Overall,

IJP is a versatile 3D printing technique for printing small sized ceramic parts, albeit with limited flexibility in complex structure design, such as the inability to print overhanging or hollow structures due to difficulties with support preparation.

Powder-based technologies

Powder-based ceramic 3D printing technologies mainly utilize powder beds normally containing loose ceramic particles as feedstock. The ceramic particles are bonded either by spreading liquid binders or by powder fusion using thermal energy provided by a laser beam.

The 3D printing technique applies printheads to selectively jet organic liquid binder droplets onto a powder bed surface. Solid layers are formed by the solidification (i. e., gluing) of the permeating liquid binder, which encloses the powder. A new layer of powder is then supplied and spread on the previous layer to repeat the building process until the part is formed. After this, loose powder is removed to reveal the part. This technique is best suited to the fabrication of porous ceramic parts. However, limitations of the application of 3D printing in the processing of advanced ceramic materials arise as the downsides of this technique, the inferior quality in resolution, surface finish, density and mechanical performance, require extra work, including infiltration and isostatic pressing for further quality improvements.

In a selective laser sintering (SLS) process, a high-power laser beam is used to selectively irradiate the surface of the target powder bed. The powder then heats up and sintering (i. e., interparticle fusion) takes place for bulk joining. After this, a new layer of powder is spread onto the previous surface for the next run of heating and joining. In this way, the process is repeated layer by layer until the designed 3D part is fabricated. A board range of material combinations can be used with the SLS technique, allowing for a comprehensive range of applications in functional and structural ceramic fabrication. Additionally, the support-free building process of SLS enables the realization of geometrically controllable complex/porous ceramic parts. Although the disadvantages of SLS ceramic include low resolution, poor surface finish and porous microstructures within the fabricated parts, it is acceptable for most applications, particularly in the tomography-assisted fabrication of scaffolds for biomedical engineering purposes. Moreover, additional processing steps, such as infiltration and isostatic pressing, should be considered to enable the fabrication of structurally sound parts with desirable strength and little geometrical distortion.

Selective laser melting (SLM), often considered as a variant of SLS, owing to the almost same process except the one-step powder bed fusion by full melting. Compared with SLS, SLM uses laser sources with much higher energy densities and requires no secondary low-melting binder powders. Thus, SLM can produce nearly fully dense homogeneous parts without post-treatments owing to its ability to fully melt the powder into the liquid phase, ensuring rapid densification, instead of heating up the powder to a specific point where the particles are partially melted and fused together as in SLS. However, due to the generation of defects such as porous microstructures, rough surface finishes and low dimensional accuracy, densification to a non-porous and isotropic ceramic body is hard to achieve, which greatly limits the applications.

Bulk solid-based technologies

Laminated object manufacturing (LOM) process was originally developed to produce paper, plastic and metal components. The application of LOM to ceramic manufacturing was first demonstrated by Griffin and co-workers in 1994, based on tape-cast alumina and zirconia green sheets. Continuous rolling of long green ceramic sheets onto the working platform allows the procedure to be fully automated. Excess material surrounding the parts being fabricated is removed layer-wise prior to lamination of each layer. Finally, high-density parts can be obtained after further binder removal and sintering at high temperatures. Attributed to the low thermal stresses in the manufacturing process, the LOM technique could eliminate the distortion and deformation. However, the main disadvantages are also directly related to such a process, since delamination, interfacial porosities and anisotropic properties along the planar directions and the building direction are the common problems associated with weak interfacial bonding behaviors between layers. Its implementation in ceramic fabrication mostly focuses on simple structures such as gear prototypes.

The fused deposition modelling (FDM) method, also known as fused deposition of ceramics (FDC), is one of the most commonly used 3D printing techniques. According to the path controlled by computer-aided design (CAD) layered data, the hot melt nozzle is used to squeeze the hot melt semiflow material so that the material is deposited in a specified position and solidified, and the solid part is finally obtained through deposition layer by layer. At present, the FDM technology has the following problems: (1) the printing precision of parts is low, and the surface of the parts is rough; (2) compared with other AM technologies, the printing speed is slow, and it is not suitable for preparing large parts. Therefore, FDM is mainly used commercially in the preparation of small-scale customized products, and there is still some distance away from largescale applications.

(**Selected from:** Wen C. Structural Biomaterials: Properties, Characteristics, and Selection [M]. Woodhead Publishing, (2021-04-06) [2023-02-22] https://www.elsevier.com/books/structural-biomaterials/wen/978-0-12-818831-6.

Baldacchini T. Three-Dimensional Microfabrication Using Two-Photon Polymerization [M]. 2nd ed. Elsevier, 2020.

Shi Y, Yan C, Zhou Z, et al. Materials for Additive Manufacturing [M/CD]. 1st ed. Elsevier, 2021.)

New Words and Expressions

  feedstock *n.* 原料，给料（指供送入机器或加工厂的原料）
  slurry *n.* 泥浆，悬浮液
  viscosity *n.* 黏性，黏度
  stereolithography *n.* 立体光刻
  inkjet *n.* 喷墨，喷墨打印机
  laminated *adj.* 层压的，层积的，薄板状的
  ultraviolet *adj.* 紫外的，紫外线的

debinding   *n*. 脱脂，脱黏
cellular   *adj*. 蜂窝状的，蜂窝式的
scaffold   *n*. 支架
projection   *n*. 投影，投影图
nozzle   *n*. 喷嘴，管口
rheological   *adj*. 流变学的，液流学的
isostatic pressing   *n*. 等静压
tomography   *n*. X射线断层摄影术
densification   *n*. 密实化
isotropic   *adj*. 各向同性的
delamination   *n*. 分层

Notes

(1) Compared with traditional fabrication methods of ceramics, the 3D printing technology owns advantages of high efficiency, high resolution, low cost, efficient utilization of material and most significant, capability of fabricating intricate 3D structures, which is extremely useful for fabricating precise components in various fields, such as construction, medicine and aerospace.

——这是一个长难句+定语从句。
——参考译文：与传统的陶瓷制造方法相比，3D打印技术具有高效、高分辨率、低成本、材料高效利用等优点，最为重要的是它能够制造复杂的3D结构，这对制造用于如建筑、医药和航空航天等各个领域的精密部件极为有用。

(2) To date, with the latest advances in material science and computer science, virous 3D printing technologies have been specifically developed for ceramic manufacturing.

——with the latest advances in... 随着……的最新进展。
——参考译文：迄今为止，随着材料科学和计算机科学的最新进展，各种3D打印技术已被具体地开发并用于陶瓷制造业。

(3) During the SLA process, ceramic-containing slurry layers are radiated by a light source of a certain wavelength (usually in the ultraviolet range), causing the polymerization of the slurry layers with incorporation of the ceramic particles.

——be radiated by... 被……照射。
——参考译文：在立体光刻过程中，含有陶瓷的浆料层被特定波长（通常在紫外范围内）的光源照射，导致掺有陶瓷颗粒的浆料层聚合。

(4) The ultra-fast light switching and integral projection allow the DLP 3D printing process time to be dramatically reduced as it is much faster than the conventional SLA point-line-layer scanning process.

——参考译文：超快的光切换和集成投影使得数字光处理3D打印过程时间大大缩短，因为它比传统的立体光刻点-线-层扫描过程快得多。

(5) The higher efficiency of DLP over the conventional SLA process makes it a promising 3D printing technique for ceramic fabrication.

— 参考译文：与传统立体光刻工艺相比，数字光处理的更高效率使其成为制造陶瓷的（更）有前景的 3D 打印技术。

(6) The ceramic particles are bonded either by spreading liquid binders or by powder fusion using thermal energy provided by a laser beam.

— 参考译文：陶瓷颗粒通过散布液体黏结剂或使用激光束提供的热能进行粉末熔接来黏结。

Exercises

1. Questions for discussion

(1) Please identify 3D-printing.

(2) What is the advantage of 3D-printing ceramics?

(3) What is the difference between SLA and 3D-printing?

(4) Please illustrate the first application of LOM.

2. Translate the following into Chinese

(1) However, the main disadvantages are also directly related to such a process, since delamination, interfacial porosities and anisotropic properties along the planar directions and the building direction are the common problems associated with weak interfacial bonding behaviors between layers.

(2) After this, a new layer of powder is spread onto the previous surface for the next run of heating and joining.

(3) Moreover, additional processing steps, such as infiltration and isostatic pressing, should be considered to enable the fabrication of structurally sound parts with desirable strength and little geometrical distortion.

3. Translate the following into English

(1) 选择性激光熔化　　　(2) 叠层实体制造

(3) 利用率　　　　　　　(4) 允许，使……有可能

(5) 然后提供一层新粉末并将其散布在前一层上以重复构建过程，直到形成零件。

(6) 因此，SLM 可以在不进行后处理的情况下生产几乎完全致密的均匀零件，因为它能够将粉末完全熔化到液相，确保快速致密化，而不是像 SLS 那样将粉末加热到特定点，在该点上颗粒部分熔化并熔融在一起。

(7) 由于制造过程中的热应力较低，LOM 技术可以消除扭曲和变形。

(8) 然而，由于多孔微结构、粗糙表面光洁度和低尺寸精度等缺陷的产生，SLM 难以实现非多孔各向同性陶瓷体的致密化，这极大地限制了其应用。

(9) 在层压每一层之前，逐层去除正在制造的部件周围的多余材料。

(10) 它在陶瓷制造中的应用场景主要集中在简单的结构上，例如齿轮原型。

Unit 3.4

Text

Ceramics solid electrolytes

这篇课文主要介绍用于锂金属电池的固态陶瓷电解质材料，主要包括氧化物、硫化物、磷酸盐以及硼酸盐。课文简述了常见的固态陶瓷电解质材料的特点、组分、制备方法以及性能。

Lithium-metal batteries (LMBs) with a large capacity or a high operating voltage can deliver better performance in comparison with commercial lithium-ion batteries. However, the large-scale application of LMBs is seriously hampered by safety issues associated with the liquid electrolyte. The replacement of liquid electrolytes with solid-state electrolytes has been considered as a promising solution. Inorganic ceramics electrolytes exhibit a high Li^+ conductivity that is close to that of liquid organic electrolytes, good thermal and electrochemical stability and excellent mechanical strength, and thus have been considered one of the most promising candidate electrolytes for solid-sate LMBs in the future.

Oxide-based solid electrolytes

Oxide-based solid electrolytes have received an increasing attention owing to their high chemical stability and a wide electrochemical stability domain. Among the family of oxides, Li-containing perovskites are of particular interest. The peculiar mechanical features of the ABO_3 perovskite structure can withstand a variety of distortions depending on the change of ionic substitution, pressure and temperature. For example, $Li_{3-x}La_{2/3-x}TiO_3$ (LLTO) shows a value of ionic conductivity as high as 10^{-3} S·cm^{-1}. Nevertheless, crystalline LLTO suffers from an instability towards metallic lithium due to a low lithium intercalation potential (less than 1.8 V vs. Li/Li^+) and an increased electronic conductivity caused by the reduction of Ti^{4+} to Ti^{3+}. Additionally, high grain boundary resistance would further diminish the global conductivity below 10^{-3} S·cm^{-1}.

Garnet-type $Li_7La_3Zr_2O_{12}$ (LLZO) has also been intensively investigated as a promising ceramic solid electrolyte owing to its relatively high ionic conductivity (about 10^{-3} S·cm^{-1}), broad potential window (over 6 V), mechanical stiffness and hardness. However, the tetragonal phase of LLZO with a low conductivity is thermodynamically more stable than the cubic phase counterpart with a high conductivity. As a result, the introduction of foreign doping elements (e.g. Ga or Al) in the framework is usually needed to ensure the stabilization of cubic phase at lower temperatures. Partial substitution of Zr with Nb or Ta is favorable to increasing ionic conductivity and improving chemical stability window with no side reactions occurred between 0 and 9 V vs. Li/Li^+.

Another major issue for garnet materials is its high interfacial resistances caused by a poor interfacial wetting, which hinders a uniform contact with cathode materials and Li metal and leads to uneven current distribution. Utilizing high external pressure to increase the surface contact, appropriately adjusting electrolyte particle size and constructing artificial interlayers to improve the superficial wetting have been proposed to overcome the above-mentioned challenge.

Boron-silicate Li_2O-B_2O_3-SiO_2 glass can be used as an additive to cubic LLZO. The introduction of 1 wt% glass could result in a twofold increase in conductivity at room temperature. Nevertheless, further glass introduction would cause a drop in conductivity and a rise in activation energy, possibly be due to the high resistance of the glassy phase.

In addition to traditional perovskites and garnet oxides, new oxide structures with a composition of $Li_{29}Zr_9Nb_3O_{40}$ (LZNO) was also explored as solid electrolytes. The undoped and Al-doped LZNO were synthesized via a sol-gel process followed by annealing. The precursor was dissolved in deionized water containing a 20mol% excess of Li_2CO_3 and different amounts of $Al(NO_3) \cdot 9H_2O$. After dehydration, the obtained gel was first heated to 300℃ to remove the organic component and then annealed at 700℃ for 5 hours. Last, the cold-pressed pellets were sintered at 700℃ for 5 h followed by 900℃ for 40 min. LZNO phase displays a distorted rock-salt orthorhombic structure with Li, Zr and Nb ions located at octahedral sites. The Al-doped LZNO with $x=0.5$ exhibits a high ionic conductivity of 2.41×10^{-4} S·cm^{-1} at room temperature, which is much higher than undoped LZNO (5.90×10^{-5} S·cm^{-1}). Nevertheless, further increase in Al content would lead to a decrement in conductivity, indicating an optimal Al content of $x=0.5$. The simulation investigations reveal that the ionic migration pathways follow an O-T-O diffusion mechanism via face-shared tetrahedral sites, which is in accordance with the typical diffusion in rock-salt structures.

Sulfide-based solid electrolytes

Sulfide-based solid electrolytes have also been intensively studied due to their high ionic conductivity. For example, oxysulfide amorphous glasses in the system Li_2S-SiS_2-Li_4SiO_4 obtained by melt-quenching process have been proven to display high conductivity over 10^{-3} S·cm^{-1}. Moreover, sulfide-based ion conductors generally exhibit a high mechanical strength and low grain boundary resistance. Unfortunately, these materials suffer from a poor compatibility with Li-metal at high voltages, low chemical stability and a high cost. To improve the stability of sulfide-based electrolytes, the partial inclusion of oxygen by introduction of metal oxides as H_2S absorbers or partial replacement of Li^+ ions by other cations with a similar ionic radius have been proposed. For example, the investigation on the effects of partial oxygen-doping in the $Li_{6.15}Al_{0.15}Si_{1.35}S_{6-x}O_x$ conductor suggests that the anion substitution significantly improve the performance in comparison with the undoped analogous counterpart. The value of ionic conductivity is up to 1.23×10^{-3} S·cm^{-1} when x is 0.6. Moreover, the evaluated stability window spans from 0 to 5 V without obvious side reactions occurred, suggesting that the oxygen introduction may be an effective solution to enhance the chemical stability of sulfides.

Recently, the superionic conductor $Li_{3.85}Sn_{0.85}Sb_{0.15}S_4$ has been explored as solid electrolyte. It has a Li_4SnS_4 or thorombic phase consisting of a series of isolated SnS_4 or SbS_4

tetrahedra which includes four different Li$^+$ allocation sites. Li$_{3.84}$Sn$_{0.85}$Sb$_{0.15}$S$_4$ not only exhibits a high ionic conductivity of 8.5×10^{-2} S·cm^{-1} but also shows the formation of hydrate phases and negligible H$_2$S evolution upon exposure to humid environment. The partial substitution of Sn with Sb in the lattice reduces the overall occupation of octahedral Li$^+$ sites and induces the formation of Li vacancies. Li$^+$ can migrate from tetrahedral to octahedral sites along four different diffusion pathways and the introduction of Sb may be responsible for the steep increase in conductivity.

Phosphate-based solid electrolytes

Compared with garnet LLZO, phosphates exhibit better stability to moist air and a lower volumetric density (3 g/cm^3). However, they generally show moderate ionic conductivities, in the order of $10^{-5}-10^{-4}$ S·cm^{-1} for Li$_{1+x}$Al$_x$Ge$_{2-x}$(PO$_4$)$_3$ (LAGP) and Li$_{1+x}$Al$_x$Ti$_{2-x}$(PO$_4$)$_3$ (LATP) or about 10^{-6} S·cm^{-1} for lithium phosphorus oxynitride LiPON. Additionally, LAGP and LATP also suffer from non-stability towards Li metal, leading to the reduction of Ti^{4+} and Ge^{4+} cations and a rapid increase of interface resistance. A thorough investigation of the Li interface behavior in LATP indicates that a phase transition would occur during lithium intercalation, enabling a Li-rich phase at Li interface. This phase is responsible for a rapid increase from 3×10^{-9} S·cm^{-1} to 2.9×10^{-6} S·cm^{-1} in electronic conductivity after cycling. This increased electronic conductivity may result in a local potential gradient that can facilitate lithium concentration in specific sites, and thus leading to dendrite evolution and formation of electronic-conductive byproducts. To address this issue, the introduction of an electronic insulating layer at the lithium interface is necessary to prevent the undesired side reactions.

LiPON has advantages of broad potential window up to 5.5 V and compatibility with Li metal. A solid-state thin film battery with LiPON exhibits a good capacity retention of 90% over 10000 cycles, together with a negligible electrolyte degradation against metallic lithium, which can be attributed to the uniform morphology of LiPON obtained by the sputtering process. In addition to LiPON, a new Li-ion conductive material LiTa$_2$PO$_8$ has also been explored. The LiTa$_2$PO$_8$ can be synthesized by a solid-state reaction between Li$_2$CO$_3$ (10 mol% excess), Ta$_2$O$_5$ and (NH$_4$)$_2$HPO$_4$. LiTa$_2$PO$_8$ has a monoclinic crystal lattice, which is consistent with the C2/c space group. The crystal unit consists of three Ta atoms, three Li atoms and eight O atoms. The Ta atoms located in octahedral sites generate two different TaO$_6$ octahedra, which are connected to form a 2D grid layer corresponding to [Ta$_3$O$_{18/2}$]$^{3-}$. While another TaO$_6$ is linked to PO$_4$ tetrahedra and form the trimer [TaP$_2$O$_{14/2}$]$^+$. The [Ta$_3$O$_{18/2}$]$^{3-}$ 2D grid and the trimers assemble together to construct a 3D framework, in which void spaces are occupied by Li atoms in the three discrete sites Li(1), Li(2) and Li(3). LiTa$_2$PO$_8$ exhibit a relatively high total ionic conductivity of 2.5×10^{-4} S·cm^{-1} but a low electronic conductivity of 4.1×0^{-9} S·cm^{-1} at room temperature, comparable with those of LATP and cubic LLZO.

Borate-based solid electrolytes

Lithium borate (Li_3BO_3) could be used as a sintering additive to decrease the preparation temperature of LLZO. The LLZO-Li_3BO_3 composite can be sintered at 900℃, exhibiting an ionic conductivity of 10^{-4} S·cm^{-1} at 30℃. The amorphous borate within the composite can act as a sintering agent and generate a thins layer at the grain boundaries, not only decreasing the boundary resistance but also improving the total ionic conductivity of the composite. Recently, a series of Li_3BO_3-Li_2SO_4 glass and glass-ceramic materials prepared via mechanical milling and subsequent thermal treatment have been reported. Raman spectroscopy investigation shows the presence of characteristic bands related to SO_4^{2-} and BO_3^{3-}, suggesting the formation of a net structure. After annealing at 400℃, the average values of ionic conductivities for glass-ceramic product is about two orders of magnitude higher than that for un-treated glass counterpart. Among the prepared glass ceramics, $90Li_3BO_3 \cdot 10Li_2SO_4$ (mol% composition) exhibits a conductivity of 1.4×10^{-5} S·cm^{-1} at room temperature, together with good mechanical properties.

Borohydrides also represent a promising alternative to sulfides in solid electrolytes. They not only exhibit a remarkable stability in both lithium and sodium environments (broad electrochemical stability windows can exceed 5 V vs. Li/Li^+) but also show a facile deformation property. These features allow them to create a good interface with active electrodes. The lithium borohydride ($LiBH_4$) would undergo an orthorombic-to-hexagonal phase transition above 380 K, with the latter phase showing an ionic conductivity of the order of 10^{-3} S·cm^{-1} due to Li superionic conduction. The introduction of lithium halides (e.g., LiI, LiBr and LiCl) can reduce the structural transition temperature. Nevertheless, the $LiBH_4$ used in cell would react with lithium cobaltoxide (LCO) cathode and exhibit a severe capacity decay. Introducing an intermediate Li_3PO_4 layer between $LiBH_4$ and LCO has been demonstrated as an effective strategy in decreasing the interface resistance.

Recently, a mixed amide $Li_2BH_4NH_2$-borohydride $Li_4(BH_4)(NH_2)_3$ phases with a low-temperature ionic conductivity in the order of 10^{-4} S·cm^{-1} has been reported. In a typical preparation, the mixture of $LiBH_4$ and $LiNH_2$ is mechanically milled and then heated at 333 K for $Li_2BH_4NH_2$ and 373 K for $Li_4(BH_4)(NH_2)_3$ in argon atmosphere. $Li_2BH_4NH_2$ displays a trigonal structure, while $Li_4(BH_4)(NH_2)_3$ presents a cubic structure, both of which are distinctively different from the crystal structures of $LiBH_4$ and $LiNH_2$. These complex borohydrides show distinct Li^+ sites within the crystal framework.

(**Selected from:** Campanella D, Belanger D, Paolella A. Beyond garnets, Phosphates and Phosphosulfides Solid Electrolytes: New Ceramic Perspectives for All Solid Lithium Metal Batteries [J]. Journal of Power Sources, 2021, 482: 228949.

Liu Y, Xu B, Zhang W, et al. Composition Modulation and Structure Design of Inorganic-in-Polymer Composite Solid Electrolytes for Advanced Lithium Batteries [J]. Small, 2020, 16 (15): 1902813.)

New Words and Expressions

electrolyte   *n.* 电解质

perovskite　*n*. 钙钛矿
distortion　*n*. 扭曲，变形
intercalation　*n*. 夹入，插入
diminish　*vt*. 减少，减损，贬低
garnet　*n*. 石榴石
tetragonal　*adj*. 四角形的，四边形的，四棱的，四方相
superficial　*adj*. 表面的，肤浅的
dehydration　*n*. 脱水
inclusion　*n*. 包含
orthorombic　*adj*. 正交晶系的；斜方晶系的
volumetric　*adj*. 体积的，容积的
oxynitride　*n*. 氮氧化合物
dendrite　*n*. 树突，树枝状结晶
borate　*n*. 硼酸盐
halide　*n*. 卤化物

Notes

(1) Inorganic ceramics electrolytes exhibit a high Li^+ conductivity that is close to that of liquid organic electrolytes, good thermal and electrochemical stability and excellent mechanical strength, and thus have been considered one of the most promising candidate electrolytes for solid-sate LMBs in the future.

——参考译文：无机陶瓷电解质具有接近液态有机电解质的高锂离子电导率、良好的热稳定性和电化学稳定性以及优异的机械强度，因此被认为是未来固态锂金属电池最有希望的候选电解质之一。

(2) The peculiar mechanical features of the ABO_3 perovskite structure can withstand a variety of distortions depending on the change of ionic substitution, pressure and temperature.

——参考译文：具有独特力学特性的 ABO_3 钙钛矿结构，可以承受随着离子取代、压力和温度的变化而发生的各种变形。

(3) Partial substitution of Zr with Nb or Ta is favorable to increasing ionic conductivity and improving chemical stability window with no side reactions occurred between 0 and 9 V vs. Li/Li^+.

——参考译文：用 Nb 或 Ta 部分取代 Zr 能够提高离子电导率，改善化学稳定性窗口，并且在 0—9 V（vs. Li/Li^+）之间没有发生副反应。

(4) A solid-state thin film battery with LiPON exhibits a good capacity retention of 90% over 10000 cycles, together with a negligible electrolyte degradation against metallic lithium, which can be attributed to the uniform morphology of LiPON obtained by the sputtering process.

——参考译文：用 LiPON 制备的固态薄膜电池表现出良好的循环性能（10000 次循环后，容量保持率为 90%），并且在金属锂上的电解液降解可以忽略不计，这可以归因于通过溅射工艺获得的均匀 LiPON。

Chapter 4

Polymers

【本章导读】

本章介绍聚合物材料，包括五个单元，共有五篇课文和四篇阅读材料。五篇课文分别介绍聚合物的基本概念、合成方法、结构特征、食品包装应用和光控微流体的新技术。四篇阅读材料分别讲解了聚合物的结晶度、聚合物逐步聚合合成方法、聚合物材料的发展历史、生物聚合物材料及其应用。

Unit 4.1

Text

Introduction to polymers

这篇课文讲述了聚合物的发展历程与现状，讨论了聚合物的定义和一些典型结构。文中还介绍了聚合物的一种经典分类方法，该方法将聚合物分为热塑性聚合物、弹性体和热固性聚合物三类，并讨论了各类聚合物的主要特点。

Polymers have existed in natural form since life began, and those such as DNA, RNA, proteins and polysaccharides play crucial roles in plant and animal life. From the earliest times, man has exploited naturally-occurring polymers as materials for providing clothing, decoration, shelter, tools, weapons, writing materials and other requirements. However, the origins of today's polymer industry commonly are accepted as being in the nineteenth century when important discoveries were made concerning the modification of certain natural polymers.

In 1820, Thomas Hancock discovered that when masticated (i. e. subjected repeatedly to high shear forces), natural rubber becomes more fluid making it easier to blend with additives and to mould. Some years later, in 1839, Charles Goodyear found that the elastic properties of natural rubber could be improved, and its tackiness eliminated, by heating with sulphur. Patents for this discovery were issued in 1844 to Goodyear, and slightly earlier to Hancock, who christened the process vulcanization. In 1851, Nelson Goodyear, Charles' brother, patented the vulcanization of natural rubber with large amounts of sulphur to produce a hard material more commonly known as hard rubber, ebonite or vulcanite.

Cellulose nitrate, also called nitrocellulose or gun cotton, first became prominent after Christian Schönbein prepared it in 1846. He was quick to recognize the commercial value of this material as an explosive, and within a year gun cotton was being manufactured. However, more important to the rise of the polymer industry, cellulose nitrate was found to be a hard elastic material which was soluble and could be moulded into different shapes by the application of heat and pressure. Alexander Parkes was the first to take advantage of this combination of properties and in 1862 he exhibited articles made from Parkesine, a form of plasticized cellulose nitrate. In 1870, John and Isaiah Hyatt patented a similar but more easily processed material, named celluloid, which was prepared using camphor as the plasticizer. Unlike Parkesine, celluloid was a great commercial success.

In 1892, Charles Cross, Edward Bevan and Clayton Beadle patented the "viscose process" for dissolving and then regenerating cellulose. The process was first used to produce viscose rayon textile fibers, and subsequently for the production of cellophane film.

The polymeric materials described so far are semi-synthetic since they are produced from natural polymers. Leo Baekeland's phenol-formaldehyde "Bakelite" resins have the distinction of being the first fully synthetic polymers to be commercialized, their production beginning in 1910. The first synthetic rubber to be manufactured, known as methyl rubber, was produced from 2,3-dimethylbutadiene in Germany during World War I as a substitute, albeit a poor one, for natural rubber.

Although the polymer industry was now firmly established, its growth was restricted by the considerable lack of understanding of the nature of polymers. For over a century, scientists had been reporting the unusual properties of polymers, and by 1920, the common belief was that they consisted of physically-associated aggregates of small molecules. Few scientists gave credence to the viewpoint so passionately believed by Hermann Staudinger, that polymers were composed of very large molecules containing long sequences of simple chemical units linked together by covalent bonds. Staudinger introduced the word "macromolecule" to describe polymers, and during the 1920s, vigorously set about proving his hypothesis to be correct. Particularly important were his studies of the synthesis, structure and properties of polyoxymethylene and of polystyrene, the results from which left little doubt as to the validity of the macromolecular viewpoint. Staudinger's hypothesis was further substantiated by the crystallographic studies of natural polymers reported by Herman Mark and Kurt Meyer, and by the classic work of Wallace Carothers on the preparation of polyamides and polyesters. Thus by the early 1930s, most scientists were convinced of the macromolecular structure of polymers. During the following 20 years, work on polymers increased enormously: the first journals devoted solely to their study were published and most of the fundamental principles of Polymer Science were established. The theoretical and experimental work of Paul Flory was prominent in this period, and for his long and substantial contribution to Polymer Science, he was awarded the Nobel Prize for Chemistry in 1974. In 1953, Staudinger had received the same accolade in recognition of his pioneering work.

Not surprisingly, as the science of macromolecules emerged, a large number of synthetic polymers went into commercial production for the first time. These include polystyrene, poly (methyl methacrylate), nylon 6.6, polyethylene, poly (vinyl chloride), styrene-butadiene rubber, silicones and polytetrafluoroethylene, as well as many others. From the 1950s onwards, regular advances, too numerous to mention here, have continued to stimulate both scientific and industrial progress, and as the discipline of polymer science progresses into the twenty-first century there is increasing emphasis on the development of more specialized, functional polymers for biomedical, optical and electronic applications.

Whilst polymer science undoubtedly is now a mature subject, its breadth and importance continue to increase and there remain many demanding challenges awaiting scientists who venture into this fascinating multidisciplinary science.

Several important terms and concepts must be understood in order to discuss fully the synthesis, characterization, structure and properties of polymers. Some are of such fundamental importance that they must be defined at the outset.

In strict terms, a polymer is a substance composed of molecules which have long sequences of one or more species of atoms or groups of atoms linked to each other by primary, usually covalent, bonds. The emphasis upon substance in this definition is to highlight that although the words polymer and macromolecule are used interchangeably, the latter strictly defines the molecules of which the former is composed.

Macromolecules are formed by linking together monomer molecules through chemical reactions, the process by which this is achieved being known as polymerization. For example, polymerization of ethylene yields polyethylene, a typical sample of which may contain molecules with 50,000 carbon atoms linked together in a chain. It is this long chain nature which sets polymers apart from other materials and gives rise to their characteristic properties.

The definition of macromolecules presented up to this point implies that they have a linear skeletal structure which may be represented by a chain with two ends. Whilst this is true for many macromolecules, there are also many with non-linear skeletal structures of the type shown in **Figure 4.1.1**.

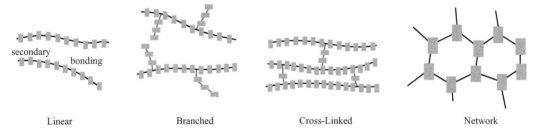

Figure 4.1.1　Skeletal structures representative of linear, and non-linear polymers.

Cyclic polymers (ring polymers) have no chain ends and show properties that are quite different to their linear counterparts. Branched polymers have side chains, or branches, of significant length which are bonded to the main chain at branch points (also known as junction points), and are characterized in terms of the number and size of the branches. Network polymers have three dimensional structures in which each chain is connected to all others by a sequence of junction points and other chains. Such polymers are said to be crosslinked and are characterized by their crosslink density, or degree of crosslinking, which is related directly to the number of junction points per unit volume.

Branched and network polymers may be formed by polymerization, or can be prepared by linking together (i. e. crosslinking) pre-existing chains. These variations in skeletal structure give rise to major differences in properties. For example, linear polyethylene has a melting point about 20℃ higher than that of branched polyethylene. Unlike linear and branched polymers, network polymers do not melt upon heating and will not dissolve, though they may swell considerably in compatible solvents. The importance of crosslink density has already been described in terms of the vulcanization (i. e. sulphur-crosslinking) of natural rubber. With low crosslink densities (i. e. low levels of sulphur) the product is a flexible elastomer, whereas it is a rigid material when the crosslink density is high.

In addition to these more conventional skeletal structures, there has been growing interest in more elaborate skeletal forms of macromolecules. Of particular interest are dendrimers, which are highly branched polymers of well-defined structure and molar mass, and hyperbranched polymers, which are similar to dendrimers but have a much less well-defined structure and molar mass. Simple depiction of the skeletal structure of the dendrimer polymer is shown in **Figure 4.1.2**. Research into these types of polymers intensified during the 1990s and they are now beginning to find applications which take advantage of their unusual properties. For example, because of their high level of branching, they are extremely crowded but as a consequence have voids and channels within the molecule and have a large number of end groups around their periphery that can be functionalized, leading to therapeutic applications such as in targeted drug delivery.

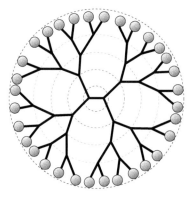

Figure 4.1.2　Skeletal structure of the dendrimer polymer.

The most common way of classifying polymers is outlined in **Figure 4.1.3** where they are first separated into three groups: *thermoplastics*, *elastomers* and *thermosets*. Thermoplastics are then further separated into those which are crystalline and those which are amorphous (i. e. non-crystalline). This method of classification has an advantage in comparison to others since it is based essentially upon the underlying molecular structure of the polymers.

Thermoplastics, often referred to just as plastics, are linear or branched polymers which become liquid upon the application of heat. They can be moulded (and remoulded) into virtually any shape using processing techniques such as injection moulding and extrusion, and now constitute by far the largest proportion of the polymers in commercial production.

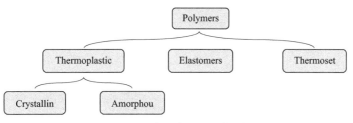

Figure 4.1.3 Classification of polymers.

Generally, thermoplastics do not crystallize easily upon cooling to the solid state because this requires considerable ordering of the highly coiled and entangled macromolecules present in the liquid state. Those which do crystallize invariably do not form perfectly crystalline materials but instead are semi-crystalline with both crystalline and amorphous regions. The crystalline phases of such polymers are characterized by their melting temperature T_m, above which such polymers can be converted into artefacts by conventional polymer-processing techniques such as extrusion, injection moulding and compression moulding.

Many thermoplastics are, however, completely amorphous and incapable of crystallization, even upon annealing. Amorphous polymers (and amorphous phases of semi-crystalline polymers) are characterized by their glass transition temperature T_g, the temperature at which they transform abruptly from the glassy state (hard) to the rubbery state (soft). This transition corresponds to the onset of chain motion; below T_g the polymer chains are unable to move and are "frozen" in position. Both T_m and T_g increase with increasing chain stiffness and increasing forces of intermolecular attraction.

It is a common misnomer that completely amorphous polymers "melt"; they do not (because they have no ordered phases, there is nothing to melt!) and may simply be considered as reducing steadily in viscosity as temperature increases above T_g until the viscosity becomes low enough for so-called melt processing.

Elastomers are crosslinked rubbery polymers (i.e. rubbery networks) that can be stretched easily to high extensions (e.g. $3\times$ to $10\times$ their original dimensions) and which rapidly recover their original dimensions when the applied stress is released. This extremely important and useful property is a reflection of their molecular structure in which the network is of low crosslink density. The rubbery polymer chains become extended upon deformation but are prevented from permanent flow by the crosslinks, and driven by entropy, spring back to their original positions on removal of the stress. The word rubber, often used in place of elastomer, preferably should be used for describing rubbery polymers that are not crosslinked.

Thermosets normally are rigid materials and are network polymers in which chain motion is greatly restricted by a high degree of crosslinking. As for elastomers, they are intractable once formed and degrade rather than become fluid upon the application of heat. Hence, their processing into artefacts is often done using processes, such as compression moulding, that require minimum amounts of flow.

(**Selected from:** Young R J, Lovell P A. Introduction to Polymers [M]. 3rd ed. CRC Press, 2011.
Callister W D, Rethwisch D G. Materials Science and Engineering - An Introduction: vol 1 [M]. 10th ed. Wiley, 2018.)

New Words and Expressions

polysaccharide   *n*. 多糖，多聚糖
masticate   *v*. 咀嚼，粉碎，撕捏
mould   *n*. 模具，铸模；*v*. 使……成形，用模子制作，浇铸
tackiness   *n*. 黏着性
christen   *v*. 为……命名
vulcanization   *n*. 橡胶的硫化
ebonite   *n*. 硬橡胶，硬化橡皮，硬橡皮
vulcanite   *n*. 硬橡皮（经硫化处理的橡皮），硬橡胶
nitrocellulose   *n*. 硝化纤维（素），火棉
prominent   *adj*. 重要的，著名的
viscose   *n*. 纤维胶，人造丝
distinction   *n*. 差别，区分，杰出，卓越
albeit   *conj*. 虽然，尽管
crystallographic   *adj*. 结晶的
substantiate   *v*. 证实，使实体化
substantial   *adj*. 大量的，基本的，实质性的
convince   *v*. 使确信，使信服，说服，劝服
many others   许多其他的
at the outset   开始，起初
interchangeable   *adj*. 可互换的，可交换的，可交替的
monomer   *n*. 单体，单元结构
compatible   *adj*. 兼容的，可共存的
elaborate   *adj*. 复杂的，详尽的；*v*. 详细说明，详尽阐述
dendrimer   *n*. 聚合物，树形分子
well-defined   *adj*. 界限清楚的，定义明确的
periphery   *n*. 外围，边缘，圆周
therapeutic   *adj*. 治疗的；*n*. 疗法，治疗药物
extrusion   *n*. 挤出，推出，喷出
coil   *n*. 卷，匝，线圈；*v*. 卷，盘绕
entangle   *vt*. 使纠缠，卷入，使混乱
invariably   *adv*. 始终如一地，一贯地
artefact   *n*. 手工制品，人造物
abruptly   *adv*. 突然地，陡然

Notes

（1）Polymers have existed in natural form since life began，and those such as DNA，RNA，proteins and polysaccharides play crucial roles in plant and animal life.

—play roles in 在……中发挥作用。

—参考译文：自生命诞生以来，聚合物就以天然的形式存在，如 DNA、RNA、蛋白质和多糖（等聚合物）在动植物的生命中起着至关重要的作用。

（2）From the 1950s onwards, regular advances, too numerous to mention here, have continued to stimulate both scientific and industrial progress.

—too numerous to mention 不胜枚举。

—参考译文：自 20 世纪 50 年代起，（聚合物材料）不胜枚举的进展持续地激励着科学和工业的进步。

（3）Some are of such fundamental importance that they must be defined at the outset.

—of importance 相当于 important，为形容词。

—参考译文：一些术语具有根本的重要性，必须在一开始就将它们加以定义。

（4）These variations in skeletal structure give rise to major differences in properties.

—give rise to 导致，类似于 lead to，result in 等短语。

—参考译文：骨架结构上的这些变化导致了它们性能上的巨大差异。

（5）The polymeric materials described so far are semi-synthetic since they are produced from natural polymers.

—参考译文：迄今为止提及的聚合物材料都是基于天然聚合物制备而成的，因此它们被认为是半合成的（聚合物）。

（6）Particularly important were his studies of the synthesis, structure and properties of polyoxymethylene and of polystyrene, the results from which left little doubt as to the validity of the macromolecular viewpoint.

—参考译文：他对聚甲醛和聚苯乙烯的合成、结构和性质的研究尤为重要，其研究结果使得大分子观点的有效性毋庸置疑。

Exercises

1. Question for discussion

（1）What is a polymer?

（2）What is the difference between the polymer and macromolecule?

（3）What is the polymerization?

（4）What is the linear polymer? cyclic polymer? branched polymer? and network polymer?

（5）What is the classification of polymers?

2. Translate the following into Chinese

（1）However, the origins of today's polymer industry commonly are accepted as being in the nineteenth century when important discoveries were made concerning the modification of certain natural polymers.

(2) Of particular interest are dendrimers, which are highly branched polymers of well-defined structure and molar mass, and hyperbranched polymers, which are similar to dendrimers but have a much less well-defined structure and molar mass.

(3) Generally, thermoplastics do not crystallize easily upon cooling to the solid state because this requires considerable ordering of the highly coiled and entangled macromolecules present in the liquid state.

3. Translate the following into English

(1) 甲基　　　　　　　　　　(2) 信心
(3) 精神旺盛地，活泼地　　　(4) 大分子
(5) 聚合作用　　　　　　　　(6) 骨骼的
(7) 药物输送　　　　　　　　(8) 膨胀
(9) 一些术语具有根本的重要性，必须在一开始就将它们加以定义。
(10) 随着大分子科学的出现，大量合成聚合物首次进入商业生产。

Reading Material
Polymer crystallinity

这篇课文介绍了聚合物结晶度的相关知识。课文阐述了聚合物结晶度的概念和计算方法，分析了影响结晶度的因素，并展示了一些典型的晶态聚合物的形态。

The crystalline state may exist in polymeric materials. However, because it involves molecules instead of just atoms or ions, as with metals and ceramics, the atomic arrangements will be more complex for polymers. We think of polymer crystallinity as the packing of molecular chains to produce an ordered atomic array. Crystal structures may be specified in terms of unit cells, which are often quite complex. For example, **Figure 4.1.4** shows the unit cell for polyethylene and its relationship to the molecular chain structure; this unit cell has orthorhombic geometry. Of course, the chain molecules also extend beyond the unit cell shown in the figure.

Molecular substances having small molecules (e.g., water and methane) are normally either totally crystalline (as solids) or totally amorphous (as liquids). As a consequence of their size and often complexity, polymer molecules are often only partially crystalline (or semicrystalline), having crystalline regions dispersed within the remaining amorphous material. Any chain disorder or misalignment will result in an amorphous region, a condition that is fairly common, because twisting, kinking, and coiling of the chains prevent the strict ordering of every segment of every chain. Other structural effects are also influential in determining the extent of crystallinity.

The degree of crystallinity may range from completely amorphous to almost entirely (up to about 95%) crystalline; in contrast, metal specimens are almost always entirely crystalline, whereas many ceramics are either totally crystalline or totally noncrystalline. Semicrystalline polymers are, in a sense, analogous to two-phase metal alloys.

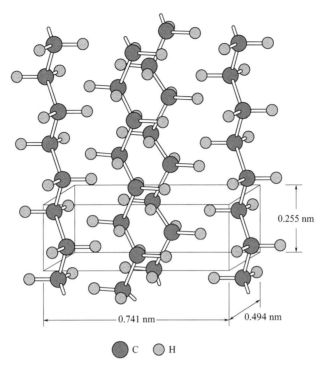

Figure 4.1.4 Arrangement of molecular chains in a unit cell for polyethylene.

The density of a crystalline polymer will be greater than an amorphous one of the same material and molecular weight because the chains are more closely packed together for the crystalline structure. The degree of crystallinity by weight may be determined from accurate density measurements, according to

$$\%\text{Crystallinity} = \frac{\rho_c (\rho_s - \rho_a)}{\rho_s (\rho_c - \rho_a)} \times 100\% \qquad (4.1.1)$$

where ρ_s is the density of a specimen for which the percent crystallinity is to be determined, ρ_a is the density of the totally amorphous polymer, and ρ_c is the density of the perfectly crystalline polymer. The values of ρ_a and ρ_c must be measured by other experimental means.

The degree of crystallinity of a polymer depends on the rate of cooling during solidification as well as on the chain configuration. During crystallization upon cooling through the melting temperature, the chains, which are highly random and entangled in the viscous liquid, must assume an ordered configuration. For this to occur, sufficient time must be allowed for the chains to move and align themselves.

The molecular chemistry as well as chain configuration also influence the ability of a polymer to crystallize. Crystallization is not favored in polymers that are composed of chemically complex repeat units (e. g., polyisoprene). However, crystallization is not easily prevented in chemically simple polymers such as polyethylene and polytetrafluoroethylene, even for very rapid cooling rates.

For linear polymers, crystallization is easily accomplished because there are few restrictions to prevent chain alignment. Any side branches interfere with crystallization, such that branched polymers are never highly crystalline; in fact, excessive branching may prevent any crystallization whatsoever. Most network and crosslinked polymers are almost totally amorphous because the crosslinks prevent the polymer chains from rearranging and aligning into a crystalline structure. A few crosslinked polymers are partially crystalline. With regard to the stereoisomers, atactic polymers are difficult to crystallize; however, isotactic and syndiotactic polymers crystallize much more easily because the regularity of the geometry of the side groups facilitates the process of fitting together adjacent chains. Also, the bulkier or larger the side-bonded groups of atoms, the less is the tendency for crystallization.

For copolymers, as a general rule, the more irregular and random the repeat unit arrangements, the greater is the tendency for the development of noncrystallinity. For alternating and block copolymers there is some likelihood of crystallization. However, random and graft copolymers are normally amorphous.

To some extent, the physical properties of polymeric materials are influenced by the degree of crystallinity. Crystalline polymers are usually stronger and more resistant to dissolution and softening by heat.

It has been proposed that a semicrystalline polymer consists of small crystalline regions (crystallites), each having a precise alignment, which are interspersed with amorphous regions composed of randomly oriented molecules. The structure of the crystalline regions may be deduced by examination of polymer single crystals, which may be grown from dilute solutions. These crystals are regularly shaped, thin platelets (or lamellae) approximately 10 to 20 nm thick, and on the order of 10 μm long, as shown in **Figure 4.1.5**.

Many bulk polymers that are crystallized from a melt are semicrystalline and form a spherulite structure. As implied by the name, each spherulite may grow to be roughly spherical in shape; one of them, as found in natural rubber, is shown in the transmission electron micrograph in the margin photograph (**Figure 4.1.6**). The spherulite consists of an aggregate of ribbonlike chain-folded crystallites (lamellae) approximately 10 nm thick that radiate outward from a single nucleation site in the center. In this electron micrograph, these lamellae appear as thin white lines. The detailed structure of a spherulite is illustrated schematically in **Figure 4.1.7**. Shown here are the individual chain-folded lamellar crystals that are separated by amorphous material. Tie-chain molecules that act as connecting links between adjacent lamellae pass through these amorphous regions.

As the crystallization of a spherulitic structure nears completion, the extremities of adjacent spherulites begin to impinge on one another, forming more or less planar boundaries; prior to this time, they maintain their spherical shape. Spherulites are considered to be the polymer analogue of grains in polycrystalline metals and ceramics. However, as discussed earlier, each spherulite is really composed of many different lamellar crystals and, in addition, some amorphous material. Polyethylene, polypropylene, poly (vinyl chloride), polytetrafluoroethylene, and nylon form a spherulitic structure when they crystallize from a melt.

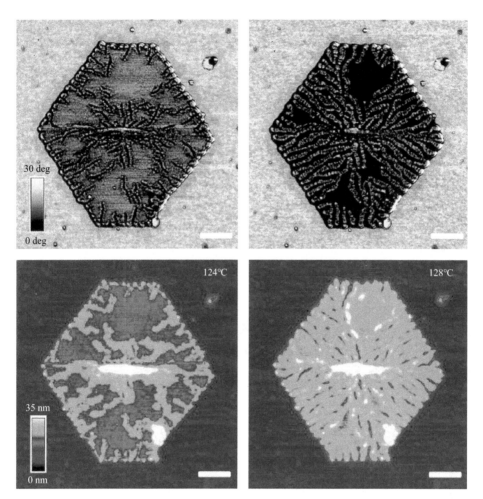

Figure 4.1.5　Morphology of a polyethylene single crystal with in situ annealing. The scale bar is 500 nm.

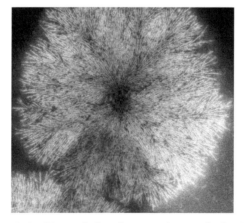

Figure 4.1.6　Transmission electron micrograph showing the spherulite structure in a natural rubber specimen.

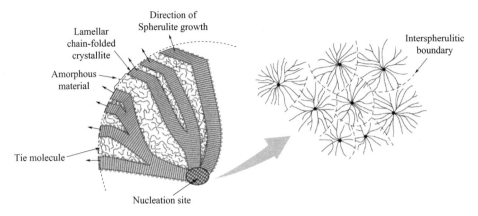

Figure 4.1.7　Schematic representation of the detailed structure of a spherulite.

New Words and Expressions

crystallinity　*n.* 结晶度
unit cell　*n.* 晶胞
orthorhombic　*adj.* 正交晶的，斜方晶系的
geometry　*n.* 几何，几何学，几何形状
kink　*n.* 扭结；*v.* （使）扭结
specimen　*n.* 样品，标本
whereas　*conj.* 但是，然而
viscous　*adj.* 黏性的，黏的
accomplish　*v.* 完成，实现
whatsoever　*adj.* 任何，无论什么
stereoisomer　*n.* 立体异构体
atactic　*adj.* 不规则的，协同不能的
isotactic　*adj.* 全同立构的，等规立构的
syndiotactic　*adj.* 间同的，间规的
likelihood　*n.* 可能，可能性
intersperse　*v.* 散布，在……中夹杂某物
lamellae　*n.* 片晶，薄片
aptly　*adv.* 恰当地，贴切地
spherulite　*n.* 球粒
aggregate　*n.* 总数，合计；*adj.* 合计的；*v.* 集合，聚集
impinge　*v.* 对……有明显作用（或影响），妨碍，侵犯

Notes

（1）Semicrystalline polymers are, in a sense, analogous to two-phase metal alloys.

　　—in a sense　在某种意义上。
　　—参考译文：在某种意义上来说，半结晶聚合物类似于两相金属合金。

(2) The degree of crystallinity of a polymer depends on the rate of cooling during solidification as well as on the chain configuration.

—参考译文：聚合物的结晶度取决于凝固过程中的冷却速率和聚合物链的构型。

(3) Also, the bulkier or larger the side-bonded groups of atoms, the less is the tendency for crystallization.

—参考译文：此外，侧键原子基团越大或体积越大，（聚合物）结晶的倾向就越小。

(4) The degree of crystallinity may range from completely amorphous to almost entirely (up to about 95%) crystalline; in contrast, metal specimens are almost always entirely crystalline, whereas many ceramics are either totally crystalline or totally noncrystalline.

—in contrast 与此相反，相比之下。
—参考译文：（聚合物）结晶度的范围可以从完全无定形到几乎完全（高达约95%）结晶；相比之下，金属样品几乎总是完全结晶，而许多陶瓷要么完全结晶，要么完全非晶。

(5) The molecular chemistry as well as chain configuration also influence the ability of a polymer to crystallize.

—as well as 以及，既……又……。
—参考译文：分子化学和链构型也影响着聚合物的结晶能力。

(6) To some extent, the physical properties of polymeric materials are influenced by the degree of crystallinity.

—to some extent 在某种程度上，在一定程度上。be influenced by … 由……影响。
—参考译文：在一定程度上，聚合物材料的物理性质受结晶度的影响。

(7) As the crystallization of a spherulitic structure nears completion, the extremities of adjacent spherulites begin to impinge on one another, forming more or less planar boundaries; prior to this time, they maintain their spherical shape.

—more or less 或多或少。prior to 在……之前。
—参考译文：当球晶结构的结晶完成之前，它们保持球形；当球晶结构的结晶接近完成时，相邻球晶的端部开始相互碰撞，形成或多或少的平面边界。

Unit 4.2

Text

Methods for synthesis of polymers

这篇课文主要介绍了聚合物的合成方法。课文阐述了聚合物合成的基本要求，分别介绍了缩合聚合、加成聚合、逐步聚合和连锁聚合的概念，并分析了单体对聚合物性质的影响。

The most basic requirement for polymerization is that each molecule of monomer must be capable of being linked to two (or more) other molecules of monomer by chemical reaction, i. e. monomers must have a functionality of two (or higher). Given this relatively simple requirement, there are a multitude of chemical reactions and associated monomer types that can be used to effect polymerization. Consequently, the number of different synthetic polymers that have been prepared is extremely large and many can be formed by more than one type of polymerization. Hence, the number of individually different polymerization reactions that have been reported is extraordinarily large. To discuss each of these would be an enormous task which fortunately is not necessary since it is possible to categorize most polymerization reactions into a relatively small number of classes of polymerization, each class having distinctive characteristics.

The purpose here is to set out some of the most important guiding principles of polymerization. The classification of polymerization reactions used in the formative years of polymer science was due to Carothers and is based upon comparison of the molecular formula of a polymer with that of the monomer (s) from which it was formed. Condensation polymerizations are those which yield polymers with repeat units having fewer atoms than are present in the monomers from which they are formed. This usually arises from chemical reactions which involve the elimination of a small molecule (e. g. H_2O, HCl). Addition polymerizations are those which yield polymers with repeat units having identical molecular formulae to those of the monomers from which they are formed.

Carothers' method of classification was found to be unsatisfactory when it was recognized that certain condensation polymerizations have the characteristic features of typical addition polymerizations and that some addition polymerizations have features characteristic of typical condensation polymerizations. A better basis for classification is provided by considering the underlying polymerization mechanisms, of which there are two general types. Polymerizations in which the polymer chains grow step-wise by reactions that can occur between any two molecular species are known as step-growth polymerizations. Polymerizations in which a polymer chain grows only by reaction of monomer with a reactive end-group on the growing chain are known as chain-growth polymerizations, and usually require an initial reaction between the monomer and an initiator to start the growth of the chain. The modern preference is to simplify these names to step polymerization and chain polymerization, and this practice will be used here.

In step polymerizations the degree of polymerization increases steadily throughout the reaction, but the monomer is rapidly consumed in its early stages (e. g. when $\bar{x}_n=10$ only 1% of the monomer remains unreacted). All the polymer chains continue to grow throughout the reaction as the conversion of functional groups into chain links increases. Carothers was quick to recognize that in order to attain even moderately high degrees of polymerization ($\bar{x}_n>100$), the extent of reaction of functional groups needs to be extremely high (greater than 99.9%), something that he set about achieving by careful design of laboratory apparatus. This key feature of step polymerizations also highlights the importance of using clean reactions in which contributions from side reactions are completely absent or negligibly small.

By contrast, in chain polymerizations, high degrees of polymerization are attained at low monomer conversions, the monomer being consumed steadily throughout the reaction. After its growth has been initiated, each polymer chain forms rapidly by successive additions of molecules of monomer to the reactive site at the chain end. In many chain polymerizations, more than 1000 repeat units are added to a single propagating chain in less than a second and the activity of the chain is lost (i. e. its activity dies and it can no longer propagate) after only a fraction of a second or a few seconds of chain growth. As the percentage conversion of monomer into polymer increases, it is simply the number of polymer molecules formed that increases; the degree of polymerization of those polymer molecules already formed does not change.

The functionality of a monomer is best defined as the number of chain links it can give rise to, because it is not necessarily equal to the number of functional groups present in the monomer, i. e. it is not always immediately obvious from the chemical structure of the monomer.

If a monomer has a functionality greater than two, then this will lead to the formation of branches and possibly to the formation of a network polymer, depending on the particular polymerization and reactant stoichiometry. The effect of monomer functionality is illustrated schematically in **Figure 4. 2. 1**. Although this depiction of the effect of monomer functionality is very much an oversimplification, it nevertheless demonstrates the importance of functionality. The formation of dendrimers and hyperbranched polymers is achieved using monomers with functionalities of three or higher, but in a way that is controlled so that a network polymer cannot form.

Chemical reactions proceed as a consequence of collisions during encounters between mutually-reactive functional groups. At each encounter, the functional groups collide repeatedly until they either diffuse apart or, far more rarely, react. Under normal circumstances, the reactivity of a functional group depends upon its collision frequency and not upon the collision frequency of the molecule to which it is attached. As molecular size increases, the rate of molecular diffusion decreases, leading to larger time intervals between encounters (i. e. to fewer encounters per unit time). This effect is compensated by the greater duration of each encounter giving rise to a larger number of functional group collisions per encounter. Hence the reactivity of a functional group can be expected to be approximately independent of molecular size.

Mathematical analysis of polymerizations is simplified greatly by assuming that the intrinsic reactivity of a functional group is independent of molecular size and unaffected by the reaction of other functional group (s) in the molecule of monomer from which it is derived. This principle of equal reactivity of functional groups was proposed by Flory who demonstrated its validity for functional groups in many step polymerizations by examining the kinetics of model reactions. Similarly, analysis of the kinetics of chain polymerizations shows that it is reasonable to assume that the reactivity of the active species at the chain end is independent of the degree of polymerization.

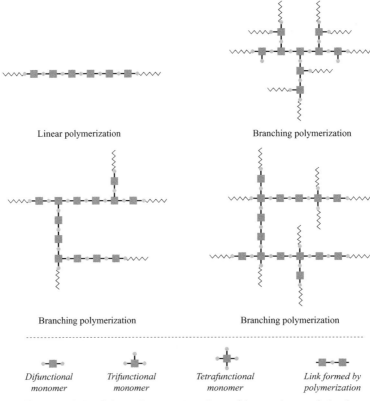

Figure 4.2.1 Schematic representations of how polymer skeletal structure is affected by monomer functionality.

(**Selected from:** Young R J, Lovell P A. Introduction to Polymers [M]. 3rd ed. CRC Press, 2011.)

New Words and Expressions

synthetic   *adj.* 合成的，人造的
extraordinarily   *adv.* 极其地，极端地
categorize   *v.* 将……分类，把……列作
distinctive   *adj.* 独特的，与众不同的
set out   出发，开始
principle   *n.* 准则，原则，基本原则，原理，定律
classification   *n.* 分类，分级，类别，级别
formative   *adj.* 形成的，格式化的
elimination   *n.* 消除，排除，淘汰
initiator   *n.* 引发剂，发起人，创始人
apparatus   *n.* 设备
nevertheless   *adv.* 然而，不过
interval   *n.* 间隔，间隙
compensate   *v.* 赔偿，偿付
kinetics   *n.* 动力学

Notes

(1) Chemical reactions proceed as a consequence of collisions during encounters between mutually-reactive functional groups.

—as a consequence of... 由于……的结果。
—参考译文：化学反应是相互反应的官能团相遇时发生碰撞的结果。

(2) Mathematical analysis of polymerizations is simplified greatly by assuming that the intrinsic reactivity of a functional group is independent of molecular size and unaffected by the reaction of other functional group (s) in the molecule of monomer from which it is derived.

—is independent of... 不依赖于……。derive from... 衍生于……。
—参考译文：假设一个官能团的固有反应活性与分子大小无关，且不受其衍生的单体分子中其他官能团反应的影响，这将大幅简化聚合反应的数学分析。

(3) The most basic requirement for polymerization is that each molecule of monomer must be capable of being linked to two (or more) other molecules of monomer by chemical reaction, i.e. monomers must have a functionality of two (or higher).

—be linked to... 与……连接。
—参考译文：聚合的最基本要求是每个单体分子必须能够通过化学反应与两个（或两个以上）其他单体分子连接，即单体必须具有两个（或两个以上）的官能度。

(4) Consequently, the number of different synthetic polymers that have been prepared is extremely large and many can be formed by more than one type of polymerization.

—参考译文：因此，已经制备出的合成聚合物数量非常多，并且许多（聚合物）可以通过不止一种类型的聚合反应来制备。

(5) Hence, the number of individually different polymerization reactions that have been reported is extraordinarily large.

—参考译文：因此，已报道的不同的聚合反应数量非常多。

(6) Polymerizations in which the polymer chains grow step-wise by reactions that can occur between any two molecular species are known as step-growth polymerizations.

—be known as... 被称为……。
—参考译文：聚合物链通过任何两种分子之间的反应逐步增长的聚合称为逐步聚合。

(7) This key feature of step polymerizations also highlights the importance of using clean reactions in which contributions from side reactions are completely absent or negligibly small.

—highlight 用作动词，表示强调、突出。
—参考译文：逐步聚合的这一关键特征也突出了使用清洁反应的重要性，在这种反应中，副反应的贡献完全不存在或小到可以忽略不计。

Exercises

1. Question for discussion

(1) What is the most basic requirement for polymerization?

(2) What is the condensation polymerization?

(3) What is the addition polymerization?

(4) Why did the scientists propose the classification of polymerization by considering the underlying polymerization mechanisms?

(5) What is the step-growth polymerization? And the chain-growth polymerization?

2. Translate the following into Chinese

(1) To discuss each of these would be an enormous task which fortunately is not necessary since it is possible to categorize most polymerization reactions into a relatively small number of classes of polymerization, each class having distinctive characteristics.

(2) The classification of polymerization reactions used in the formative years of polymer science was due to Carothers and is based upon comparison of the molecular formula of a polymer with that of the monomer (s) from which it was formed.

(3) The functionality of a monomer is best defined as the number of chain links it can give rise to, because it is not necessarily equal to the number of functional groups present in the monomer, i. e. it is not always immediately obvious from the chemical structure of the monomer.

3. Translate the following into English

(1) 副反应；副作用　　　　　(2) 传播，繁殖

(3) 官能团　　　　　　　　　(4) 极其

(5) 相比之下　　　　　　　　(6) 大约

(7) 实验室仪器

(8) 相比之下，在链式聚合中，单体转化率低，聚合度高，单体在整个反应过程中稳定消耗。

(9) 单体的官能度最好定义为它能产生的聚合物链的链接数，因为它不一定等于单体中存在的官能团数，也就是说，单体的官能度并不能简单的从单体的化学结构中观察得到。

(10) 根据链聚合动力学分析，我们可以合理地假设链端活性物质的反应性与聚合度无关。

Reading Material
Step-growth polymerization

这篇课文介绍了逐步聚合反应。文中从逐步聚合的概念出发，阐述了逐步聚合的过程和分子量的分布统计控制。

Step-growth polymerization is a very important method for the preparation of some of the most important engineering and specialty polymers. Polymers such as polyamides, poly (ethylene terephthalate), polycarbonates, polyurethanes, polysiloxanes, polyimides, phenol polymers and resins, urea, and melamine formaldehyde polymers can be obtained by step-growth polymerization through different types of reactions such as esterification, polyamidation, formylation, substitution, and hydrolysis.

In step-growth polymerization, only one kind of reaction is involved in the formation of a polymer and the reaction proceeds step by step. The main feature of this type of reactions is that two monomers, which bear different functionalities, can react with each other, or with a polymer of any size, through the same kind of reactions. In this case, the individual polymer molecules can grow over the course of the whole process; each reaction step of a polymer molecule implies that the reactive end of a monomer or polymer encounters another species with which it can form a link. The functional group at the end of a monomer is usually assumed to have the same reactivity as that on a polymer chain of any size. The polymerizations proceed by the stepwise reaction between the functional groups of reactants. The size of the polymer molecules increases at a relatively slow pace in polymerizations that proceed from monomer to dimer, trimer, tetramer, pentamer, and so on.

As a consequence of this mechanism, it is expected that the molecular weight grows in a slow manner at early stages of the reaction, where the reactions of oligomerization are the predominant ones. However, considering that the reactivity of the functional groups is the same during the formation of oligomers and during later stages of the reaction to form high polymer, the evolution of the average molecular weight with conversion follows the behavior shown in **Figure 4.2.2**. For this reason, a reaction with high conversion of functional groups is required to produce polymers with high molecular weight. Therefore, if polymers with high molecular weight are required, high conversions must be reached and side reactions must be avoided; in this case, the purity of the monomers plays an important role.

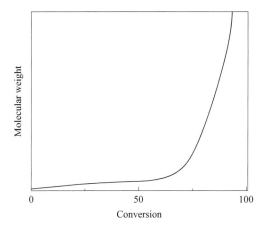

Figure 4.2.2 Profile of molecular weight versus conversion in a step-growth polymerization.

Many step-growth polymerizations involve an equilibrium between reactants and products, the latter comprising macromolecular species and (usually) eliminated small molecules.

The course of these polymerizations and of the distribution of the molecular weights is statistically controlled. High polymers cannot coexist with much monomer in a system in equilibrium. Step-growth polymerizations are evidently reversible and also involve interchange reactions in which terminal functional groups in a growing chain react with linking units of other molecules producing changes in the molecular weight distributions. There are step-growth polymerization reactions in which a small molecule is not produced (e. g., the reaction between a diol and a diisocyanate); these reactions are considered irreversible and are usually very fast, leading to high degrees of polymerization.

As mentioned before, the main characteristic of the step-growth polymerization is that it proceeds stepwise, according to the reactivity of the two functionalities involved in the formation of the new linkage. The average functionality f_{av} is the average number of functional groups per monomer molecule and it is defined by **Equation (4.2.1)**.

$$f_{av} \equiv \frac{\sum N_i f_i}{\sum N_i} \qquad (4.2.1)$$

where N_i is the number of molecules of the species i and f_i is the functionality of monomeric species i. This equation is valid when the opposite functionality is present in equal concentration and in the absence of side reactions.

Consider a polymerization that forms AB links and in which $n_A < n_B$, where n_i is the number of equivalents of the functional groups of type i. In this case, the number of B equivalents that can react cannot exceed n_A, and **Equation (4.2.1)** adopts the form of **Equation (4.2.2)**.

$$f_{av} = \frac{2n_A}{\sum N_i} = \frac{2n_A}{N_A + N_B} \qquad (4.2.2)$$

The number-average degree of polymerization in the reaction mixture (X_n) is defined as the average number of structural units (or monomer units) per polymer molecule. A structural unit equals a monomer unit, that is, the residue of each monomer in the polymer. For an AB polymerization, a repeating unit is made of two structural units. This differs from what some authors term the average degree of polymerization (D_p, i. e., the average number of repeating units per polymer molecule). In the step growth polymerization of a single molecule that bears two functionalities and can form a polymer, $X_n = D_p$, and when two monomers are involved, $X_n = 2D_p$, and is defined as

$$X_n = \frac{N_0}{\frac{2N_0 - N_0 p f_{av}}{2}} = \frac{2}{2 - p f_{av}} \qquad (4.2.3)$$

where N_0 is the initial number of monomer molecules and p is the extent of reaction, which is equal to the fraction of functional groups that have reacted $0 \leqslant p \leqslant 1$. An example of the use of the number average degree of polymerization concept is the reaction between 1,2-ethanediol with 1,6-hexanedioic acid (adipic acid). In this system, an esterification reaction takes place through the two functionalities, alcohol and acid, in the monomers and a polymer is formed. The repeating unit is $[OCH_2-CH_2-O-C(O)-(CH_2)_4-C(O)]$, which has a molecular weight of 172 g/mol. If the monomers are in equivalent concentrations, then $f_{av} = 2$ and $X_n =$

$2/(2-2p)$. When the conversion is 90%, the number-average degree of polymerization is $X_n=2/(2-2\times 0.9)=10$ and the molecular weight is 860 g/mol. On the other hand, when the conversion is almost complete (99.5%), $X_n=2/(2-2\times 0.995)=200$ and the molecular weight is 17200 g/mol. From this concept, the effect of conversion on molecular weight in step-growth polymerizations is clear.

Now consider the situation in which the 1,2-ethanediol monomer has a 99% (w/w) purity and contains ethanol in a concentration of 1% (w/w), as shown in **Table 4.2.1**. When 146 g of adipic acid (1 mol) is reacted with 62 g of 1,2-ethanediol (0.99 mol, considering the purity reported), at nearly complete (99.5%) conversion, the number-average degree of polymerization is $X_n=2/(2-0.995\times 1.9935)=121$ and the molecular weight obtained is 10440 g/mol. This result clearly contrasts with the previous data found of $X_n=200$ and molecular weight of 17200 g/mol for the same polymerization and conversion, but with pure monomer. This example also clarifies the enormous effect of the monomers' purity on the course of a step-growth polymerization.

Table 4.2.1 Polyester polymerization of 1,2-ethanediol with adipic acid

Reagent	Reagent concentration/wt%	Moles	Functionality	Equivalents
1,2-ethanediol	99	0.99	2	1.98
Ethanol	1*	0.0135	1	0.0135
Adipic acid	100	1	2	2
Total	—	2.0035	—	3.9935

* impurity in ethanediol.

The prediction of the molecular weight distribution for a step-growth polymerization assumes that both the probability of the reaction and the reaction rate of two functional groups are independent of the sizes of the involved molecules (monomers or growing chains).

For a polymerization system involving functional groups, A, reacting with another functional group (say B), the probability of finding by random selection a growing molecule with i monomer units is given by the probability that $i-1$ A groups have reacted (p^{i-1}) multiplied by the probability that the last functional group has not reacted, which is $1-p$ (since the probabilities that a given functional group has reacted or not must add up to 1). The resulting probability is evidently $p^{i-1}(1-p)$. Then, the probability that a randomly selected molecule will be an i-mer equals the mole fraction x_i of i-mers in the reaction mixture and is calculated as

$$x_i=(1-p)p^{i-1} \quad (4.2.4)$$

On the other hand, the total number N of molecules remaining at an extent of reaction p is

$$N=N_0(1-p) \quad (4.2.5)$$

Therefore, the number N_i of i-mer molecules is given by $N_{xi}=N_0(1-p)^2 p^{i-1}$, and the molecular weight of an i-mer is iM_0.

Since the total weight of all molecules equals $N_0 M_0$ (neglecting unreacted ends), the weight fraction w_i of i-mers is

$$w_i = \frac{N_0 i M_0 (1-p)^2 p^{i-1}}{N_0 M_0}$$

$$w_i = i(1-p)^2 p^{i-1} \tag{4.2.6}$$

Equation (4.2.6), together with **Equation** (4.2.5), describes a random distribution of molecular sizes; this distribution is also known as the Flory-Schulz distribution or the most probable distribution. Recently, Wutz and Kricheldorf proposed a model describing the frequency distribution (f_i) and formulated the weight distribution (w_i) of linear chains in step-growth polymerizations considering the cyclization reaction, which is one of the most important side reactions in step-growth polymerization.

The number-average molecular weight M_n, defined as the total weight of a polymer sample divided by the total number of molecules, is given by

$$M_n = M_0 X_n = \frac{M_0}{1-p} \tag{4.2.7}$$

where M_0 is the formula weight of the repeating unit.

(**Selected from:** Saldivar-Guerra E, Vivaldo-Lima E. Handbook of Polymer Synthesis, Characterization, and Processing [M]. 1st ed. John Wiley & Sons, 2013.)

New Words and Expressions

specialty   *n.* 专业，专长

polyamide   *n.* 聚酰胺（尼龙）

polycarbonate   *n.* 聚碳酸酯

polyurethane   *n.* 聚氨酯，聚亚安酯

polysiloxane   *n.* 聚硅氧烷，聚砂氧烷

polyimide   *n.* 聚酰亚胺

resin   *n.* 树脂，松香

urea   *n.* 尿素

melamine formaldehyde   三聚氰胺甲醛

esterification   *n.* 酯化作用

polyamidation   *n.* 聚酰胺化

formylation   *n.* 甲酰化作用

hydrolysis   *n.* 水解作用

individual   *adj.* 单独的，个人的

species   *n.* 种，物种，种类

dimer   *n.* 二聚物

trimer   *n.* 三聚物

tetramer   *n.* 四聚物

pentamer   *n.* 五节聚化物

oligomerization   *n.* 低聚

reversible   *adj.* 可逆的

terminal *n.* 终点站，终端
stepwise *adj.* 逐步的，逐渐的
functionality *n.* 功能
diisocyanate *n.* 二异氰酸盐
adopt *v.* 收养，采取，采纳
residue *n.* 残渣，剩余，滤渣
adipic acid 己二酸
molecular weight 分子量
enormous *adj.* 巨大的，极大的
cyclization reaction 环合反应，环化反应

Notes

(1) In step-growth polymerization, only one kind of reaction is involved in the formation of a polymer and the reaction proceeds step by step.

—is involved in... 包括……中，涉及……。

—参考译文：在逐步聚合反应中，聚合物的形成只涉及一种反应，而且反应是逐步进行的。

(2) There are step-growth polymerization reactions in which a small molecule is not produced (e. g., the reaction between a diol and a diisocyanate); these reactions are considered irreversible and are usually very fast, leading to high degrees of polymerization.

—参考译文：有些逐步聚合反应（例如，二醇和二异氰酸酯之间的反应）没有小分子生成，这些反应被认为是不可逆的并且通常很快，会导致高聚合度。

(3) In this case, the individual polymer molecules can grow over the course of the whole process; each reaction step of a polymer molecule implies that the reactive end of a monomer or polymer encounters another species with which it can form a link.

—参考译文：在这种情况下，单个聚合物分子可以在整个过程中生长；一个聚合物分子的每个反应步骤都意味着单体或聚合物的反应末端遇到另一个可以与之形成链接的个体。

(4) Many step-growth polymerizations involve an equilibrium between reactants and products, the latter comprising macromolecular species and (usually) eliminated small molecules.

—参考译文：许多逐步聚合反应都涉及反应物和产物之间的化学平衡，后者包括大分子和（通常）被消除的小分子。

(5) This differs from what some authors term the average degree of polymerization.

—参考译文：这不同于一些作者所说的平均聚合度。

(6) In this system, an esterification reaction takes place through the two functionalities, alcohol and acid, in the monomers and a polymer is formed.

—参考译文：在该体系中，单体通过醇和酸两种官能团发生酯化反应，形成聚合物。

(7) A structural unit equals a monomer unit, that is, the residue of each monomer in the

polymer.

—参考译文：一个结构单元等于一个单体单元，即每个单体在聚合物中的残留物。

Unit 4.3

Text

Structures and properties of polymers

这篇课文主要介绍了聚合物的结构及性质。课文从单体的种类、聚合物链的线性度及链长三个方面，详细介绍了聚合物的结构特征，说明了聚合物的性质与组成和结构之间的错综复杂关系，文中还讨论了聚合物的机械、热和电性能。

Polymer structure

A polymer consists of a long chain backbone composed of repeat subunits called monomers, the length scale of which ranges from sub-nm up to macro-meter. Hence, to understand the structure of a polymer, we will firstly discuss the identity of the repeat monomers. The structure of a polymer also refers to the physical arrangement monomer residues along the backbone of a single chain. Polymers may form different phases with different arrangements, for instance, through crystallization, the glass transition or microphase separation. These features have a significant impact on the physical and chemical properties of a polymer.

Monomer identity

The identity of the monomers comprising a polymer determines the most important attribute of a polymer. Generally, most polymers are organic hydrocarbons, that is, they are composed of hydrogen and carbon via covalent bonds. Polymer nomenclature is generally based upon the type of monomer residues. A homopolymer is a polymer which only contains a single type of repeat units, while a polymer containing two or more types of repeat units is known as a copolymer. For example, polystyrene only consists of repeat styrene units, while ethylene-vinyl acetate is composed of several types of repeat units. In particular, a polyelectrolyte or ionomer refers to a polymer containing ionizable subunits (high, low fraction respectively).

Biological polymers, such as polynucleotides, proteins, and starch, are composed of a variety of structurally relevant monomer residues. Polynucleotides (for example, deoxyribonucleic acid) are composed of four types of nucleotide subunits, while proteins comprise long chains amino acid residues and the repeat units in starch are glucose monomers.

Chain linearity

In a polymer, monomer residues are physically arranged along the backbone of the long chain. Depending on the arrangement of monomer residues and reaction conditions, a variety of polymer structures can be produced. The simplest structure of polymers is a straight chain composed of a single unbranched chain, for example, unbranched polyethylene, which can be characterized and recorded using an atomic force microscopy under liquid medium. Polymers could also be composed of branched macromolecules with a main chain and one or more side chains. Particularly, a grafted polymer refers to a polymer containing a side chain which has different composition or configuration compared with the main chain. If four or more distinct chains emanate from a branch point in a polymer, this polymer is termed a cross-link polymer.

Depending on the chain linearity and the degree of cross-linking, polymers exhibit distinct physical and thermal properties. Unbranched and slightly branched polymers are generally thermoplastics, while close-meshed crosslinked polymers are thermosets. In the solid state, linear macromolecules can be semi-crystalline, while highly branched polymers are amorphous with molecules randomly interacting with each other.

Chain length

During the polymerization process, a number of monomers are incorporated together, forming a long polymer chain. More monomers incorporated result in higher degree of polymerization, and higher molecular weight. Therefore, the length of a chain relates to the degree of polymerization, and can also be expressed in terms of molecular weight. However, almost all polymerization techniques produce a statistical distribution of chain lengths. This leads to the expression of molecular weight in terms of weighted averages, namely, the number average molecular weight (M_n) and weight average molecular weight (M_w). The ratio of these two values is commonly used to describe the width of the molecular weight distribution, termed the polydispersity.

The length of polymer chains (or the molecular weight) has a strong influence on the viscosity, strength and toughness of the polymer. For example, if M_w of the polymer is above the entanglement molecular weight, increasing the polymer chain length would result in increase in its viscosity in the melt. Moreover, increasing chain length would decrease its mobility, hence increasing its strength, toughness. This can be attributed to the increased chain interactions which tend to confine the chains more strongly and prevent deformations and matrix breakup under higher stresses due to the increased chain length.

Polymer properties

Since the first synthesis of polymers from small molecules, the field of materials has been virtually revolutionized by the advent of synthetic polymers. In some fields, polymers have been used as alternatives to replace their counterparts since synthesized polymers exhibit satisfactory properties (even are superior to their counterparts) and can be produced inexpensively. Furthermore, modern scientific research tools have made possible the determination of polymer components and structures to synthesize polymers with particular characteristics. The properties of polymers express how polymers behave as bulk materials under different conditions, which are intricately related to their components and structures. In this section, we will discuss the mechanical, thermal, and electrical properties of polymers.

Mechanical properties

The mechanical properties of a polymer describe how it behaves under stress on a macroscopic scale, such as how much it can be stretched, how much it can be bent, how hard or soft it is and so on. Hence, it is of great importance to study basic mechanical properties of polymers before their practical applications.

The most important parameters to evaluate the mechanical properties of polymers are strength, Young's modulus of elasticity and viscoelasticity. The strength quantifies how much stress a polymer can endure before breakup. Young's modulus expresses the elasticity of the polymer. It is defined as the ratio of stress to the strain for small strains.

These two parameters are very important in polymer applications that requires high strength and elasticity. Viscoelasticity refers to a complicated time-dependent elastic response, which will exhibit hysteresis in the stress-strain curve.

Glass transition and crystallization

As aforementioned, polymers can be either semi-crystalline or amorphous due to their various structures. The thermal properties of polymers describe how polymers behave under increasing temperature.

At low temperature, polymer molecules are in frozen state and can only vibrate slightly, making the polymers brittle and rigid analogous to glass. Hence this state is termed the glassy state. As temperature increases, polymer chains are capable of wiggling. This makes the polymers rubbery and viscous. Both semi-crystalline and amorphous polymers may go through this glass transition, and the temperature at which the transition occurs is called glass transition temperature (T_g). If temperature further increases, semi-crystalline polymers will undergo crystallization and melting transitions. This temperature is referred as melting temperature (T_m), at which ordered polymer phase turns to disordered phase.

The glass transition temperature is a crucial physical parameter for polymer manufacturing, processing, and applications. It can be influenced by many factors, such as chain stiffness, chain length, intermolecular forces, cross-linking, plasticizers and so on.

Electrical properties

Most conventional polymers are electrical insulators. However, polymer-based semiconductors can be formed by using π-conjugated molecules as building blocks, which is also known as organic semiconductors. This is because charge carriers in semiconductors are injected from the electrodes or generated by doping.

Organic semiconductors have attracted considerable attentions because they possess many excellent properties, such as low density, cheap production, easy processing, and mechanical flexibility. These unique properties make them superior to their conventional inorganic counterparts. Particularly, organic semiconductors have been successfully commercialized in fabricating active-matrix organic light-emitting diodes for high-end displays. Thanks to the tremendous efforts dedicated by researchers, it has also been demonstrated that organic semiconductors possess great potential for the fabrication of solar cells, transistors, and photodetectors.

New Words and Expressions

macromolecule *n*. 高分子，大分子
monomer *n*. 单体，（病毒）单壳粒
backbone *n*. 脊梁骨，脊柱，支柱
organic hydrocarbon 有机烃
nomenclature *n*.（尤指某学科的）命名法
residue *n*. 残留，残渣，残留物
ethylene-vinyl acetate *n*. 乙烯醋酸乙烯酯
polyelectrolyte *n*. 高分子电解质，聚电解质
ionomer *n*.（高分子）离聚物，离子交联聚合物
polynucleotides *n*. 多聚核苷酸
deoxyribonucleic acid *n*. 脱氧核糖核酸
emanate *vt*. 散发，产生，表现，显示
close-meshed *adj*. 密网眼的
virtually *adv*. 实际上，几乎，事实上，虚拟
alternative *adj*. 可供替代的，另类的，非传统的；*n*. 可供选择的事物
intricately *adj*. 错综复杂的，复杂的
macroscopic *adj*. 宏观的，肉眼可见的
bent *v*. 弯曲
stretch *v*. 伸展，拉长
breakup *n*. 破裂，终结，分裂，解体
endure *v*. 忍耐，忍受，持续，持久
viscoelasticity *n*. 黏弹性
hysteresis *n*. 磁滞，滞后，滞变，平衡阻碍
aforementioned *adj*. 前面提到的，上述的
amorphous *adj*. 无定形，无固定形状的，不规则的，无组织的
vibrate *v*. 振动，（使）颤动，摆动
rigid *adj*. 固执的，死板的，坚硬的，一成不变的，不弯曲的，僵直的
term *v*. 把……称为，把……叫作
wiggling *v*.（使）扭动，摆动，摇动，起伏
plasticizer *n*. 增塑剂，塑化剂
tremendous *adj*. 巨大的，极好的，了不起的

Notes

(1) A homopolymer is a polymer which only contains a single type of repeat units, while a polymer containing two or more types of repeat units is known as a copolymer.

——which 引导的从句修饰前面的 polymer。

——参考译文：均聚物是仅含有一种重复单元的聚合物，而含有两种或多种重复单元的聚合物称为共聚物。

(2) Since the first synthesis of polymers from small molecules, the field of materials has been virtually revolutionized by the advent of synthetic polymers.

—be revolutionized by 被彻底改变。by the advent of... 随着……的出现。

—参考译文：自从第一次由小分子合成聚合物以来，合成聚合物的出现几乎彻底改变了材料领域。

(3) Furthermore, modern scientific research tools have made possible the determination of polymer components and structures to synthesize polymers with particular characteristics.

—参考译文：此外，现代科学研究工具使得通过确定聚合物成分和结构来合成具有特定特性的聚合物成为可能。

(4) The properties of polymers express how polymers behave as bulk materials under different conditions, which are intricately related to their components and structures.

—参考译文：聚合物的性质体现了聚合物作为块材料在不同条件下的表现，这与它们的成分和结构密切相关。

(5) Hence, it is of great importance to study basic mechanical properties of polymers before their practical applications.

—it is of great importance 这是非常重要的。

—参考译文：因此，在实际应用之前对聚合物的基本力学性质进行研究十分重要。

(6) Viscoelasticity refers to a complicated time-dependent elastic response, which will exhibit hysteresis in the stress-strain curve.

—参考译文：黏弹性涉及复杂的时间依赖的弹性响应，它会在应力-应变曲线中表现出滞回现象。

(7) As aforementioned, polymers can be either semi-crystalline or amorphous due to their various structures. The thermal properties of polymers describe how polymers behave under increasing temperature.

—参考译文：如前所述，由于结构的差异，聚合物可以是半结晶的或无定形的。聚合物的热学性质描述了聚合物在温度升高时的行为。

Exercises

1. Question for discussion

(1) What determines the most important attribute of a polymer?
(2) Which are the repeat units in starch?
(3) Which is the simplest structure of polymers?
(4) Please identify the polydispersity.
(5) Please list the most important parameters to evaluate the mechanical properties of polymers.
(6) Please define Young's modulus.

(7) Please define glass transition temperature.

2. Translate the following into Chinese

(1) More monomers incorporated result in higher degree of polymerization, and higher molecular weight.

(2) Polymers could also be composed of branched macromolecules with a main chain and one or more side chains.

(3) At low temperature, polymer molecules are in frozen state and can only vibrate slightly, making the polymers brittle and rigid analogous to glass.

(4) If temperature further increases, semi-crystalline polymers will undergo crystallization and melting transitions.

(5) The glass transition temperature is a crucial physical parameter for polymer manufacturing, processing, and applications.

(6) This is because charge carriers in semiconductors are injected from the electrodes or generated by doping.

3. Translate the following into English

(1) 因此，要了解聚合物的结构，我们将首先讨论重复单体的特性。

(2) 聚合物的最简单结构是由单个无支链结构组成的直链，例如无支链聚乙烯，可以在液体介质下使用原子力显微镜对其进行表征和记录。

(3) 增加链长会降低聚合物链的流动性，从而增加其强度、韧性。

(4) 这可归因于增强的链相互的作用倾向于更强烈地限制链，防止由于链长的增加而造成在更高应力下产生的变形和基质破裂。

(5) 从那时起的许多进步，例如通过更有效的工艺、用更便宜的原材料合成单体、更先进的表征技术和对聚合物性质的更好理解，持续激励着21世纪聚合物科学和工业的进步。

(6) 这两个值的比率通常用于描述分子量分布的宽度，称为多分散性。

(7) 特别是有机半导体已成功商业化应用，用于制造高端显示器中的有源矩阵有机发光二极管。

(8) 这些独特的特性使它们优于传统的无机对应物。

(9) 有机半导体因具有密度低、生产成本低、易于加工和机械柔韧性等许多优良特性而引起了人们的广泛关注。

(10) 以π共轭分子为构建单元可以形成聚合物基半导体，也被称为有机半导体。

(11) 玻璃化转变温度受链刚度、链长、分子间作用力、交联、增塑剂等多种因素的影响。

(12) 有机半导体被证明在制造太阳能电池、晶体管和光电探测器方面具有巨大潜力。

Reading Material
History of polymers and their typical synthesis approaches

这篇课文简述了聚合物的历史及发展进程中出现的问题，详细介绍了聚合物合成的三种主要途径，即实验室合成、自然界的生物合成以及通过氧化、交联和封端等方法对天然聚合物进行改性。

History

Polymers are naturally existing in plants and animals, such as wood, rubber, cotton, and silk, which are of importance in our daily life. Natural polymers, such as proteins, enzymes, DNA and cellulose, also paly fundamental roles in biological and physiological processes. It is not until 100 years ago that human started synthesizing polymers via chemical methods, which are usually called "plastics", although at this moment the nature of polymers was not understood. Initially, natural polymers were used to synthesise polymeric materials, which were also known as semi-synthetic polymers. Vulcanized rubber was prepared by heating natural rubber with sulphur to improve its durability in the 19th century, which was regarded as the first popularized semi-synthetic polymer. Another commercially important example is cellulose nitrate, which was formed by reacting nitric acid and cellulose. The first commercialized completely synthetic polymer is Bakelite, which was fabricated in 1907 by Baekeland by reacting phenol and formaldehyde at defined temperature and pressure.

Despite these successes in synthesis of polymers, the lack of correct theoretical knowledge prevented the further development of polymers until the 1920s. Initially, according to the association theory proposed by Thomas Graham, most scientists believed that polymers were colloidal aggregates of small molecules held together by unknown forces. However, Hermann Staudinger proposed a controversial work that polymers in fact consisted of long chains of covalently bonded atoms in 1920s. This was eventually accepted, and he was awarded the Nobel Prize in 1953.

Due to the discovery of macromolecules science and the strong economical drive, many new types of cheap polymers were discovered, synthesized, and given commercial applications, including polystyrene, nylon, polyethylene, poly (vinyl chloride), silicones and polytetrafluoroethylene and so on. Since then, many advances, such as synthesis of monomers from cheaper raw materials via more efficient processes, advanced techniques for characterisation and better understanding of the nature of polymers, have continued to stimulate both scientific and industrial progress in the 21st century. It is of importance to develop more specialized, functional polymers for biomedical, optical, and electronic applications.

Synthesis approaches

The process of polymer synthesis is generally termed polymerization, during which small repeat units (or monomer residues) are covalently incorporated into a long chain or network. Some chemical groups may be lost from each monomer due to the polymerization reaction.

Laboratory synthesis

Depending on the nature of monomer to polymer chain incorporation, laboratory synthetic methods can classified into step growth polymerization and chain growth polymerization. During chain growth polymerization, monomers can only be added onto the chain at limited active sites one at a time, whereas in step growth polymerization, monomer chains may directly combine with other chains. Step growth polymerization can be further divided into polycondensation and polyaddition depending on whether small molecules are yielded as by-products. Although catalysts are not necessarily required in polymerization reactions, they can significantly increase the efficiency of the polymerization process. It has been demonstrated that biopolymers, for example, polypeptides, can be formed via laboratory synthetic methods.

Biological synthesis

In nature, there are three main categories of natural biopolymers, namely, polysaccharides, polypeptides, and polynucleotides, which can be synthesized via enzyme-mediated processes in living cells. For example, the incorporation of nucleotides into DNA is catalysed by DNA polymerase. The synthesis of proteins is more complicated and involves multiple enzyme-mediated processes to combine amino acids into the formation of specific proteins. Other natural biopolymers, such as rubber, cellulose and starch, have been widely used and modified for practical applications.

Modification of natural polymers

As discussed before, natural polymers, such as cotton, starch, and rubber, have widely been used to produce commercially important polymers for years before polymers are synthesised via chemical methods. Methods for the modification of natural polymers include oxidation, cross-linking, and end-capping.

New Words and Expressions

macromolecule *n*. 高分子，大分子
monomer *n*. 单体，（病毒）单壳粒
durability *n*. 耐久性，耐久（用）性，经（持）久性
synthesis *n*. （人工的）合成
semi-synthetic *adj*. 半合成的
vulcanize *v*. 硫化，硬化，热补
sulphur *n*. 硫黄，硫
cellulose nitrate 硝酸纤维素
nitric acid *n*. 硝酸
phenol *n*. 酚，苯酚
formaldehyde *n*. 甲醛，福尔马林
association *n*. 协会，社团，联盟，联合
colloidal *adj*. 胶体的，胶质的
aggregate *adj*. 总数的，总计的；*n*. 合计，总数，（可成混凝土或修路等用的）骨料，集料；*vt*. 合计，总计
polytetrafluoroethylene *n*. 聚四氟乙烯
covalently *adv*. 共价地
classify *vt*. 分类，划分

polycondensation　*n*. 缩聚，缩聚反应
polyaddition　*n*. 加聚
by-product　*n*. 副产品
catalysts　*n*. 催化剂
polypeptide　*n*. 多肽，肽类，肽链
via　*prep*. 通过，经由
category　*n*. 类别，（人或事物的）种类
amino　*n*. 氨基
cross-linking　*n*. 交联
end-capping　*n*. 封端

Notes

（1）Polymers are naturally existing in plants and animals, such as wood, rubber, cotton, and silk, which are of importance in our daily life.

——这是一个 which 引导的从句，修饰前面的 wood, rubber, cotton, and silk。
——参考译文：聚合物天然存在于动植物中，例如木材、橡胶、棉花和丝绸，它们在我们的日常生活中非常重要。

（2）Natural polymers, such as proteins, enzymes, DNA and cellulose, also play fundamental roles in biological and physiological processes.

——play fundamental roles in... 在……中发挥基本作用。
——参考译文：天然聚合物，如蛋白质、酶、DNA 和纤维素，也在生物和生理过程中发挥基本作用。

（3）Initially, natural polymers were used to synthesise polymeric materials, which were also known as semi-synthetic polymers.

——参考译文：最初，天然聚合物被用于合成聚合物材料，（它们）也被称为半合成聚合物。

（4）Despite these successes in synthesis of polymers, the lack of correct theoretical knowledge prevented the further development of polymers until the 1920s.

——参考译文：尽管在聚合物合成方面取得了这些成功，但 19 世纪 20 年代之前，正确理论知识的缺乏仍然阻碍着聚合物的进一步发展。

（5）During chain growth polymerization, monomers can only be added onto the chain at limited active sites one at a time, whereas in step growth polymerization, monomer chains may directly combine with other chains.

——be added onto... 被添加到……上。
——参考译文：在链增长聚合过程中，单体一次只能在有限的活性位点上添加到链上，而在逐步增长聚合中，单体链可以直接与其他链结合。

（6）Step growth polymerization can be further divided into polycondensation and polyaddition depending on whether small molecules are yielded as by-products.

—be divided into... 被分类为……。
—参考译文：根据是否产生小分子副产物，逐步生长聚合可进一步分为缩聚和加聚。

(7) It has been demonstrated that biopolymers, for example, polypeptides, can be formed via laboratory synthetic methods.

—It is demonstrated that... 事实证明……。
—参考译文：已经证明生物聚合物，例如多肽，可以通过实验室合成方法制备。

(8) For example, the incorporation of nucleotides into DNA is catalysed by DNA polymerase.

—参考译文：例如，核苷酸与 DNA 的融合是由 DNA 聚合酶催化的。

(9) The synthesis of proteins is more complicated and involves multiple enzyme-mediated processes to combine amino acids into the formation of specific proteins.

—参考译文：蛋白质的合成更为复杂，涉及多种酶介导的过程，以将氨基酸组合形成特定蛋白质。

(10) Other natural biopolymers, such as rubber, cellulose and starch, have been widely used and modified for practical applications.

—参考译文：其他天然生物聚合物，如橡胶、纤维素和淀粉，已被广泛使用和改性以用于实际应用。

Unit 4.4

Text

Polymers for food packaging

这篇课文介绍了聚合物在食品包装中的应用。课文从食品包装的发展历程开始讲述，介绍了聚合物包装的发展现状，展望了生物可降解聚合物的广泛应用前景。

The packaging materials used in food and allied industries are highly varied. The main responsibility of these materials is to keep foods fresh and safe from the production stage all the way to consumption including the storage and distribution chains as well. Materials that have traditionally been utilized in food packaging and have worked well throughout the years include glass, metals (aluminum, foils and laminates, tin-free steel, tin plate, etc.), plastics, paper, and paperboards. In general, flexible and rigid synthetic packaging materials are used in food products.

Throughout history, the most common materials used for protecting and covering goods have been glass and paper. Similar to many other matters in day to day life, however, each of these materials has its own advantages and disadvantages.

For example, glass containers used in different applications and shapes such as glass bottles are the oldest packaging group and are still used to this day. Glass packages are most suitable for containing liquids as they are made of natural components with a nonpermeable barrier (one of their advantages). However, their biggest disadvantage is their fragility.

The paper packaging industry also offers the best quality kraft and sack paper. Kraft paper is a strong paperboard or cardboard (usually brown paper) that is produced by processing wood chemical pulp. It is mainly used for bags and/or as wrapping paper. Sack paper is a kraft paper that is permeable and sponge-like with high tear resistance and elasticity. It is widely used for packaging products that demand strength and durability. An advantage point for the paper packaging industry is that innovative solutions marketing experts can design on it for clientele in the industrial, medical, and consumer sections. The greatest disadvantage for paper packaging, however, is the absorption of water and moisture.

Over the course of history, especially after the industrial revolution in the eighteenth century, manufacturing technology revolutionized the entire concept of packaging when manufactures were pushed to develop more durable and resistant types of protection (packaging) to make the transportation of products from factory to shop and later to customers' homes possible. Soon after, the development and manufacturing of plastic materials for packing applications began in the 1860s. This was done by altering hard rubber. As science progressed, synthetic plastics and multiple compounds such as PVC and PET were gradually invented.

For a long time, petrochemical polymers (plastics) have served as the go-to packaging materials due to their economical abundance, desirable presentation, excellent barrier properties toward oxygen, aroma compounds, softness, tensile and tear strength, lightness, and transparency. An example of petrochemical polymers is polyethylene (PE) foam, which is a durable, lightweight, closed-cell material. Its main applications are in industrial and agricultural packaging. PE is an excellent material choice in these applications due to its insulation and vibration control properties. It also offers high resistance to chemicals and moisture and comes in various shapes such as pouches, sheets, bundles, die cuts, and tubing. The advantages of PE foam in pouch shape, for example, are its flexibility, lightness, surface sensitivity, low cost, shock absorbability, and resistance to mildew or mold.

In fact, plastics are polymers with additives and are based on long-chain molecules developed from alkenes. To prepare different polymers, usually one or more hydrogen atoms is replaced by a different atom. Plastics are commonly used in the packaging industry due to their easy molding and transformation. Various plastics and their different properties allow for many packaging options such as shape, color, size, weight, function, printing, etc. Today, polymers are an integral part of modern life owing to their desirable properties including stability, resilience, and ease of production. In other words, plastics are favorable for producers for the reasons that they are light in weight and malleable and/or flexible, thus, can be formed into any shape by different means such as blowing, extrusion or coextrusion, casting, lamination, etc. This makes it possible to package unique objects that are difficult to fit into normal/basic containers. Polymer plastics are useful for different parts of food packaging due to their barrier properties that help to keep products fresh, prevent contamination, and increase shelf-life. Polymer packaging can be seen as beneficial to the environment because food industry manufacturers and businesses are able to reduce waste by preserving foods for longer than in a case without polymer packaging. For example, Tajeddin et al. reviewed polymers for modified atmosphere packaging (MAP) applications. These compounds are the main materials for flexible package structures used for MAP, but they can also be used in a rigid or semirigid packaging solution such as a lidding on a tray. PEs including LDPE, linear LLDPE, HDPE; PP; PS; polyesters including PET or PETE, polycarbonate (PC), and polyethylene naphthalate; PVC; polyvinyl dichloride; polyamide (PA) or nylon; and ethylene vinyl alcohol are the main polymeric materials (plastic films) used for MAP applications. On the other hand, researchers are still working to correct the structure of some of these polymers for different purposes. For instance, Kobayashi and Saito worked on the structural evolution of blends of PC and poly (methyl methacrylate) by simultaneous biaxial stretching. Since the mechanical and gas barrier properties of polymer films can be improved using the biaxial stretching process, these biaxially stretched films are greatly utilized as packaging materials for industrial and food products. In another example, Han et al. synthesized polyvinyl formal (PVF) through the reaction of polyvinyl alcohol (PVA) and formaldehyde. The synthesized PVF showed a higher decomposition and glass transition temperature (T_g), and a lower melting point compared to PVA. The synthesized PVF could be melted and processed at much lower temperatures than PVA. Polypropylene carbonates (PPC) and the synthesized PVF were melted and blended in a Haake mixer machine. The PPC/PVF blends presented a higher Vicat softening temperature and tensile strength compared to those of pure PPC. Thermogravimetric analysis findings showed that the thermal stabilities of the PPC/PVF blends were reduced as the PVF content was increased. Observations obtained from scanning electron microscopy revealed that the interfacial compatibility of the PPC and PVA matrix was worse than that of the PPC and PVF matrix. The PPC/PVF blends indicated superior broad properties compared to those of pure PPC, which presents as a feasible method to expand the application of PPC copolymers.

As previously mentioned, polymers (plastic materials) are the most used materials in the packaging industry. For example, in an analyzed packaging categories, it is estimated that 95%—99% of plastics use in North American packaging. Comparative to other packaging substances such as aluminum, glass, paper, steel, etc., plastic-based packaging materials make up 39%—100% of total North American market request for the packaging categories analyzed in this study. However, the most important problem with plastics is that the production and processing of plastics are energy used processes, leading to the production of greenhouse gases and contributing enormously to global warming. Furthermore, Smith, as cited by Yadav et al., stated that on burning, plastics release noxious emissions such as CO_2, chlorine, hydrochloric acid, dioxin, furans, amines, benzene, 1,3-butadiene, and acetaldehyde, which threaten the environment and public health. Created wastes such as of plastics are also a big challenge for many years due to their degradation resistance. They produce substantial environment pollutions, thus, hurt natural resources and wildlife when they are spread in nature. For example, the disposal of nondegradable plastic bags, to-go food containers, plastic straws, and plastic bottles for bottled water has a major and direct negative impact on sea life and can ultimately disrupt the natural food chain. It is estimated that by 2050, there will be more plastic waste in our oceans than fish! Plastic bags are choking our planet and, in particular, the oceans. They expose children to lung problems and can increase the risk of prostate cancer in men. Every year, more than 500 billion plastic bags are disposed of in landfills, and less than 3% of these bags are being recycled. These plastics are mainly made of PE and can take hundreds of years to degrade without proper recycling technology and also release damaging greenhouse gases that contribute to global warming. Therefore it is better to reduce the distribution of these materials and to use other packaging materials instead.

Figure 4.4.1 shows an example of packaging materials composition for replacing plastic materials. In fact, Brandt and Pilz developed a model for a theoretical substitution of plastic packaging. The results of their study are summarized in **Figure 4.4.1**, if plastic packaging (LDPE, LLDPE, HDPE, PP, PVC, PS, expanded polystyrene, and PET in this study) were replaced by other materials (tin plate and steel packaging, aluminum, glass, corrugated board, cardboard, paper and fiber cast, paper-based composites, and wood in this study).

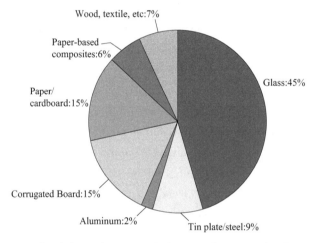

Figure 4.4.1　A case study of the replacement of plastic packaging with other packaging materials.

As shown in **Figure 4.4.1**, in addition to the possibility of replacing plastic materials with other packaging materials (which is a widespread debate and needs to be addressed elsewhere), through the use of renewable natural materials and new technologies, there may be other alternatives to substitute for polymeric materials.

Nowadays, with the arrival of new technologies such as three dimensional (3D) printing, it is possible to have a diverse, good-quality packaging material. The application of 3D printing technology in packaging industries has shifted the dynamic of packaging in many different aspects such as in the potential to have full color graphics and text for labels. 3D printing is a process that uses computer control to form layers and coatings for materials. It is convenient in terms of package design. People are able to provide color coatings and layers in almost any material regardless of shape or geometry. Although the current advantage of 3D printing technology in packaging is not clear, it requires a lot of future exploration because of its huge potential.

It should be noted that the packaging industry and researchers work on the recycling and reusing of synthetic packaging materials for waste management and control. For example, PET can be recycled repeatedly, and about 681000 metric tons of recycled PET containers and bottles are recovered each year in the United States. One method of recycling is to wash and remelt PET, and use it in new products that require PET as a component. Another method of recycling PET is to break it down with chemical processes into its raw materials. Due to this reusability, PET is highly sustainable and its sustainability is an ever-increasing pursuit for scientists and a lot of research is going into the development of facilities that are capable of transforming used PET packages and bottles into new food-grade PET containers and bottles. The only drawback to the reuse of PET is the amount of material collected. One of the most interesting uses of recycled PET resin is in the production of filament for use in 3D printers.

The waste disposal problem of petroleum materials and their nonrenewable nature have caused a spike in the level of interest in the sustainable development of biodegradable polymers, recycling, and/or environment protection. The degradation of materials causes structural and morphological transformations that can result in major changes in the properties of polymer materials. Generally, biodegradable polymers get hydrolyzed and turn into methane (CH_4), carbon dioxide (CO_2), mineral mixtures or compounds, or biomass. To create biodegradable polymers, bio-origin materials obtained from cellulose, starch, and microbial fermentations are used. This has led to their incredible success in the food packaging industry in the past few years.

(**Selected from:** Tajeddin B, Arabkhedri M. Polymers and Food Packaging [M]. Polymer Science and Innovative Applications, 2020, 525-543.)

New Words and Expressions

allied   *adj.* 结盟的，联盟的
fragility   *n.* 脆弱，易碎
kraft   *n.* 牛皮纸
sack   *n.* 麻布（或厚纸、塑料等）大袋
pulp   *n.* 纸浆
sponge-like   *adj.* 海绵样的
tear resistance   *n.* 撕裂强度
elasticity   *n.* 弹性，弹力，灵活性
clientele   *n.* 客户
invent   *v.* 发明，创造
petrochemical   *adj.* 石油化工的
foam   *n.* 泡沫，气泡，泡沫橡胶
mildew   *n.* 霉，霉病

alkene *n*. 烯烃，链烯烃
integral *adj*. 必需的，必要的，完整的
resilience *n*. 恢复力，弹力
malleable *adj*. 有延展性的，容易改变的
blowing *n*. 吹制，吹气法
lamination *n*. 层压，薄板
tray *n*. 托盘
biaxial stretching 双轴拉伸
polyester *n*. 聚酯
polycarbonate *n*. 聚碳酸酯
polyvinyl formal 聚乙烯醇缩甲醛
polyvinyl alcohol 聚乙烯醇
formaldehyde *n*. 甲醛
melting point 熔点
compatibility *n*. 共存，兼容性
greenhouse gas 温室效应气体
noxious *adj*. 有害的，有毒的
dioxin *n*. 二氧芑
furan *n*. 呋喃
amine *n*. 胺
threaten *v*. 威胁，恐吓
choking *adj*. 令人窒息的，透不过气的
prostate cancer 前列腺癌
landfill *n*. 废物填埋，垃圾填埋
pursuit *n*. 追求，寻找，继续，追逐，追捕
spike *n*. 长钉，道钉
mineral *n*. 矿物质，矿物
incredible *adj*. 不可思议的，难以置信的

Notes

(1) Over the course of history, especially after the industrial revolution in the eighteenth century, manufacturing technology revolutionized the entire concept of packaging when manufactures were pushed to develop more durable and resistant types of protection (packaging) to make the transportation of products from factory to shop and later to customers' homes possible.

> 参考译文：在历史的进程中，尤其是十八世纪工业革命之后，制造技术彻底改变了包装的全部概念，推动制造商开发更耐用且更具抵抗力的保护类型（包装），使产品从工厂到商店再到顾客家里成为可能。

(2) Polymer packaging can be seen as beneficial to the environment because food industry

manufacturers and businesses are able to reduce waste by preserving foods for longer than in a case without polymer packaging.

　　—参考译文：聚合物包装可以被视为对环境有益，这是因为相对于没有聚合物包装的情况，食品工业制造商和企业能够更长时间地保存食物以减少浪费。

（3）People are able to provide color coatings and layers in almost any material regardless of shape or geometry.

　　—be able to 可以，能够。regardless of 不管，不顾。
　　—参考译文：人们能够在几乎任何材料上提供颜色涂层而不受材料形状和结构的限制。

（4）It should be noted that the packaging industry and researchers work on the recycling and reusing of synthetic packaging materials for waste management and control.

　　—it should be noted that　值得注意的是。
　　—参考译文：值得注意的是，包装行业和研究人员致力于回收和再利用合成包装材料，以进行废物管理和控制。

（5）On the other hand, researchers are still working to correct the structure of some of these polymers for different purposes.

　　—参考译文：另一方面，研究人员仍在努力修改其中一些聚合物的结构以满足不同的用途。

（6）This makes it possible to package unique objects that are difficult to fit into normal/basic containers.

　　—make it possible to... 使……成为可能。
　　—参考译文：这使得包装难以放入普通/基本容器的特殊物体成为可能。

（7）In general, flexible and rigid synthetic packaging materials are used in food products.

　　—in general　通常，总的来说，总体上。
　　—参考译文：总的来说，柔性和刚性的合成包装材料可用于食品（的包装）。

Exercises

1. Question for discussion

（1）What are the advantages and disadvantages of glass as the food packaging material?

（2）What are the advantages and disadvantages of paper as the food packaging material?

（3）What is the percentage of the plastic-based packaging material in the North American market?

（4）What is the most important problem of the plastics to our environment?

（5）How did they reuse the PET containers and bottlesin the United States?

2. Translate the following into Chinese

（1）The application of 3D printing technology in packaging industries has shifted the dynamic of packaging in many different aspects such as in the potential to have full color graphics and text for labels.

(2) These plastics are mainly made of PE and can take hundreds of years to degrade without proper recycling technology and also release damaging greenhouse gases that contribute to global warming.

(3) The waste disposal problem of petroleum materials and their nonrenewable nature have caused a spike in the level of interest in the sustainable development of biodegradable polymers, recycling, and/or environment protection.

3. Translate the following into English

(1) 包装纸 (2) 有渗透性的
(3) 聚萘二甲酸乙二酯 (4) 聚二氯乙烯
(5) 灯丝，细丝 (6) 纤维素
(7) 聚碳酸丙烯酯 (8) 热重分析
(9) 保质期
(10) 食品和相关行业使用的包装材料种类繁多。
(11) 玻璃包装最适合盛装液体，因为它们由具有不可渗透屏障的天然成分制成，这是它们的优点之一。
(12) 聚乙烯的主要应用是工业和农业包装。

Reading Material
Biopolymers and their potential as a packaging material

这篇课文介绍了生物聚合物在包装材料方面的潜在应用。课文阐述了生物聚合物包装材料的分类，即直接从生物质中提取、通过经典聚合过程合成、由微生物或转基因细菌产生。文中还介绍了一些典型的生物聚合物的性质与应用场景。

Packaging is an integral component of the food processing sector. Food packaging is a combination of art, science and technology of enclosing a product for achieving safe transportation and distribution of the products in wholesome conditions to the users at least price. Most of the conventional packaging materials are products of petrochemicals like PVC, PET, polystyrene (PS), polypropylene (PP), polyamide (PA). The properties which make them unique for packaging of food are low cost, excellent physical properties (density, molecular weight), mechanical properties (tensile strength), transmission properties (O_2, CO_2), which not only increase the shelf-life of the product, but also add functionality in terms of convenience and attractiveness to the consumers. The only problem with synthetic polymers is their resistance to degradation in the environment. According to ASTM standards D-5488-94d, biodegradable is defined as capable of undergoing decomposition into carbon dioxide, methane, water, inorganic compounds and biomass. With increased awareness on sustainability, the packaging industries around the globe are looking for biopolymers as the replacement of synthetic polymer. Biopolymers may be defined as the polymers that are biodegradable by the enzymatic action of microbes. In last two decades, a lot of research has been done on biopolymers for food packaging applications. Based on the researches biopolymer based packaging materials may be divided into three main groups based on their origin and production.

Group 1 Thesis constitutes polymers which are directly extracted or removed from biomass. Certain polysaccharides such as starch, cellulose, and proteins (like casein and gluten) constitute represent this category. All these are, by nature, hydrophilic and somewhat crystalline and create problems while processing. Besides, further, their performances are also poor especially in relation to packaging of moist food products. However, their excellent gas barrier properties make them suitable for their utilization in food packaging industry.

Group 2 Thesis includes polymeric materials which are synthesized by a classical polymerization procedure such as aliphatic aromatic copolymers, aliphatic polyesters, poly-lactide, aliphatic copolymer (CPLA), using renewable bio-based monomers such as poly (lactic acid) and oil-based monomers like poly-caprolactones. A good example of polymer produced by classical chemical synthesis using renewable bio-based monomers is polylactic acid (PLA), a biopolyester polymerized from lactic acid monomers. The monomers themselves may be produced via fermentation of various carbohydrate feed stocks. PLA may be plasticized with its monomers or, alternatively, oligomeric lactic acid. PLA can be formed into blown film, injected mold objects and coating. Therefore, all together explaining why PLA is the first novel bio-based material produced at commercial scale.

Group 3 Polymers which are produced by microorganisms or genetically modified bacteria constitute this group. Till date, this group of bio-based polymers consists mainly of the polyhydroxy-alkanoates, but developments with bacterial cellulose and other polysaccharides are also in progress.

Starch based biopolymers

Starches are low cost polysaccharides, abundantly available and one of the cheapest groups of biodegradable polymers. It is also known hydrocolloid biopolymer. It is composed of amylose (poly-α-1, 4-D-glucopyranoside), a linear and crystalline polymer and amylopectin (poly-α-1, 4-D-glucopyranoside and α-1, 6-D-glucopyranoside), a branched and amorphous polymer. The amylose and amylopectin contents of starch ranges from about $10\% - 20\%$ and $80\% - 90\%$, respectively, depending on the source. Amylose is soluble in water and forms a helical structure. Various kinds of starches like potato, cassava, rice, corn, and tapioca are used for the preparation of biopolymers. Starch is usually used as a thermoplastic. It is plasticized through destructuration in presence of specific amounts of water or plasticizers and heat and then it is extruded. So thermoplastic starch has high sensitivity to humidity. Starches are poor resistance to moisture and their mechanical property restricts their uses. To improve these properties starches are blended with various biopolymers and certain additives.

Protein based biopolymers

Many of the proteins like gelatin, keratin, and casein consisting very interesting features of polymers such as flexural, shear strength, tensile modulus, as well as exceptional material properties including toughness, strength and elasticity. Thus, these proteins are also useful for the creation of new biodegradable polymer for various commercial applications. Protein-based biodegradable polymers have an expanding range of potential applications in formation of food and non-food packaging, and as biomaterials like reconstructive surgery, tissue engineering, etc. Therefore, the protein-based polymer can be used for the polymer reinforcement. The mechanical properties of protein polymer can be further enhanced by blending them with other protein and/or non-protein molecules. Blending technology gives us an opportunity to develop next generation biodegradable polymer/plastics which can replace the conventional plastics from the market. In food packaging industries films made by protein polymers (like Milk proteins, Whey protein, Gelatin, Wheat gluten, Corn, Zein, Soy protein, Egg white, etc.) are used as an edible films so that they can consume along with the food. Plant proteins from soybean, wheat, and corn are readily available and films from these proteins have been investigated extensively. The employment of protein-based film concepts to edible packaging materials promises to improve barrier and mechanical properties and facilitate the effective incorporation of bioactive ingredients and other functions such as tampering resistance, a barrier from oxygen, water vapor and dust, etc. In non-food packaging polymer of keratin, casein, zein, gelatin and soy-protein, etc., could play a crucial role in the development of various commercial products like shopping bags, mulch film, flushable sanitary product, etc. Blends of protein with non-protein, natural molecules such as chitosan, cellulose, and with synthetic polymer like polypropylene, polyethylene, polyvinyl chloride, etc., were prepared to improve the plastic properties of protein-based polymer which are suitable for food and non-food packaging.

Polylactic acid (PLA)

PLA is one of the biopolymer that has gained lot of attention in recent years because of its economic and commercial viability during processing. Poly (lactic acid) (PLA) belongs to the family of aliphatic polyesters made up from alpha-hydroxyacids, including polyglycolic acid or polymandelic. The polylactic acid (PLA) is obtained from the controlled depolymerization of the lactic acid monomer obtained from the fermentation of sugar feedstock, corn, etc., which are readily biodegradable. PLA is a sustainable alternative to petrochemical-derived products, since the lactides are produced by the microbial fermentation of agricultural byproducts, mainly the carbohydrate rich substances. PLA is becoming a growing alternative as a green food packaging material because it was found that in many circumstances its performance was better than synthetic plastic materials.

PLA is usually obtained from polycondensation of D- or L-lactic acid or from ring opening polymerization of lactide, a cyclic dimer of lactic acid. Properties that make PLA a good food packaging material are their high molecular weight, water solubility resistance, good process ability, i. e., easy to process by thermoforming and biodegradability. PLA has the tensile strength modulus, flavor and odor barrier of polyethylene and PET or flexible PVC; the temperature stability and process ability of polystyrene; and the printability and grease resistance of polyethylene. PLA can be processed by several approaches which include injection molding, sheet extrusion, blow molding, thermoforming and film forming. Processed PLA comes in the form of films, containers and coatings for paper and paper boards. PLA can be further recycled by chemical conversion back to lactic acid and then re-polymerized. Although PLA seems to be potential biodegradable polymer to be utilized in packaging of various food products, it exhibits certain limitations in unmodified form, viz. it is more brittle and degrades easily at substantial temperature rise.

Polyhydroxyalkanoates (PHAs)

Polyhydroxyalkanoates (PHAs), a family of bacterial polyesters, are formed and accumulated by various bacterial species under unbalanced growth conditions. These polymers are produced in nature by bacterial fermentation of sugar and lipids. Structurally, PHAs comprise simple macromolecules composed of 3-hydroxy fatty acid monomers. PHA has a considerably low volume of the biopolymer market, somewhere around 453.59 tons per year. In 2008, approximately 55115.57 tons of PHAs were commercially produced. PHAs have thermo-mechanical properties similar to synthetic polymers such as polypropylene. PHA polyesters are biodegradable, biocompatible and can be obtained from renewable resources. They have several desirable properties such as petroleum displacement and greenhouse gas minimization apart from their fully biodegradable nature. Poly (3-hydroxybutyrate) (PHB) is one of the biodegradable PHA (polyhydroxyalkanoates) and is naturally occurring β-hydroxyacid linear polyester. The general structure of the repeating units of these polyesters is different depending on the type of bacteria and the feed, it is typically $-(CH_2)_n-CH_3$ for most naturally occurring PHAs. Applications of PHA as a biodegradable packaging include bottles, containers, sheets, films, laminates, fibers and coatings. Over 100 monomers and copolymers can be developed from PHAs. Some of the polymers used are PHB, PHV, PHBV (Metabolix), PHBO, PHBH, PHBD. PHAs exhibit good tensile strength, printability, flavor and odor barrier properties, heat sealability, grease and oil resistance, temperature stability and are easy to dye, which boosts its application in food industry. For example, Metabolix, a US-based company, produces "Metabolix PHA", which is a blend of polyhydroxybutyrate (PHB) and poly (3-hydroxyoctanoate) that has been approved by the FDA for production of food additives and making packages that maintain all the performance characteristics of non-degradable plastics. Polyhydroxybutyrate (PHB), a lipid-like polymer of 3-hydroxybutyrate accumulate as a carbon and energy reserve under unbalanced (unfavorable) growth conditions, such as nutrient limitation. In general, PHB accumulation is favored by adequate availability of a suitable carbon source and a limiting supply of nitrogen, phosphate or dissolved oxygen or certain micro components like sulfur, potassium, tin, iron or magnesium. PHB exists in the cytoplasmic fluid in the form of crystalline granules having diameters of 0.2—0.7 μm and are surrounded by a membrane coat composed of lipid and protein about 2 nm thick and can be isolated as native granules or by solvent and enzymatic extraction.

(**Selected from:** Mangaraj S, Yadav A, Bal L M, et al. Application of Biodegradable Polymers in Food Packaging Industry: A Comprehensive Review [J/OL], Journal of Packaging Technology and Research, 2019, 3: 77-96.)

New Words and Expressions

wholesome   *adj.* 有益健康的

polystyrene   *n.* 聚苯乙烯

polypropylene   *n.* 聚丙烯

biodegradable   *adj.* 可生物降解的

awareness   *n.* 认识，意识，感悟能力

biopolymer   *n.* 生物高聚物

starch   *n.* 淀粉

somewhat   *adv.* 稍微，有点

moist   *adj.* 潮润的，微湿的

aliphatic   *adj.* 脂肪族的

aliphatic polyesters   脂族聚酯

aliphatic copolymer   脂肪族共聚物

poly-lactide   *n.* 聚交酯

renewable   *adj.* 可更新的，可再生的

polylactic acid   聚乳酸

biopolyester   *n.* 生物聚酯

fermentation   *n.* 发酵

microorganism   *n.* 微生物

genetically   *adv.* 基因地，遗传地

polyhydroxy-alkanoate   *n.* 聚羟基脂肪酸酯

hydrocolloid   *n.* 水状胶质，水状胶体

amylose   *n.* 直链淀粉

amylopectin   *n.* 支链淀粉

helical   *adj.* 螺旋形的

cassava   *n.* 木薯

tapioca   *n.* 树薯粉

extrude   *vt.* 挤出，逐出；*vi.* 突出

gelatin   *n.* 明胶，动物胶

keratin   *n.* 角蛋白，角素

casein   *n.* 酪蛋白

flexural   *adj.* 弯曲的

shear strength   切变强度

tensile modulus   拉伸模量

reconstructive   *adj.* 改造的

surgery   *n.* 外科手术

tissue engineering   组织工程学

reinforcement   *n.* 加强，加固

blending technology 共混工艺
edible *adj.* 可食用的
polyvinyl chloride 聚氯乙烯
hydroxyacid *n.* 羟基酸，羟基烃酸
polyglycolic acid 聚乙醇酸
feedstock *n.* 原料
lactide *n.* 丙交酯
microbial *adj.* 微生物的
byproduct *n.* 副产品
thermoforming *n.* 加热成形，热力塑型
biodegradability *n.* 生物降解能力
grease resistance 耐油脂性
somewhere *adv.* 在某处；*n.* 某个地方
biocompatible *adj.* 生物适合的
lipid *n.* 脂质，油脂
cytoplasmic *adj.* 细胞质的
granule *n.* 颗粒

Notes

(1) The properties which make them unique for packaging of food are low cost, excellent physical properties (density, molecular weight), mechanical properties (tensile strength), transmission properties (O_2, CO_2), which not only increase the shelf-life of the product, but also add functionality in terms of convenience and attractiveness to the consumers.

——参考译文：低成本、优异的物理性质（密度、分子量）、机械性能（拉伸强度）、透气性质（O_2、CO_2）等使它们在食物的包装上有着独特的优势，（它们）不仅增加了产品的保质期，而且既方便又可以吸引消费者。

(2) Based on the researches biopolymer based packaging materials may be divided into three main groups based on their origin and production.

——参考译文：根据研究，生物聚合物包装材料根据其来源和生产方式可分为三大类。

(3) The amylose and amylopectin contents of starch ranges from about $10\% \sim 20\%$ and $80\% \sim 90\%$, respectively, depending on the source.

——参考译文：根据其来源，直链淀粉和支链淀粉在淀粉中的含量分别约为 $10\% \sim 20\%$ 和 $80\% \sim 90\%$。

(4) PLA is one of the biopolymer that has gained lot of attention in recent years because of its economic and commercial viability during processing.

——参考译文：由于在加工过程中的经济性和商业可行性，聚乳酸是近年来备受关注的生物聚合物之一。

(5) These polymers are produced in nature by bacterial fermentation of sugar and lipids.

—参考译文：这些聚合物是在自然界中通过糖和脂质的细菌发酵产生的。

(6) They have several desirable properties such as petroleum displacement and greenhouse gas minimization apart from their fully biodegradable nature.

—参考译文：它们除了具有可完全生物降解的性质外，还具有几个理想的特性，比如取代石油产品（的使用）以及减少温室气体（的产生）。

Unit 4.5

Text

Photocontrol of fluid slugs in liquid crystal polymer microactuators

这篇课文主要介绍了一种利用液晶聚合物控制微致动器中流体的光控制方法。课文从微流控及管状微致动器（TMA）的背景出发，详细介绍了新型线性液晶聚合物的设计制备方案以及由此构建的 TMA。课文重点展示了利用线性液晶聚合物及 TMA 的光控变形操纵微量液体，阐释了相应的微观机制，并展望了该技术的应用。

Background

Manipulating small amounts of liquids is of great interest in both scientific research and practical applications. Conversion of light energy to liquid motion is a new paradigm for the actuation of microfluidic systems by using optical forces (through radiation pressure and optical tweezers), light modulation of electrical actuation (optoelectrowetting and photocontrol of electroosmotic flow), or light-induced capillary forces. The last approach has advantages since it requires neither special optical set-ups nor complex microfabrication steps: it uses capillary forces generated from a light-induced wettability gradient and Marangoni effects. However, the capillary force arising from a wettability gradient is too small to overcome the effect of contact-line pinning, so the motion is limited to specific liquids over a relatively short distance, in simple linear trajectories, and at low speed ($10-50\ \mu m \cdot s^{-1}$). In addition, use of the light-induced Marangoni effect requires either local heating or the addition of photosensitive surfactants to liquids, which is undesirable for biomedical applications and undoubtedly produces sample contamination.

It is well known that a completely wetting liquid droplet confined in a conical capillary is self-propelled towards the narrower end because of the axial force arising from differing curvature pressures across its end caps. If the geometry of tubular microactuators (TMAs) could be dynamically adjusted by light, a simple and straight-forward method to manipulate liquids would be achieved; therefore, a smart material capable of photodeformation is crucial to building such TMAs. Photodeformable crosslinked liquid crystal (LC) polymers are ordered polymers that show large and reversible deformation through the orientation change of LCs and allow temporal, localized, remote and isothermal triggering and actuation. Hence, crosslinked LC polymers are good candidates for actuators for precise and direct manipulation of liquids through photodeformation. Unfortunately, TMAs have not yet been fabricated from existing photodeformable crosslinked LC polymers, since they show poor processability owing to chemical crosslinking.

Inspired by the lamellar structure of artery walls, a novel linear LC polymer (LLCP) is designed, which has a long alkyl backbone containing double bonds and azobenzene moieties in side chains acting as both mesogens and photoresponsive groups [**Figure 4.5.1(a)**]. The flexible backbones and the azobenzene mesogens can self-assemble into a nano-scaled lamellar structure [**Figure 4.5.1(b)**] due to the molecular cooperation effect of LCs. In addition, the long spacers provide enough free volume for the azobenzene mesogens to generate a highly ordered structure and undergo a fast photoresponse. In order to promote mechanical robustness, ring-opening metathesis polymerization — living polymerization that allows synthesis of a high-molecular-weight polyolefin with narrow polydispersity — was employed to prepare the LLCP. The number-average molecular weight of the LLCP reached 3.6×10^5 g · mol^{-1}, which is at least one order of magnitude larger than that of the generally used photoresponsive azobenzene LC polymers. Tensile tests show that the LLCP fiber has a moderate elastic modulus [(96 ± 19) MPa], high toughness [(319 ± 41) MJ · m^{-3}], high strength (~ 20 MPa) and a large elongation at break (or fracture strain; $2089\% \pm 275\%$). Thus the LLCP is a strong and tough material, which is ascribed to the ordered lamellar structure and the high molecular weight. Moreover, the absence of a chemical network means that the broken samples can be reshaped; a "healed" fiber with a cross-sectional area of 0.02 mm^2 can still sustain a large load, up to ~ 52 g.

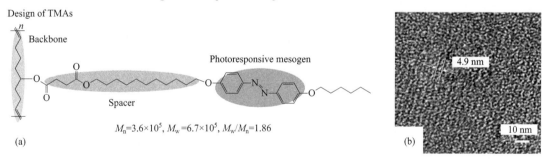

Figure 4.5.1 (a) Molecular structure of a novel LLCP. M_n, number-average molecular weight; M_w, weight-average molecular weight.
(b) TEM image showing the lamellar structure of the LLCP film.

Thanks to rational structure design and the robust mechanical properties of the LLCP, we were able, for the first time, to fabricate structurally defined and robust TMAs via a solution processing method. In the process, a glass capillary was filled with a solution of the LLCP in dichloromethane (~ 3 wt%). After evaporation of the dichloromethane at 50 ℃, the inner surface of the glass capillary was uniformly coated with the LLCP. The coated capillary was annealed at 50 ℃ for 30 min and then immersed in hydrofluoric acid to remove the glass. The produced free-standing TMA is robust enough to resist large deformation for many cycles.

Photocontrol of fluid slugs

The obtained TMAs show asymmetric geometry change upon irradiation by 470-nm light with an intensity gradient along the TMA (**Figure 4.5.2**); such irradiated TMAs can thus successfully manipulate liquid motion by light. In addition, these TMAs show unique abilities to propel a wide range of liquids spanning nonpolar to polar liquids, such as silicone oil, hexane, ethyl acetate, acetone, ethanol and water. More surprisingly, the TMAs can also propel complex fluids efficiently, such as a train of slugs, emulsion, liquid-solid fluid mixtures and even petrol, which have not yet been handled using existing light manipulation principles. Liquids widely used in biomedical engineering such as bovine serum albumin solution, phosphate buffer solution, cell culture medium and cell suspension, can also be manipulated by the TMAs, which is of great significance for biomedical analysis and micro-engineering. Moreover, it is noticed that the direction of movement of the slug can easily be controlled by varying the direction in which the intensity of the actinic light decreases. After turning off the light source, the TMA returns to its initial size due to elastic recovery of unexposed regions and entropic restoring forces imparted by the exposed region. Such reversible deformation on intermittent irradiation with 470-nm light can be repeated over 100 cycles without obvious fatigue, because the creep of the LLCP is minimized by the smectic organization of the side groups, which might act as physical crosslinks.

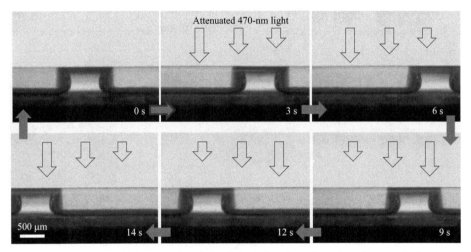

Figure 4.5.2 Lateral photographs of the light-induced motion of a silicone oil slug in a TMA. On irradiation by 470-nm light whose intensity (represented by open arrows) is attenuated increasingly from left to right (top row), the silicone oil slug is self-propelled towards the right; when the direction of attenuation is reversed (bottom row), the direction of movement of the slug is also reversed.

Photodeformation mechanism of the TMAs

All these liquid handling abilities arise from asymmetric photo-deformation of the TMAs in response to attenuated 470-nm light, which is a novel principle for inducing capillary force. It has been reported that azobenzene mesogens can be realigned along the direction perpendicular to the polarized direction of actinic linearly polarized blue light after repetition of trans-cis-trans isomerization cycles, which is known as the Weigert effect. In the case of unpolarized light, only the propagation direction is perpendicular to the polarized direction of the unpolarized light, thus the azobenzene mesogens orientate along the propagation direction of the actinic unpolarized light. When the TMAs are exposed to unpolarized 470-nm light whose actinic direction is perpendicular to the long axis of the TMAs, the azobenzene mesogens are reorientated along the propagation direction (**Figure 4.5.3**). Therefore, the tilt angles ϕ of azobenzen mesogens in the different exposed areas are different because the lamellae of the LLCP are arranged coaxially in the TMA wall.

In order to facilitate understanding of this photo-reorientation, the wall of the TMA is flattened out into a plane, as shown in **Figure 4.5.3**. According to the tilt angle of azobenzene mesogens calculated from 2D wide-angle X-ray diffraction, the azobenzene mesogens in ~70% of the exposed area are reoriented to exhibit $\phi \leqslant 65°$, which means this area expands along the y axis. The rest of the azobenzene mesogens are tilted with $65° < \phi \leqslant 90°$, leading to contraction along the y axis. In other words, the expansion of the light-exposed area is far larger than the contraction of that area. This photoinduced reorientation results in a decrease of the thickness of the TMA wall (along the x axis in **Figure 4.5.3**) and an elongation of the perimeter of the TMA (along the y axis in **Figure 4.5.3**), which together cause an increase of the cross-sectional area of the TMA. Moreover, the higher the light intensity, the larger the increase in cross-sectional area. The cross-sectional areas of the photo-deformed TMA at different positions increase with the increase of the light intensity upon irradiation by attenuated 470-nm light, whereas the cross-sectional areas at different positions without irradiation are almost the

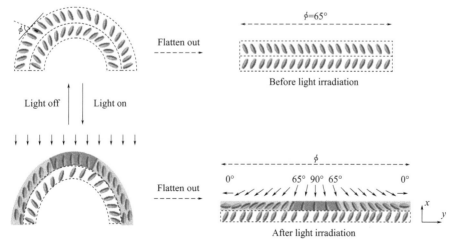

Figure 4.5.3 Schematics illustrating the reorientation of mesogens in the cross-sectional area of the TMA before and after irradiation by unpolarized 470-nm light.

same. Therefore, the TMA thus deforms to an asymmetric cone-like geometry, which generates adjustable capillary force to propel liquids in the direction of light attenuation.

To facilitate understanding the photo-reorientation, the wall is flattened out into a plane. The normal direction of the lamellae is along the x direction in the scheme. Before irradiation by the light, ϕ of all the LC mesogens is $65°$ (top). On light irradiation, the LC mesogens in the exposed surface of the TMA are realigned to the direction of the actinic light, which results in the change of ϕ in the exposed area (bottom). The orange and blue parts of the cross-sectional area respectively expand and contract along the y axis on light irradiation. This photoinduced reorientation leads to the decrease in thickness of the TMA wall (along the x axis) and the elongation of the perimeter of the TMA (along the y axis), which contributes to the increase of cross-sectional area.

Conclusion

The current TMAs present a conceptually novel way to propel liquids by capillary force arising from photo-induced asymmetric deformation, which relies on neither wettability gradients nor the Marangoni effect. They thus have considerable application as micro-pumps in microsystems technology and architecture without any aid from additional components. They are also excellent candidates for application in the fields of micro-reactors, laboratory-on-a-chip contexts and micro-optomechanical systems.

(**Selected from:** Lv J, Liu Y, Wei J, et al. Photocontrol of Fluid Slugs in Liquid Crystal Polymer Microactuators [J]. Nature, 2016, 537 (7619): 179-184.)

New Words and Expressions

paradigm *n.* 范式，范例，典范

tweezer *n.* 镊子，镊子钳

modulation *n.* 调制，调谐

optoelectrowetting *n.* 光电致湿润

electroosmotic *v.* 电渗，电渗透，电渗流

capillary force 毛细力

overcome *vt.* 克服，解决，战胜

trajectory *n.* 弹道，（射体在空中的）轨道，轨迹

photosensitive *adj.* 光敏的

surfactants *n.* 表面活性剂，表面活化剂，界面活性剂，表面活性物质

contamination *n.* 污染，污秽，（语言的）交感，（文章，故事等的）混合

droplet *n.* 液滴，小滴

curvature *n.* 弯曲，曲度，曲率

liquid crystal 液晶

isothermal trigger 等温触发器

artery *n.* 动脉

lamellar *adj.* 薄片状的，层式的，成薄层的，多层（片）的

novel linear LC polymer 新型线性液晶聚合物

azobenzene *n.* 偶氮苯

moiety *n.* 一半，二分之一，一部分

mesogen *n.* 液晶基元

photoresponsive *adj.* 光响应的，感光的

robustness *n.* 坚固性，健壮性

metathesis *n.* 易位

polyolefin *n.* 聚烯烃，聚烯烃系，聚烯烃纤维，聚烯烃树脂，聚烯

polydispersity *n.* 多分散性

elastic modulus 弹性模量

sustain *vt.* 维持（生命、生存），遭受，使保持，使稳定持续，经受，证明，支撑

dichloromethane *n.* 二氯甲烷

anneal *v.* 退火

hydrofluoric *n.* 氢氟酸

asymmetric *adj.* 不对称的

irradiate *v.* 辐照

hexane *n.* 己烷

ethyl acetate 乙酸乙酯

acetone *n.* 丙酮

actinic light 光化性光

entropic *adj.* 熵的

polarized *v.* （使）两极化，截然对立，使（光波等）偏振，使（物体）极化

Notes

（1）The last approach has advantages since it requires neither special optical set-ups nor complex microfabrication steps: it uses capillary forces generated from a light-induced wettability gradient and Marangoni effects.

—neither … nor … 既不……也不……。wettability gradient 湿度梯度。Marangoni effect 马兰戈尼效应。

—参考译文：最后一种方法具有优势，因为它既不需要特殊的光学装置，也不需要复杂的微加工步骤，它使用由光诱导的润湿性梯度和马兰戈尼效应产生的毛细力。

（2）However, the capillary force arising from a wettability gradient is too small to overcome the effect of contact-line pinning, so the motion is limited to specific liquids over a relatively short distance, in simple linear trajectories, and at low speed.

—be limited to 受限于。

—参考译文：然而，由润湿性梯度产生的毛细力太小，无法克服接触线钉扎的影响，因此运动仅限于特定的液体在简单线性轨迹上以低速运动（的情形）。

（3）In addition, use of the light-induced Marangoni effect requires either local heating or the addition of photosensitive surfactants to liquids, which is undesirable for biomedical applications and undoubtedly produces sample contamination.

——in addition 此外。
——参考译文：此外，使用光诱导的马兰戈尼效应需要局部加热或向液体中添加光敏表面活性剂，这对生物医学应用来说并不理想，而且无疑会产生样品污染。

（4）It is well known that a completely wetting liquid droplet confined in a conical capillary is self-propelled towards the narrower end because of the axial force arising from differing curvature pressures across its end caps.

——it is well known that 众所周知。
——self-propel 自动推进。
——参考译文：众所周知，由于端盖上不同的曲率压力产生轴向力，限制在锥形毛细管中的完全润湿的液滴会自行朝较窄端推进。

（5）Photodeformable crosslinked liquid crystal (LC) polymers are ordered polymers that show large and reversible deformation through the orientation change of LCs and allow temporal, localized, remote and isothermal triggering and actuation.

——Photodeformable crosslinked liquid crystal (LC) polymers 可光变形交联液晶聚合物。
——that 引导的从句，其中 that 指代前面的光变形交联液晶聚合物。
——参考译文：可光变形交联液晶（LC）聚合物是有序聚合物，通过 LC 的取向变化表现出大且可逆的变形，并允许临时、局部、远程和恒温触发和驱动。

Exercises

1. Question for discussion

（1）Please list the merits and demerits of light-induced capillary force manipulating small amounts of liquid.

（2）Please define photo-deformable crosslinked liquid crystal (LC) polymer.

（3）How can we control the direction of movement of the slug in TMAs?

（4）How can we fabricate structurally defined and robust TMAs?

（5）Please illustrate the application of TMAs.

2. Translate the following into Chinese

（1）Unfortunately, TMAs have not yet been fabricated from existing photo-deformable crosslinked LC polymers, since they show poor processability owing to chemical crosslinking.

（2）It has been reported that azobenzene mesogens can be realigned along the direction perpendicular to the polarized direction of actinic linearly polarized blue light after repetition of trans-cis-trans isomerization cycles, which is known as the Weigert effect.

（3）Inspired by the lamellar structure of artery walls, a novel linear LC polymer (LLCP) is designed, which has a long alkyl backbone containing double bonds and azobenzene moieties in side chains acting as both mesogens and photo responsive groups.

(4) Therefore, the tilt angles ϕ of azobenzen mesogens in the different exposed areas are different because the lamellae of the LLCP are arranged coaxially in the TMA wall.

3. Translate the following into English

(1) 管状微致动器　　　　　(2) 微流控
(3) 马兰戈尼效应　　　　　(4) 间歇照射
(5) 微反应堆　　　　　　　(6) 芯片实验室
(7) 微光机械系统　　　　　(8) 光形变
(9) 数均分子量　　　　　　(10) 重均分子量

(11) 可光变形交联液晶聚合物的数均分子量达到 $3.6 \times 10^5 \, g \cdot mol^{-1}$，比常用的光响应偶氮苯液晶聚合物大至少一个数量级。

(12) 因此可光变形交联液晶聚合物是一种强韧的材料，这归因于有序的层状结构和高分子量。

(13) 由于可光变形交联液晶聚合物的合理结构设计和强大的机械性能，我们能够首次通过溶液加工方法制造结构明确且坚固的管状微致动器。

(14) 牛血清白蛋白溶液、磷酸盐缓冲液、细胞培养基和细胞悬液等生物医学工程中广泛使用的液体也可以通过管状微致动器进行操作，这对生物医学分析具有重要意义。

Chapter 5

Biomaterials

【本章导读】
本章介绍生物材料,包括三个单元,共有三篇课文和三篇阅读材料。三篇课文分别介绍生物材料和生物材料科学的基础知识、传统和新型生物材料的特点以及生物材料在医学上的应用。三篇阅读材料分别讲解聚合物生物材料、3D打印生物材料和生物材料在组织修复上的应用。

Unit 5.1

Text

Biomaterials and biomaterials science

这篇课文主要介绍生物材料的定义、意义及应用。课文说明了术语"biomaterials"与"biological materials"的区别,介绍了生物材料在遗传疾病及药物递送领域的应用潜力。

The definitions of biomaterials include "materials of synthetic as well as of natural origin in contact with tissue, blood, and biological fluids, intended for use for prosthetic, diagnostic, therapeutic, and storage applications without adversely affecting the living organism and its components" and "any substance (other than drugs) or combination of substances, synthetic or natural in origin, which can be used for any period of time, as a whole or as a part of a system which treats, augments, or replaces any tissue, organ, or function of the body". Since a definition of the term "biomaterials" is difficult to formulate, the more widely accepted working definitions include: "A biomaterial is any material, natural or man-made, that comprises whole or part of a living structure or biomedical device which performs, augments, or replaces a natural function". The term "biomaterials" should not be confused with "biological materials". In general, a biological material is a material such as skin or artery, produced by a biological system.

Admittedly, any current definition of biomaterials is neither perfect nor complete, but already provides an excellent reference or starting point for discussion. Such definitions inevitably need to be updated to reflect the evolution and revolution in the dynamic biomedical field. Synthetic materials currently employed for biomedical applications include metals and alloys, polymers and ceramics. Due to the various structures of these materials, they have different properties and therefore have different uses *in vivo*. Next, this chapter will give a review of these three types of materials.

The study of biomaterials is known as "biomaterials science", which incorporates elements of medicine, biology, chemistry, tissue engineering, and materials science. A number of factors, including the aging of the population, the trend toward younger and middle-aged patients undergoing surgery, improvements in technology and lifestyle, better understanding of body function, improved aesthetics, and the need for better function, have led to a tremendous growth in biomaterials science, which should be an ongoing process. For example, there is a growing emphasis on developing non-traditional clinical approaches, such as the prevention and treatment of major genetic diseases. These trends in medicine pose unique challenges to the field of biomaterials. Applications such as the controlled delivery of pharmaceuticals (drugs and vaccines), viral and non-viral mediated delivery agents for gene therapy, and engineering functional tissues require visionary, non-traditional thinking and novel design approaches. Most importantly, to successfully address current and future biomaterials challenges, we need materials scientists and engineers who are familiar and sensitive to cellular, biochemical, molecular, and genetic issues and who work effectively in specialized teams that include molecular biologists, biochemists, geneticists, physicians, and surgeons.

(**Selected from:** Barbucci R. Integrated Biomaterials Science[M]. Springer Science & Business Media, 2002.)

New Words and Expressions

tissue   *n.* 纸巾，面巾纸，（动物或植物的细胞）组织
prosthetic   *adj.* 假体的，非阮基的
diagnostic   *adj.* 诊断的，判断的
therapeutic   *n.* 疗法，治疗药物；*adj.* 治疗的，有疗效的
organism   *n.* 生物，有机体，有机组织，有机体系
artery   *n.* 动脉，干线，干道，干流
esthetics   *n.* 美学
pharmaceutical   *adj.* 制药的；*n.* 药
viral   *adj.* 病毒性的
non-viral   *adj.* 没有病毒性的
cellular   *adj.* 由细胞组成的，细胞的
biochemist   *n.* 生物化学家
geneticist   *n.* 遗传学家
physician   *n.* 医生，（尤指）内科医生
surgeon   *n.* 外科医生
sensitive   *adj.* 敏感的，体贴的
clinical   *adj.* 临床的，临床诊断的
address   *v.* 设法解决，应对

Notes

(1) The definitions of biomaterials include "materials of synthetic as well as of natural

origin in contact with tissue, blood, and biological fluids, intended for use for prosthetic, diagnostic, therapeutic, and storage applications without adversely affecting the living organism and its components" and "any substance (other than drugs) or combination of substances, synthetic or natural in origin, which can be used for any period of time, as a whole or as a part of a system which treats, augments, or replaces any tissue, organ, or function of the body".

—materials of synthetic as well as of natural origin 意为对材料的来源进行分类，可分为合成类和自然类。

—prosthetic, diagnostic, therapeutic, and storage applications 是对医用材料应用的一个概括，指医用材料主要用于"修复、诊断、治疗和储存"四个方面。

—参考译文：生物材料的定义包括"与组织、血液和生物体液接触的合成材料和天然材料，用于修复、诊断、治疗和储存，而不会对生物体及其组织产生不利影响"和"任何合成的或天然的物质（药物除外）或物质的组合，作为一个整体或系统的一部分，可在任何时期用于治疗、增强或替代身体的任何组织、器官或功能"。

(2) Admittedly, any current definition of biomaterials is neither perfect nor complete but has provided an excellent reference or starting point for discussion.

—参考译文：诚然，目前任何生物材料的定义既不完美也不完整，但却为讨论提供了极好的参考或出发点。

(3) The study of biomaterials is called "Biomaterials Science" which encompasses the elements of medicine, biology, chemistry, tissue engineering, and materials science.

—be called 被称作。tissue engineering 组织工程。

—参考译文：对生物材料的研究被称为"生物材料科学"，它包含了医学、生物学、化学、组织工程和材料科学等领域。

(4) For example, there is a growing emphasis on developing non-traditional clinical approaches, such as the prevention and treatment of major genetic diseases.

—参考译文：例如，（人们）越来越重视发展非传统的临床方法，例如预防和治疗重大遗传疾病。

(5) A number of factors, including the aging population, increasing tendency for younger and middle-aged candidates to undergo surgery, improvements in the technology and life style, better understanding of body functionality, improved esthetics and need for better function have led to a tremendous growth in biomaterials science, and this should be an ongoing process.

—这一句对主语进行了大量的铺陈，读者确立主语为 a number of factors，谓语为 have led to，即可理顺句子的含义。

—参考译文：人口老龄化、年轻人和中年人接受手术治疗趋势的增加、技术的提高和生活方式的改善、对身体功能更深的理解、审美的提升和对更好功能的需求等一系列因素推动了生物材料科学的巨大发展，并且这应该是一个持续的过程。

Exercises

1. Question for discussion

(1) What is biomaterial?

(2) What areas of research does the study of biomaterials cover?

2. Translate the following into Chinese

(1) The definitions of biomaterials include "materials of synthetic as well as of natural origin in contact with tissue, blood, and biological fluids, intended for use for prosthetic, diagnostic, therapeutic, and storage applications without adversely affecting the living organism and its components" and "any substance (other than drugs) or combination of substances, synthetic or natural in origin, which can be used for any period of time, as a whole or as a part of a system which treats, augments, or replaces any tissue, organ, or function of the body".

(2) These trends in medicine pose unique challenges to the field of biomaterials. Applications such as the controlled delivery of pharmaceuticals (drugs and vaccines), viral and non-viral mediated delivery agents for gene therapy, and engineering functional tissues require visionary, non-traditional thinking and novel design approaches.

3. Translate the following into English

(1) 修复　　　　　　　　　(2) 合成

(3) 药品　　　　　　　　　(4) 增加，提高

(5) 代替　　　　　　　　　(6) 器官

(7) 强调　　　　　　　　　(8) 作用

Reading Material
Make better, safer biomaterials

这篇课文主要介绍用于医学领域中聚合物的种类、性能以及制备方法。课文说明了聚合物的分子结构特点及其在医学领域的应用优势，随后介绍了不同生物材料的制备方法并概述了相应的结构表征方法及性能评估标准。

Polymers have a wide range of physical and mechanical properties suited to many purposes in medicine. For example, poly (methyl methacrylate) (PMMA), which resembles bones and teeth, has been used since the 1930s for dental implants and hip replacements. Poly (2-hydroxyethyl methacrylate) has been used since the early 1960s for soft contact lenses because it is transparent, flexible and stays swollen and wet. Strong yet bendable polyurethanes have been used for heart valves for decades.

But once in the body, polymers can cause side effects. These might be triggered by components left over from the polymerization process, such as monomers, reaction initiators or catalysts. For example, residual methyl methacrylate monomer in PMMA damages cells, irritates eyes and skin and disturbs the nervous system. Certain silicones in breast implants can cause infections. Poly (ethylene terephthalate), often used to make vascular grafts, traps proteins on its surface that can disturb blood flow and induce clots.

Clinical approval of new materials remains difficult and expensive. Rounds of extensive toxicological studies are followed by tough clinical trials to assess the safety and efficacy of a proposed device. These hurdles mean that repurposing old materials is easier than introducing new ones. But promising new options abound.

What's needed is a more integrated approach to designing and regulating polymers in biomedicine. From the start, designers need to address all the components that may render a material toxic or capable of causing cancer, birth defects, genetic mutations or blood clots. Below we outline the sort of standardized testing platforms—experimental and computational—that are needed to evaluate biocompatibility.

Advanced polymers

Progress over the past two decades holds promise for designing new biomaterials. For example, macromolecular structures can be designed and fabricated with precision. Techniques such as reversible-deactivation radical polymerization attach and detach small active molecular groups (radicals) to block undesirable steps during the reaction that forms the polymer. The range of molecular weights of the polymer chains is controlled and little catalyst is left behind (just a few parts per million). The biocompatibility of the polymers can be enhanced by further purification and by using aqueous solvents and non-metal catalysts.

Another breakthrough is "click chemistry", which builds polymers and molecules in a modular way through a series of reactions. Different sorts of polymer can be linked together, vastly broadening the range of surface chemistries possible for biomedicine. Click reactions are efficient, have high yields and few by-products. They need only mild conditions and benign or removable solvents. Click chemistry has been used to make gels, including patterned forms of hydrogel, where different areas perform different biochemical functions. Unfavourable copper catalysts and azides are being phased out through, for example, carrying out reactions with greater precision and without catalysts.

Another emerging area is assembly through physical interactions between molecules. For example, hydrophilic and -phobic groups arrange themselves differently in polar or non-polar liquids. They can self-assemble into thin plates, aggregates and three-dimensional structures. Biopolymers made from DNA and proteins are increasingly used to make materials. Polymeric materials shaped as nanotubes, nanospheres, fibres and tapes have been prepared by self-assembling peptides or macromolecules.

Surfaces can be modified to control interactions. For example, some hydrogels repel proteins electrostatically, which avoids immune reactions or the biosurface becoming fouled. Repellent coatings on medical devices such as catheters can be based on slippery, liquid-infused, porous surfaces (SLIPS) to prevent thrombosis.

Approval process

However, it is difficult to get clinical approval for new polymeric systems. Toxicological studies and clinical trials require more money and equipment than a standard academic laboratory can muster. An industrial setting is a must.

The regulatory process for multifunctional medical devices is complicated. For example, a single product such as a heart stent that slowly releases a drug can have several components, including the drug, a polymer coating and metallic frame. The US Food and Drug Administration evaluates the effectiveness and safety of either the primary use or of independent uses, depending on what the product is mainly meant to do. Thus it is easier to use established polymers in new applications than to get new ones approved.

Addressing these challenges requires various stakeholders to work together, including academia, industry and regulatory agencies. They should evaluate the biomaterial's design earlier in the research phase, based on regulatory needs and the performance specifications required for each application.

Evaluation systems

To predict how human tissues will respond to new materials before they are tested in clinical trials we must develop standardized in vitro and in vivo evaluation platforms. Toxicity and inflammatory response are particularly important to assess because these human reactions cannot be faithfully reproduced in animal models. Several options need research.

"Organ-on-a-chip" systems capture aspects of human physiology using miniaturized human tissues. Most existing platforms focus on metabolic and barrier tissues, such as liver and lung. Polymeric materials will require systems capable of testing human blood and the immune system. Variations in patients must be considered. Organs-on-a-chip could contain cells derived from specific patients.

Modelling and simulations can be used to understand and predict the behaviour of body systems, from a single cell to the whole organism. These computational tools can also be used to interpret, analyse and predict the underlying response mechanisms to new substances. For example, a mathematical method known as physiologically based pharmacokinetic modelling (PBPK) is used to determine exposure doses that can lead to toxicity. PBPK parameters use data from studies done in vitro, in vivo and in silico. The increased risk of human cancer from vinyl chloride was evaluated using such a model. Although based on data from animals, its estimates are consistent with human epidemiological data.

High-throughput screening can be used to test the safety of libraries of new polymers, while lowering the cost and reducing animal testing. To screen hundreds of thousands of chemicals, rapid throughput microarrays have been used in vitro and in vivo. Automatic systems that contain small vertebrates to conduct high-throughput pharmacological tests in vivo have been developed. Similar tools would be useful in assessing new polymers.

Integrating all these platforms into the design process would accelerate the clinical translation of biomaterials.

(**Selected from:** Peppas N A, Khademhosseini A. Make better, Safer Biomaterials [J]. Nature, 2016, 540 (7633): 335-337.)

New Words and Expressions

resemble　v. 像，与……相似
transparent　adj. 透明的，清澈的
bendable　adj. 可弯曲的
heart valves　心脏瓣膜
trigger　v. 引发，激发，起动，触发
polymerization　n. 聚合，聚合作用
initiator　n. 引发剂，发起人，创始人
residual　adj. 剩余的，残留的
irritate　v. 刺激，使疼痛，使兴奋
vascular graft　人工血管
toxicological　adj. 毒理学的
hurdle　n. 障碍，难关
biocompatibility　n. 生物相容性，生物适合性
modular　adj. 模块化的，有标准组件的
benign　adj. 良性的
azides　n. 叠氮化物
hydrophilic　adj. 亲水的
polar　adj.（液体，尤指溶剂）极性的，（固体）离子的
peptide　n. 肽
organ-on-a-chip　n. 器官芯片
pharmacokinetic　adj. 药物代谢动力学的
high-throughput screening　高通量筛选
vertebrate　n. 脊椎动物

Notes

（1）From the start, designers need to address all the components that may render a material toxic or capable of causing cancer, birth defects, genetic mutations or blood clots.

——参考译文：从一开始，设计者就需要处理可能导致材料毒性或可能导致癌症、出生缺陷、基因突变或血栓的所有成分。

（2）The biocompatibility of the polymers can be enhanced by further purification and by using aqueous solvents and non-metal catalysts.

——参考译文：聚合物的生物相容性可以通过进一步提纯和使用水溶剂和非金属催化剂来提高。

（3）Click chemistry has been used to make gels, including patterned forms of hydrogel, where different areas perform different biochemical functions.

——click chemistry　译为点击化学，是一类常用于一些具有生物相容性的小分子的生物

偶联反应，将所选底物与特定生物分子结合。

——参考译文：点击化学已被用来制备凝胶，包括图案化的水凝胶，其中不同的区域表现出不同的生化功能。

(4) Toxicity and inflammatory response are particularly important to assess because these human reactions cannot be faithfully reproduced in animal models.

——参考译文：对于毒性和炎症反应的评估尤其重要，因为这些人体反应不能在动物模型中真实再现。

Unit 5.2

Text

Traditional and new biomaterials

这篇课文主要介绍金属、陶瓷、聚合物及复合材料在生物材料领域的应用。课文介绍了以上四种材料的性能和应用领域，同时分析了四种材料的优缺点并对其未来在生物材料领域的应用进行了展望。

Metals

Metallic implant materials have long gained great clinical importance in the medical field. Many of the metals and metal alloys used here include stainless steel (316L), titanium and alloys (Cp-Ti, Ti_6Al_4V), cobalt-chromium alloys (Co-Cr), zirconium-niobium, aluminum alloys, and tungsten heavy alloys. The rapid growth and development of the biomaterials sector has created scope for many medical products made from metals, such as dental implants, cranial plates and screws; components of artificial hearts, pacemakers, clips, valves, balloon catheters, medical devices and equipment; and bone fixation devices, dental materials, prosthetics and orthotics for biomedical applications. For example, NiTi alloy with TiN nanocoating is used to fabricate a left atrial appendage (LAA) occlusion device for interventional treatment (**Figure 5.2.1**). While there are other classes of materials that can be used as biomaterials, engineers prefer metals as the key material for designing the desired biomaterial. The main criteria for selecting metal-based materials for biomedical applications are their excellent biocompatibility, convenient mechanical properties, excellent corrosion resistance and low cost.

The type of metal used in biomedical applications depends on the function of the implant and the biological environment. 316L type stainless steel (316L SS) is the mostly used alloy in all implants ranging from cardiovascular to otorhinology. The mechanical properties of metals are significant when designing load-bearing dental and orthopedic implants. However, when the implant requires high wear resistance such as artificial joints, Co-Cr-Mo alloys are used to serve the purpose. The high tensile strength and fatigue limit properties of metals make it possible to design implants that can withstand mechanical loads better than ceramics and polymers. In

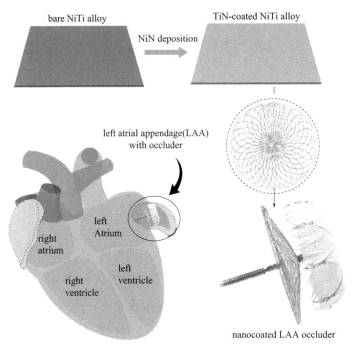

Figure 5.2.1 Schematic illustration of the left atrial appendage (LAA) occlusion and the construct of the LAA occlude.

comparison to polymers, metals have higher ultimate tensile strength and elastic modulus but lower strains at failure. However, in comparison to ceramics, metals have lower strengths and elastic modulus with higher strains to failure. It should be noted that biomedical metals used for implants are conventional, homogeneous materials compared to the nanophase, composite nature of tissues such as bone. This mismatch of mechanical properties can cause "stress shielding", a condition characterized by bone resorption (loss of bone) in the vicinity of implants. This clinical complication arises because preferential distribution of mechanical loading through the metallic prosthesis deprives bone of the mechanical stimulation needed to maintain homeostasis. The mechanical properties of a specific implant depend not only on the type of metal but also on the processes used to fabricate the material and device. Thermal and mechanical processing conditions can change the microstructure of a material. For example, when "cold working" metals, such as by rolling or forging, the resulting deformation makes the material stronger and harder. Unfortunately, as the metal becomes harder and stronger it also becomes less ductile (undergoes less deformation before failure) and more chemically reactive.

In biological media, when metal-based biomaterials are implanted, the surface of the material changes and degrades, releasing some by-products. As a result of the release process, interactions between the metal implant surface and the cells or tissues occur. This factor has stimulated today's researchers to give high priority to understand the surface properties of metal products in order to develop biocompatible materials.

Polymers

There are a large number of polymeric materials that have been served as implants. The current applications of them include cardiac valves, artificial hearts, vascular grafts, breast prosthesis, dental materials, contact and intraocular lenses, fixtures of extracorporeal oxygenators, dialysis and plasmapheresis systems, coating materials for medical products, surgical materials, tissue adhesives, etc. The composition, structure, and organization of constituent macromolecules specify the properties of polymers. Further, the versatility in diverse application requires the production of polymers that are prepared in different structures and compositions with appropriate physicochemical, interfacial, and biomimetic properties to meet specific purpose. The advantages of the polymeric biomaterials over other classes of materials are (i) ease to manufacture, (ii) ease of secondary processability, (iii) availability with desired mechanical and physical properties, and (iv) reasonable cost. Polymers for biomedical applications can be classified into two categories namely synthetic and natural. The synthetic polymeric systems include acrylics, polyamides, polyesters, polyethylene, polysiloxanes, polyurethane, etc. Although the processability of synthetic polymers is easy, the main drawback of these synthetic polymers is the general lack of biocompatibility in the majority of cases and therefore their utilization is often associated with inflammatory reactions. This problem can be avoided by the usage of natural polymers. The natural polymers such as chitosan, carrageenan, and alginate are used in biomedical applications such as tissue regeneration and drug delivery systems. Percutaneous or transcutaneous devices are important and unique, and the corresponding biological sealing at the skin-implant interface is the key to their long-term success. **Figure 5.2.2** illustrates polytetrafluoroethylene (PTFE) polymer used as an artificial heart valve, which not only has excellent mechanical durability but also enhanced hemodynamic function like a biological valve.

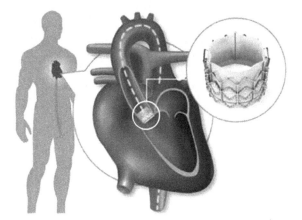

Figure 5.2.2 Schematic diagram of the polytetrafluoroethylene (PTFE) sewing ring in a polymer prosthetic valve.

The most pressing need for the use of polymeric biomaterials is their biocompatibility and degradation properties. Biocompatibility is generally defined as the ability of a biomaterial, prosthesis, or medical device to perform with an appropriate host response during a specific application. All materials intended for application in humans as biomaterials, medical devices, or prostheses undergo tissue responses when implanted into living tissue. "Appropriate host response" includes lack of blood clotting, resistance of bacterial colonization, and normal healing process. For a biomaterial implant to function properly in the patient's body, the implant should be biocompatible. Degradation of polymers requires disruption of their macromolecular structure and can occur by either alteration of the covalent interatomic bonds in the chains or alteration of the intermolecular interactions between chains. The former can occur by chain scission (cleavage of chains) or cross-linking (joining together of adjacent chains), an unlikely occurrence under physiological conditions. The latter can occur by incorporation (absorption) or loss (leaching) of low-molecular-weight compounds. Chemical reactions, such as oxidation and hydrolysis, can also change the properties of implanted polymers.

Ceramics

Ceramics are another class of materials used in the design of biomaterials. The motivation for using ceramics is their inertness *in vivo*, the ability to be machined into various shapes and pores, high compressive strength and excellent wear properties. Ceramics are used as parts of the musculoskeletal system, hip prostheses, artificial knees, bone grafts, dental and orthopedic implants, orbital and middle ear implants, cardiac valves, and coatings to improve the biocompatibility of metallic implants. **Figure 5.2.3** exhibits the wide use of ceramics in the human body. Although ceramics are used to design biomaterials, they are less popular than metals or polymers. In some cases, the application of ceramics is severely limited due to their brittleness and poor tensile strength. However, silicon-based bioceramics are widely used to make desirable biomaterials due to their high biocompatibility and osseointegration.

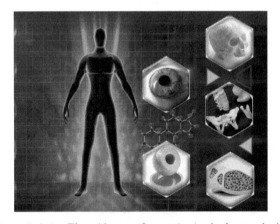

Figure 5.2.3　The wide use of ceramics in the human body.

Although they can have outstanding strength when loaded in compression, ceramics and glasses fail at low stress when loaded in tension or bending. Among biomedical ceramics, alumina has the highest mechanical properties, but its tensile properties are still below those of metallic biomaterials. Additional advantageous properties of alumina are its low coefficient of friction and wear rate. Because of these properties, alumina has been used as a bearing surface in joint replacements. The mechanical properties of calcium phosphates and bioactive glasses make them unsuitable as load-bearing implants. Clinically, hydroxyapatite has been used as filler for bone defects and as an implant in load-free anatomic sites such as nasal septal bone and middle ear. It is also used to develop bio-eye hydroxyapatite orbital implants and hydroxyapatite block ceramic. In addition to these applications, hydroxyapatite has been used as a coating material for stainless steels, titanium and its alloys-based implants, and onmetallic orthopedic and dental implants to promote their fixation in bone. Delamination of the ceramic layer from the metal surface, however, can create serious problems and lead to implant failure.

Composites

Composites are engineering materials which contain two or more physical and/or chemical distinct, properly arranged or distributed constituent materials that have different physical properties than those of individual constituent materials. Composites have a continuous body phase called the matrix and one or more discrete phases called the reinforcement, which usually have superior properties to the matrix. Separately, there is a third phase named as interphase between matrix and reinforced phases. Composites have unique properties and are usually stronger than any of the single materials from which they are made, hence are applied to some difficult problems where tissue in-growth is necessary. Composite scaffolds with porous structure tailored from combinations of bioglass particles and biodegradable polymers with mechanical properties that are close to cancellous bone are potentially in use. Hard-tissue applications such as skull reconstruction, bone fracture repair, total knee, ankle, dental, hip, and other joint replacement applications are possible with fiber-reinforced composite materials. In recent years, scientific research has also explored the development of new biomedical composites as alternative solutions for tumor treatment and tissue regeneration. For example, **Figure 5.2.4** shows an injectable thermosensitive hydrogel with nanosized black titanium dioxide nanoparticles integrated in a chitosan matrix for tumor treatment and wound healing.

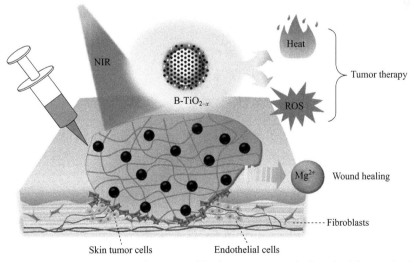

Figure 5.2.4 Illustration of an injectable thermosensitive hydrogel with nanosized black titanium dioxide nanoparticles integrated in a chitosan matrix for tumor treatment and wound healing.

The main advantage of composites is that they offer alternatives to improve many undesirable properties of homogeneous materials (metals or ceramics) despite the disadvantages of individual metal or ceramic materials, such as the low biocompatibility and corrosion of metals and the brittleness and low fracture strength of ceramic materials. The properties of the constituent materials have significant influence on composite biomaterials. One of these factors, "linear expansion" plays a key role in the design of composite biomaterials. Typically, composite materials are composed of components with similar linear expansion constants. If the constituent materials have different linear expansion constants, then the contact zone (interface) between the reinforcement and matrix materials can create large voids through the contact surface, thus obscuring the purpose of the implant. Therefore, bone tissue engineers need to be more careful in the selection of individual components when processing composite biomaterials.

(**Selected from:** Shen Y, Zhang W, Xie Y, et al. Surface Modification to Enhance Cell Migration on Biomaterials and its Combination with 3D Structural Design of Occluders to Improve Interventional Treatment of Heart Diseases[J]. Biomaterials, 2021, 279: 121208.)

New Words and Expressions

stainless steel　*n*. 不锈钢
aluminum alloy　铝合金
zirconium　*n*. 锆
niobium　*n*. 铌
tungsten　*n*. 钨
scope　*n*. 范围，领域；*v*. 评估，调查
craniofacial　*adj*. 颅面的
screw　*n*. 螺丝（钉），螺杆，螺丝杆；*v*. 用螺丝固定；拧紧，旋紧
artificial　*adj*. 人造的，人工的，人为的，不真挚的，矫揉造作的
pacemaker　*n*. 起搏器，领跑者，标兵
valve　*n*. 阀，活门；*v*. 装阀于，以活门调节
ballooncatheter　*n*. 囊导管，气泡式导管
orthodontic　*adj*. 畸齿矫正的，齿列矫正的
tensile strength　拉伸强度
fatigue　*n*. 疲乏，厌倦；*v*. 使疲劳，使劳累；*adj*. 疲劳的
withstand　*v*. 顶住，抵住，经受住，承受住（攻击、批评等），反抗
elastic modulus　弹性模量
resorption　*n*. 再吸收，再吞
vicinity　*n*. 周围地区，邻近地区，附近
by-product　*n*. 副产品，附带产生的结果，意外收获
biocompatible　*adj*. 生物适合的，不会引起排斥的
intraocular　*adj*. 眼内的
extracorporeal　*a*. 身体外面的
dialysis　*n*. 透析

plasmapheresis　*n*. 血浆除去法

polyamide　*n*. 聚酰胺（尼龙）

polyester　*n*. 聚酯

polyethylene　*n*. 聚乙烯

polysiloxane　*n*. 聚硅氧烷，聚砂氧烷

polyurethane　*n*. 聚氨酯，聚亚安酯

inflammatory　*adj*. 炎症性的，煽动性的，激动的

chitosan　*n*. 壳聚糖，脱乙酰几丁质，脱乙酰壳多糖，聚氨基葡萄糖

carrageenan　*n*. 角叉菜胶

alginate　*n*. 海藻酸盐

percutaneous　*adj*. 经皮的，经由皮肤的

transcutaneous　*adj*. 经皮的，由皮的

colonization　*n*. 定殖

macromolecular　*adj*. 大分子的

scission　*n*. 切断，分离，断开

hydrolysis　*n*. 水解作用

orthopedic　*adj*. 骨科的，矫形的

orbital　*adj*. 轨道的

hydroxyapatite　*n*. 羟基磷灰石

anatomic　*adj*. 组织的，解剖学上的，结构上的

nasal　*adj*. 鼻的；*n*. 鼻音

septal　*adj*. 中隔的，氏族的，隔膜的

delamination　*n*. 分层，层离

discontinuous　*adj*. 不连续的，间断的

reinforcement　*n*. 援兵，增援部队，加强，加固

skull　*n*. 颅骨，头骨

titania　*n*. 二氧化钛，人造金红石

wound healing　创伤修复，愈合，伤口愈合

Notes

（1） The rapid growth and development in biomaterial field has created scope to develop many medical products made of metal, such as dental implants, craniofacial plates and screws; component of artificial hearts, pacemakers, clips, valves, balloon catheters, medical devices and equipment; and bone fixation devices, dental materials, prosthetic and orthodontic devices for biomedical applications.

——此句阐述了医用材料发展所创造的新格局。create a scope to 中 scope 原意为范围，在这里可以形象化理解为"创造了空间"。

——参考译文：生物材料领域的快速发展为开发许多金属医疗产品创造了空间，例如口腔种植体、颅面板和螺钉；人造心脏、起搏器、夹子、瓣膜、球囊导管、医疗器械的组成部分；以及用于生物医学应用的骨固定装置、口腔材料、假肢和正畸装置。

（2）This clinical complication arises because preferential distribution of mechanical loading through the metallic prosthesis deprives bone of the mechanical stimulation needed to maintain homeostasis.

——deprive ... of ... 是一个固定搭配，是指剥夺了某人的某物。
——homeostasis 意为稳态，是指生物体内相互依赖的因素之间保持平衡的一种趋势。
——参考译文：因为机械载荷优先负载在金属义肢上，使骨骼失去了维持平衡所需的机械刺激，造成了这种临床并发症的产生。

（3）This factor has stimulated today's researchers to give high priority to understanding the surface properties of metal products in order to develop biocompatible materials.

——这一句中有两个固定搭配：stimulate ... to ... 和 give high priority to，分别意为刺激某人做了某事和高度重视。
——参考译文：这一因素促使当今的研究人员高度重视理解金属产品的表面特性，以开发生物相容性材料。

（4）The current applications of them include cardiac valves, artificial hearts, vascular grafts, breast prosthesis, dental materials, contact and intraocular lenses, fixtures of extracorporeal oxygenators, dialysis and plasmapheresis systems, coating materials for medical products, surgical materials, tissue adhesives, etc.

——参考译文：它们（聚合物）目前的应用包括心脏瓣膜、人工心脏、血管移植物、乳房假体、口腔材料、隐形眼镜和人工晶状体、体外氧合器的固定装置、透析和血浆置换系统、医疗产品的涂层材料、手术材料、组织粘合剂等。

（5）Though the processability is easy in case of synthetic polymers, the main disadvantage of these synthetic polymers is the general lack of biocompatibility in the majority of cases and hence their utility is often associated with inflammatory reactions.

——这一段中出现了很多学科名词，读者可以学习这些名词来加深印象。在这一句中hence意为因此，可以与and直接连用而不用断句。
——参考译文：尽管合成聚合物易于加工，但在大多数情况下这些材料的主要缺点是普遍缺乏生物相容性，因此使用此类材料通常伴随着炎症反应。

（6）Biocompatibility is generally defined as the ability of a biomaterial, prosthesis, or medical device to perform with an appropriate host response during a specific application.

——这一句是对生物相容性的定义，也是本段最为核心的概念。
——参考译文：生物相容性通常被定义为生物材料、假体或医疗设备在特定应用中具有适当的宿主反应的能力。

（7）Degradation of polymers requires disruption of their macromolecular structure and can occur by either alteration of the covalent interatomic bonds in the chains or alteration of the intermolecular interactions between chains.

——参考译文：聚合物的降解需要破坏大分子结构，可以通过改变链中原子间的共价键或改变链之间的分子间相互作用来实现。

(8) Clinically, hydroxyapatite has been used as filler for bone defects and as an implant in load-free anatomic sites such as nasal septal bone and middle ear.

—hydroxyapatite 羟基磷灰石，是一种常用的医用材料。
—load-free 是指无负载，这是因为羟基磷灰石的机械性能较差。
—参考译文：在临床上，羟基磷灰石已被用作骨缺损的填充物，以及在鼻中隔骨和中耳等无载荷解剖部位的植入体。

(9) Composites are engineering materials which contain two or more physical and/or chemical distinct, properly arranged or distributed constituent materials that have different physical properties than those of individual constituent materials.

—这一句是对复合材料的定义，应引起注意。
—参考译文：复合材料是指包含两种或两种以上物理和/或化学（性质）不同的、组成材料以适当形式排列或分布的工程材料，其物理特性与单一组成材料不同。

Exercises

1. Question for discussion

(1) What are the advantages and disadvantages of metals?
(2) What are the advantages and disadvantages of polymers?
(3) What are the advantages and disadvantages of ceramics?
(4) What are the advantages and disadvantages of composites?
(5) What are the main criteria for selection of metal-based materials for biomedical applications?

2. Translate the following into Chinese

(1) Although they can have outstanding strength when loaded in compression, ceramics and glasses fail at low stress when loaded in tension or bending.

(2) The most pressing need for the use of polymeric biomaterials is their biocompatibility and degradation properties. Biocompatibility is generally defined as the ability of a biomaterial, prosthesis, or medical device to perform with an appropriate host response during a specific application.

(3) It should be noted that biomedical metals used for implants are conventional, homogeneous materials compared to the nanophase, composite nature of tissues such as bone.

3. Translate the following into English

(1) 羟基磷灰石　　　　(2) 肿瘤治疗
(3) 伤口愈合　　　　　(4) 踝关节
(5) 复合材料是包含两种或多种物理和/或化学性质不同的、组成材料以适当形式排列或分布的工程材料。
(6) 聚合物材料目前的应用包括心脏瓣膜、人造心脏、血管移植物、乳房假体、牙科材料、隐形眼镜和人工晶状体、体外氧合器的固定装置、透析和血浆置换系统、医疗产品的涂层材料、手术材料、组织粘合剂等。

(7) 复合材料具有独特的性能，通常比构成它们的任何单一材料都更坚固，因此适用于解决一些需要长期植入的难题。

Reading Material
3D printing metallic implants: Technologies available and the future of the industry

这篇课文主要介绍3D打印技术在生物医疗领域的应用。课文分析了制造医疗植入体的艰难性，指出其中最难克服的问题之一是生物相容性，提出使用3D打印医疗植入物来克服这一难题，并详细讲解3D打印的制造方法及其在医疗植入体中的应用。

Medical implants are an ideal field for 3D printing, requiring the rapid manufacture of highly intricate and customizable products (**Figure 5.2.5**). Research in additive manufacturing continues to improve the quality of these implants and increases the number of applications 3D printing has in the medical industry.

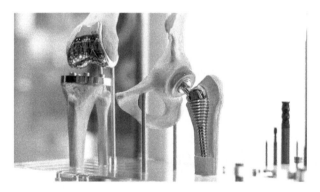

Figure 5.2.5 Artificial joints prepared by 3D printing technology for hip and knee replacement. (Image Credit: *Monstar Studio/Shutterstock.com.*)

Creating medical implants is a difficult process, as it involves the collaboration of multiple intensive and vastly different bodies of research. These products must be analyzed from a medical, biomechanical, and material science perspective.

One of the most difficult issues to overcome in medical implants is biocompatibility. This issue refers to the implant's ability to behave and perform adequately in the body over long periods of time. Incorrectly manufactured synthetic implants can emit ions, negatively affecting the tissue around the implant, or the immune system can reject the implant as a foreign body altogether. Toxicology, immune system response, and surface treatment are all essential fields to consider when ensuring biocompatibility.

Beyond biological considerations, implants are also difficult to design from a mechanical perspective. Biomechanics of body parts can be vast and complex: single components can vary in properties as small as collagen content and lamellae orientation across their bodies, giving distinct and essential mechanical properties. Joints can move countless cycles for decades with little to no wear. Recreating these biomechanical properties synthetically is incredibly difficult.

3D printing medical implants

The geometry of bones and tissue differs from patient to patient. The nature and extent of the damage that the patient has endured is also entirely unique to them. Because of this, implants must be custom-made and uniquely designed to fit their patient.

In almost all circumstances where small batches of uniquely designed products are manufactured, additive manufacturing becomes the most attractive, cost-effective method. Building complex pieces with intricate internal structures and minute but essential variations is much simpler to manufacture layer by layer, straight from a computer-aided design (CAD) drawing.

Additive manufacturing is not a new concept in the medical implant industry. Knee implants, for example, are 3D-printed from CAD models. These models are extracted from 3D scans of the patient's knee, which prevents the invasive surgeries that would traditionally take place. 3D printing has been used for other implants across the entire human body for over a decade.

Advancements in 3D printing technologies — from a more expansive range of filament materials to increased precision — mean this manufacturing method is likely to be increasingly implemented for medical implants.

Recent technological advancements in 3D printed implants: Nitinol arterial stents

Advancements in materials 3D printing — made as recently as 2020 — have led to monumental strides in the medical field. Research at the Australian government's scientific research agency CSIRO concluded in the past year that Nitinol can be 3D printed, allowing mass production of arterial stents for circulation disease.

Nitinol, a titanium and nickel alloy, is a highly desirable material for peripheral artery disease (PAD) due to its shape memory retaining properties. PAD is caused by fat deposits in the arteries in the legs or arms, causing a reduction in blood flow to the limbs. Stents placed in these defective arteries must be able to deform but retain their shape as the user moves their limbs. Therefore, Nitinol is a beneficial option when creating these stents.

CSIRO research has now concluded that Nitinol stents can be 3D printed. This is a notable advancement in the metallurgy of additive manufacturing (nitinol is rarely used in 3D printing) and an improvement in the geometries of these stents. These 3D-printed stents have a complex mesh geometry, allowing them to expand and contract more efficiently than traditional stents.

With this research comes all the benefits of additive manufacturing for PAD sufferers. Arterial stents can be custom-made to the user and manufactured rapidly. Rapid manufacture also leads to wider accessibility of the product.

Manufacturing methods of 3D printing and applications in medical implants

To truly understand the potential of 3D printing in the medical industry, it is important to note how expansive the field of 3D printing is. These two previous examples, the arterial stents and the knee implants, are manufactured using one method of 3D printing: powder bed fusion. Other techniques are available, all of which allow different properties to their products.

Powder bed fusion refers to the fusion of substances to form layers of solid material from powder using a heat source such as a laser. The powder is then placed on the previously fused cross-section, and the process is repeated, building the product up layer by layer.

This process is one of the most widely used, as it allows for precise and complex structures to be made easily. In the case of the electron beam, the precision of this method is in the range of a few microns. The process also allows for a wide range of materials to be used.

There are downsides to this method, however. Like all additive manufacturing methods, powder bed fusion creates products with rough surfaces, requiring treatment to smooth them. This can be an issue with delicate components. Furthermore, using lasers in powder bed fusion can lead to crack propagation due to small fluctuations in temperature. Finally, some powders may be hard to source. The nitinol stents are an example of this as the nitinol powder has to be custom-made.

The other most common method for 3D printing is extrusion. Material extrusion, rather than melting a sheet of powder, will have the powder (or filament) in a specialized nozzle. This nozzle, which can move across all 3 axes, heats and feeds the material onto a hotbed, forming the layers of the product. Extrusion printers are what most people think of when they hear "3D printing".

Material extrusion is more cost-effective than powder bed fusion and has less waste. However, it suffers the same issues of rough surfaces and is unable to compete with powder bed fusion's precision when it comes to certain biological factors. For example, powder bed fusion has been shown to efficiently mimic the porosity of human bone. These minuscule pores are essential in recreating the mechanical properties of biological material and are a circumstance in which powder bed fusion overcomes material extrusion methods.

3D printing is essential to the advancement of medical implants, and a mixture of these two main manufacturing techniques should be implemented to continue this. As the technology develops, so too do the possible applications for this printing method.

(**Selected from:** Ventre T. 3D Printing Metallic Implants: Technologies Available and the Future of the Industry [J/OL] 2021, [02-22 2023]. https://www.azom.com/article.aspx? ArticleID= 20542.)

New Words and Expressions

metallic   *adj.* 金属的，含金属的
implant   *n.* 植入体   *v.* 移植
manufacture   *n.* 工业品，制造业
intricate   *adj.* 错综复杂的，难以理解的
customizable   *adj.* 可定制的
collaboration   *n.* 协同，合作，协作

vastly *adv.* 非常，极大地
synthetic *adj.* 合成的，人造的
immune system 免疫系统
perspective *n.* 视角，观点
joints *n.* 关节
small batches 小批量
filament *n.* 灯丝，细丝
arterial *adj.* 动脉的，干线的
stent *n.* 支架，导管
monumental *adj.* 巨大的，伟大的
peripheral artery disease 外周动脉疾病
mesh *n.* 网眼，网丝，网格
powderbed 粉末床，粉态床
rough *adj.* 粗糙的，不平的
delicate *adj.* 易碎的，精巧的，柔和的
crack propagation 裂纹扩展
extrusion *n.* 挤出，推出
nozzle *n.* 喷嘴，排气口，鼻
cost-effective *adj.* 划算的，成本效益好的
porosity *n.* 有孔性，对孔性
mimic *v.* 模仿，模拟

Notes

（1）Incorrectly manufactured synthetic implants can emit ions, negatively affecting the tissue around the implant, or the immune system can reject the implant as a foreign body altogether.

—negatively affecting 为 ions 的同位语，对其产生的影响做了进一步解释。
—参考译文：制造不当的合成植入体会释放出离子，对植入体周围的组织产生负面影响，或者可能被当作异物被免疫系统排斥。

（2）Biomechanics of body parts can be vast and complex: single components can vary in properties as small as collagen content and lamellae orientation across their bodies, giving distinct and essential mechanical properties.

—lamellae orientation 表示蛋白质的结构取向，比如 α 螺旋和 β 折叠以及空间结构。
—参考译文：身体部位的生物力学是庞杂的：单个组成结构的特性可能不同——小到胶原蛋白含量和整个身体的蛋白质结构取向（均可能存在差异），从而赋予（身体部件）基本但独特的力学特性。

（3）Building complex pieces with intricate internal structures and minute but essential variations is much simpler to manufacture layer by layer, straight from a computer-aided design (CAD) drawing.

——minute but essential 细微且重要。
——straight from 译为直接借助。
——句子主干为：Building ... is much simpler。
——参考译文：利用计算机辅助设计（CAD）逐层地制造具有复杂内部结构和细微但具有关键性差异的部件，可以大大简化制造过程。

（4）These minuscule pores are essential in recreating the mechanical properties of biological material and are a circumstance in which powder bed fusion overcomes material extrusion methods.

——are essential 和 are a circumstance 为并列成分，共同主语 pores。
——in which... 为修饰 circumstance 的定语从句。
——意译为：这些微小的孔隙对于重塑生物材料的机械性能是至关重要的，这种情况下粉末床熔融技术优于材料挤出成型（技术）。

Unit 5.3

Text

Medical applications of biomaterials

这篇课文主要介绍生物材料在临床上的应用，包括缝合线、防粘连材料、药物递送系统、组织工程和可植入器件等，文末概述了可植入生物材料的优缺点并展望其在生物工程领域的应用前景。

Sutures and adhesives

Surgical suture can be treated as a medical device and used to hold body tissues together after an injury or surgery. Suture thread is made from numerous materials. The original sutures were made from biological materials, such as catgut suture and silk. Most modern sutures are synthetic, including the absorbable polyglycolic acid, polylactic acid, Monocryl and polydioxanone as well as the non-absorbable nylon, polyester, PVDF and polypropylene. Triclosan-coated sutures were first approved by the U. S. Food and Drug Administration (FDA) in 2002. They have been shown to reduce the chance of wound infection. Sutures come in very specific sizes and may be either absorbable (naturally biodegradable in the body) or non-absorbable. Sutures must be strong enough to hold tissue securely but flexible enough to be knotted. They must be hypoallergenic and avoid the "wick effect" that would allow fluids and thus infection to penetrate the body along the suture tract. All sutures are classified as either absorbable or non-absorbable depending on whether the body will naturally degrade and absorb the suture material over time. Absorbable suture materials include the original catgut as well as newer synthetic materials such as polyethylene glycol, polylactic acid, polydioxane, and polycaprolactone. These polymeric materials are based on one or more of five cyclic monomers:

glycolide, l-lactic acid, p-dioxane, trimethyl carbonate, and ε-caprolactone. They are capable of decomposition by various processes, including hydrolysis and proteolytic enzymatic degradation. Depending on the material, the process may vary from ten days to eight weeks. Non-absorbable sutures are made of special silk or synthetic polypropylene, polyester or nylon. They are used in patients who cannot return for suture removal, or in internal body tissues. Stainless steel wires are commonly used in orthopedic surgery and forsternal closure in cardiac surgery. Non-absorbable sutures are used either on skin wound closure, where the sutures can be removed after a few weeks, or in stressful internal environments where absorbable sutures will not suffice. Recently, an advanced suture with multifunction is shown in **Figure 5.3.1**. Here, thermoplastic polyglycolic acid surgical sutures are used as electrodes and injected through a syringe needle (0.33 mm inner diameter) to modulate cardiac function.

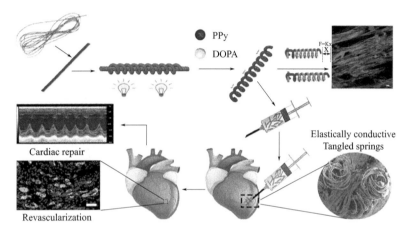

Figure 5.3.1 Schematic illustration of the design of conductive surgical suture spring.

There has been a great demand from surgeons for excellenthemostats, sealants, and adhesives because these biomaterials are very often needed in surgery. However, biomaterials that exhibit strong adhesion to tissues and are absorbed after wound healing have not yet been developed. These biomaterials should be delivered to surgeons in the form of flowable sol, viscous solution, or very pliable material and must set to a gel as quickly as possible when the sol or solution is applied to a wound tissue site. Biomaterials currently in use that nearly meet this stringent requirement include fibrin glue, microfibrillar collagen, and gelatin. All of these are of natural origin. Fibrin glue is used in the United States and Canada primarily as a surgical adhesive and sealant. This glue consists of two aqueous solutions prior to curing: fibrinogen and thrombin. The mixing of these two solutions produces a gel within ten seconds. The main problems that remain to be solved with fibrin glue are possible viral infection and poor tissue adhesion. Probably the most widely used topical hemostatic agents in the world today are sponge-type or microfibrillar collagen.

Adhesion prevention materials

After surgical operations, defective tissues often adhere to surrounding tissues and cause problems, especially in gynecological, abdominal and cardiovascular procedures. A viscous solution of hyaluronate is usually applied to the tissues that need protection, but it is not always effective in preventing tissue adhesions because of the rapid elimination. As alternatives, sheets of oxidized cellulose and derivatized hyaluronate have been applied to the damaged tissues, but tend to move from the proper site earlier than required. Human amniotic membrane has also been studied to prevent tissue adhesion after cross-linking, but the results of animal studies were not promising enough to encourage its clinical application. Inspired by the strong adhesive mechanism of biofilm and mussels, researchers recently put forward a novel dual bionic adhesive hydrogel (DBAH) based on chitosan grafted with methacrylate (CS-MA), dopamine (DA), and N-hydroxymethyl acrylamide (NMA) via a facile radical polymerization process (**Figure 5.3.2**). DBAH presents strong adhesion in water and is able to withstand high blood pressure, which is significantly higher than most clinical settings. Regenerated chitin and collagen molecules have been used clinically as skin wound coverings after being made into sponge or nonwoven forms, as skin applications do not require high strength materials.

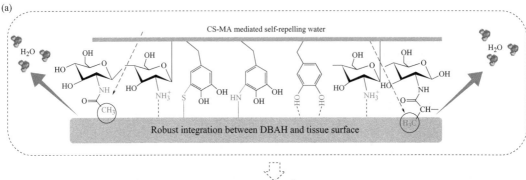

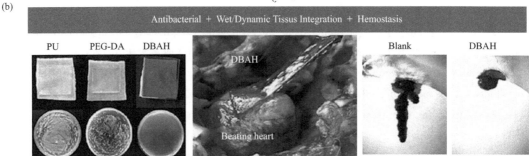

Figure 5.3.2 Schematic illustration of underwater bioinspired strong adhesion and the digital images of dual bionic hydrogel (DBAH) for antimicrobial activity analysis, tissue integration and hemostasis testing.

Drug delivery systems (DDS)

In some DDS, drug carriers made of polymers are needed for controlled release of drugs. In addition to synthetic polymers, biological materials have been explored extensively as candidates for drug carriers in the water soluble or water-insoluble state. They include collagen, gelatin, serum albumin, fibrin, starch, cellulose derivatives, dextran, pullulan, hydroxyapatite, and calcium carbonate. Although drug carriers do not require high mechanical strength, appropriate absorbability is necessary. Gelatin is the most important material for soft and hard capsulation of drugs for oral administration, while acidic cellulose derivatives are used for enteric coating of drugs. Nanomaterials-based drug delivery systems have attracted a great deal of attention in treatment of cancer because nanoparticle-based nanomedicines may have improved solubility, reduced non-specificity toxicity, altered pharmacokinetics and biodistributions compared with small-molecule drugs (**Figure 5.3.3**).

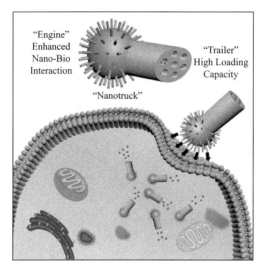

Figure 5.3.3 Schematic illustration of an engine-enhanced "nano-truck" with nano-bio interactions and cellular-level drug delivery.

Tissue engineering

Tissue engineering is a biomedical engineering discipline that uses a combination of cells, engineering, materials methods, and suitable biochemical and physicochemical factors to restore, maintain, improve, or replace different types of biological tissues. Tissue engineering often involves the use of cells placed on tissue scaffolds in the formation of new viable tissue for a medical purpose but is not limited to applications involving cells and tissue scaffolds. While it was once categorized as a sub-field of biomaterials, having grown in scope and importance it can be considered as a field in its own.

Formation of new biological tissues by tissue engineering can be achieved both in vitro and in vivo. It is seen that an artificial extracellular matrix is generally required for tissue regeneration by tissue engineering, because cell proliferation and differentiation, resulting in tissue regeneration, would be difficult unless such a matrix that functions as a cell scaffold is provided. Since this artificial extracellular matrix should disappear through absorption into the body when a new tissue is regenerated, materials for the matrix should be prepared from biodegradable polymers. This requirement as well as adequate cell adhesion onto the matrix surface make biological materials attractive in tissue engineering. In fact, collagenous, porous materials have been widely used for the scaffolding of cells. If a biological tissue is to be constructed from allo- or xenogeneic cells, the newly constructed tissue should be immunologically protected from the host self-defending system since the constructed tissue is used in direct contact with the blood or tissues of patients. An effective method for this immunological protection is to separate the heterogeneous cells from the host immune system using a membrane. This material has to supply oxygen and nutrients to the cells but prevent the attack of immunological proteins of the host against the heterogeneous cells. This membrane is termed "immunoisolation membrane" and is currently prepared mostly from biomaterials such as agarose and a combination of alginate with poly (L-lysin). Chinese researchers have developed the world's first biodegradable coated coronary stent as shown in **Figure 5.3.4**. Based on this, the first fully biodegradable stent in China, "Xinsorb", was further developed and the related fully biodegradable vascular drug stent implantation procedure was completed.

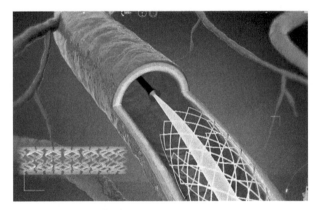

Figure 5.3.4 China's first fully degradable bio-scaffold.

Biomaterials for implantable devices

There is no doubt that important progress has been made in the clinical use of medical implants and other devices. Currently, the emphasis is on designing active biomaterials, i. e., materials that elicit specific, desired and timely responses from surrounding cells and tissues. Medical research continues to explore new scientific frontiers to diagnose, treat, cure and prevent disease at the molecular/genetic level. With this newfound knowledge, there will be a further need for innovative formulations and/or modifications of existing materials, for novel materials and non-traditional applications of biomaterials, such as tissue engineering. Promising developments include bioinspired chemical and topographic modifications of materials surfaces, current-conducting polymers, and nanophase materials.

Medical devices vary in their intended use and indications for use. For example, they range from simple, low-risk devices such as tongue depressors, medical thermometers, disposable gloves, and commodes, to complex, high-risk implants and life-sustaining devices. Examples of high-risk devices are those with embedded software such as pacemakers and prostheses. The design of medical devices constitutes a major part of the biomedical engineering field.

The ability to monitor blood flow is critical to patient recovery and patient outcomes after complex reconstructive surgery. Clinically available wired implantable monitoring techniques require careful fixation for accurate detection and need to be removed after use. A recent study was reported on the design of a pressure sensor, made entirely of biodegradable materials and based on fringe-field capacitor technology, for measuring arterial blood flow in both contact and non-contact modes (**Figure 5.3.5**). In addition to new challenges and opportunities, a number of unresolved issues from the past and present (mainly biocompatibility) will need to be addressed in the future.

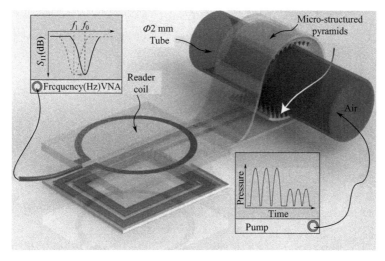

Figure 5.3.5 Schematic illustration of a biodegradable sensor for measuring arterial blood flow.

(**Selected from:** Boutry C M, Beker L, Kaizawa Y, et al. Biodegradable and Flexible Arterial-Pulse Sensor for the Wireless Monitoring of Blood Flow [J]. Nature Biomedical Engineering, 2019, 3 (1): 47-57.)

New Words and Expressions

  catgut *n.* 肠线，弦线

  polydioxanone *n.* 聚二氧六环酮

  nylon *n.* 尼龙，聚酰胺纤维，尼龙袜

  polypropylene *n.* 聚丙烯

  hypoallergenic *adj.* 低过敏性

  wick *n.* 灯芯，蜡烛心

  polyglycolic *n.* 聚乙醇酸

  glycolide *n.* 乙交酯

  l-lactide *n.* 左旋乳酸

  p-dioxanone *n.* 对二氧环己酮

  trimethylene *n.* 三甲烯，环丙烷

  carbonate *n.* 碳酸盐

  ε-caprolactone *n.* ε-己内酯

  sternal *adj.* 胸骨的，近胸骨的

hemostat n. 止血剂，止血钳子
flowable adj. 可流动的，易流动的
viscous adj. 黏性的，黏的
pliable adj. 柔韧的，柔软的，圆滑的，易曲折的
stringent adj. 严格的，严厉的，紧缩的，短缺的
fibrin n. 纤维蛋白，血纤蛋白
collagen n. 胶原，胶原质
gelatin n. 明胶，动物胶，胶制品
fibrinogen n. 纤维蛋白原，纤朊原
bionic adj. 仿生学的，利用仿生学的
methacrylate n. 甲基丙烯酸酯，异丁烯酸甲酯
dopamine n. 多巴胺（一种治脑神经病的药物）
chitin n. 壳质，几丁质，角素，甲壳素
serum albumin n. 血清白蛋白
starch n. 淀粉，含淀粉的食物；v. 给……上浆
cellulose n. 纤维素，(植物的)细胞膜质
derivative n. 派生物，衍生物；adj. 非独创的，模仿他人的
extracellular adj. (位于或发生于)细胞外的
proliferation n. 增殖，扩散，分芽繁殖
agarose n. 琼脂糖
newfound adj. 新发现的，新得到的

Notes

(1) These biomaterials should be delivered to surgeons in the form of flowable sol, viscous solution, or very pliable material and must set to a gel as quickly as possible when the sol or solution is applied to a wound tissue site.

—本段话为两个并列句，主要说明了生物材料的使用形式。
—in the form of... 以……的形式。
—参考译文：这些生物材料应以可流动溶胶、黏性溶液或非常柔韧的材料形式提供给外科医生，并且作用于伤口组织部位的溶胶或溶液须尽快定型成凝胶。

(2) Regenerated chitin and collagen molecules have been used clinically as skin wound covers, after being fabricated into a sponge or nonwoven fabric form, since skin application does not require high-strength materials.

—此句是一个复杂句，句子前半部分时态为现在完成时，句子末尾部分为原因状语从句，since 是因为的意思。
—参考译文：因为应用于皮肤的材料不需要高强度，再生甲壳素和胶原蛋白分子制成海绵或无纺布后已被临床用于包扎皮肤伤口。

(3) Nanomaterials-based drug delivery systems have attracted a great deal of attention in treatment of cancer, because nanoparticle-based nanomedicines may have improved solubility,

reduced non-specificity toxicity, altered pharmacokinetics and bio-distributions compared with small-molecule drugs.

—参考译文：在癌症治疗中，基于纳米材料的药物递送系统受到极大的关注，因为与小分子药物相比，基于纳米颗粒的纳米药物可能具有更高的溶解度、更低的非特异性毒性，（而且可以）改变药物代谢动力学和生物分布情况。

（4）Tissue engineering often involves the use of cells placed on tissue scaffolds in the formation of new viable tissue for a medical purpose, but is not limited to applications involving cells and tissue scaffolds.

—这是一个 but 引导的从句，句子主要对 tissue engineering 的作用作了进一步说明。
—参考译文：组织工程通常涉及使用放置在组织支架上的细胞来形成用于医疗目的的新组织，但不限于涉及细胞和组织支架的应用。

（5）All sutures are classified as either absorbable or non-absorbable depending on whether the body will naturally degrade and absorb the suture material over time.

—be classified as 被分为。
—参考译文：根据是否会随着时间的推移自然降解和吸收，所有缝合线被分为可吸收和不可吸收（两类）。

（6）Absorbable suture materials include the original catgut as well as newer synthetic materials such as polyethylene glycol, polylactic acid, polydioxane, and polycaprolactone.

—参考译文：可吸收缝合材料包括最初的羊肠线以及新的合成材料，例如聚乙醇酸、聚乳酸、聚二氧烷酮和聚己内酯。

（7）As alternatives, sheets of oxidized cellulose and derivatizedhyaluronate have been applied to the damaged tissues, but tend to move from the proper site earlier than required.

—be applied to 被应用于。
—参考译文：作为替代品，氧化纤维素片和衍生的透明质酸片已被应用于受损组织，但往往过早地从适当的位置移开。

（8）A recent study was reported on the design of a pressure sensor, made entirely of biodegradable materials and based on fringe-field capacitor technology, for measuring arterial blood flow in both contact and non-contact modes.

—参考译文：最近的一项研究报道了一种基于边缘场电容技术的压力传感器，完全由可生物降解的材料制成，可用于测量接触和非接触模式下的动脉血流量。

Exercises

1. Question for discussion

（1）What are the main medical applications of biomaterials?

（2）Where is the current focus for clinical applications of medical implants and other devices?

2. Translate the following into Chinese

(1) They are broken down by various processes including hydrolysis (polyglycolic acid) and proteolytic enzymatic degradation. Depending on the material, the process can be from ten days to eight weeks.

(2) After surgical operations, defective tissues tend to adhere to surrounding tissues and cause problems, especially in gynecological, abdominal and cardiovascular procedures.

(3) Unquestionably, important advances have been made in the clinical use of medical implants and other devices. Presently, emphasis is placed on the design of proactive biomaterials, that is, materials that elicit specific, desired, and timely responses from surrounding cells and tissues.

3. Translate the following into English

(1) 止血剂　　　　　　　　　(2) 聚丙烯
(3) 凝血酶　　　　　　　　　(4) 己内酯
(5) 心血管手术　　　　　　　(6) 尼龙
(7) 温度计　　　　　　　　　(8) 纤维蛋白，血纤蛋白（有凝血作用）
(9) 缝线必须足够牢固以固定组织，但又必须足够柔韧方便打结。
(10) 这些生物材料应以可流动溶胶、黏性溶液或非常柔韧的材料形式提供给外科医生，并且作用于伤口组织部位的溶胶或溶液须尽快定型成凝胶。

Reading Material
Biomaterials for tissue repair

这篇课文主要介绍了生物支架材料在组织工程中的应用及制备方法，阐述了合成生物材料支架和脱细胞支架。课文总结了两种方法的优劣势，并展望了它们在生物医学领域的发展前景。

An approach to regenerative medicine that is showing promise involves the use of biomaterials as tissue scaffolds. Biomaterials scaffolds have been used for >20 years in tissue engineering to improve the transplantation of cells and growth factors. More recently, biomaterials that can promote tissue repair and regeneration on their own without the need for delivering cells or other therapeutics have emerged as a potentially powerful paradigm for regenerative medicine.

In most diseases and injuries, the extracellular matrix (ECM), which influences all aspects of cell behavior, is damaged, altered, or lost. Regenerative medicine approaches involving delivery of cells to these regions have produced disappointing results. The diseased microenvironment does not resemble healthy ECM as it has both abnormal biochemical components and different mechanical properties. Therefore, when cells, such as stem cells, are delivered, they receive abnormal ECM cues. Even in cases where the delivered cells are intended to modulate the immune response and recruit endogenous cells, such as with delivery of mesenchymal stem cells (MSCs), infiltrating cells are also exposed to a diseased ECM.

In contrast to the cell-only regenerative medicine paradigm, a biomaterial scaffold, if appropriately designed, can create a new microenvironment in diseased tissue that mimics the original healthy ECM and/or provides cues that influence the behavior of infiltrating cells to promote tissue repair or regeneration (**Figure 5.3.6**). Synthetic biomaterials can be precisely tuned in terms of mechanical properties, architecture, and/or degradation rate. Conversely, purified naturally derived materials, such as animal-derived collagen and human-derived fibrin, already contain cell adhesion ligands and are susceptible to proteolytic degradation that enable cell infiltration and remodeling. Synthetic scaffolds have typically been used as a delivery modality for cells and other biologics or they are modified with peptides to encourage cell infiltration. However, some synthetics have reached the clinic in which the scaffold alone has been designed to encourage regeneration. Synthetic scaffolds have focused on architecture, pore size, and degradation rate as ways to modulate the host response and encourage endogenous repair. Most examples to date are in orthopedic and dental applications. For example, a nanocrystalline hydroxyapatiteporous silica gel matrix has been used in dental applications in Europe. Although such products are available, given their classification as devices (which can have limited requirements for clinical studies prior to approval), there have typically been few and only small-scale (<20 to 40 patients) randomized clinical studies, which makes it difficult to truly assess clinical efficacy across products or compared to standard of care.

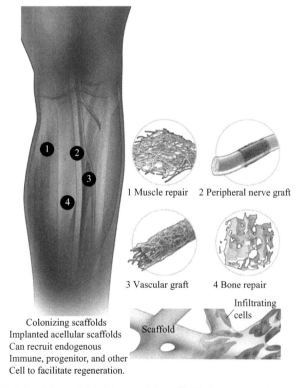

Figure 5.3.6 The cellular biomaterial scaffolds in regenerative medicine.

For orthopedics, most synthetic scaffolds for bone regeneration have incorporated hydroxyapatite, a mineral found in bone, to encourage osteogenesis. However, a purely polymeric scaffold has made it into clinical trials for cartilage repair. A poly (ethylene glycol) (PEG) diacrylate scaffold applied with a chondroitin sulfate adhesive in articular defects served as a scaffold to promote cartilage growth from MSCs released by microfracture of neighboring bone. Magnetic resonance imaging of the 15 treated patients showed increased tissue filling, decreasing water content, and increasing tissue reorganization. Reported pain also decreased compared to three control patients who only received microfracture; however, knee function was similar between the treated and control groups. Other synthetic polymer-only scaffolds are also being explored to modulate tissue repair in patients.

Another scaffold-only approach that more fully mimics the original ECM takes advantage of what nature has already produced by decellularizing xenogeneic (derived from another animal) or allogeneic (cadaver-derived) tissues. Decellularized ECM scaffolds have been used for >10 years in surgical applications, including hernia repair and treatment of diabetic foot ulcers, and traditionally regulated by the U. S. Food and Drug Administration (FDA) as 510 (k) medical devices. Decellularization has typically been performed through the use of detergents to adequately remove cellular debris such that the remaining ECM, which degrades as host cells infiltrate and remodel the scaffold, promotes a pro-remodeling versus rejection response. However, complete removal of all cellular components is never achieved, and this remains an important safety consideration in manufacturing of existing and new products given that failure of a decellularized porcine heart valve that led to the deaths of pediatric patients was attributed to inadequate decellularization. Nonetheless, decellularized biomaterials are being investigated for tissue regeneration in several forms, including intact, implantable ECM and injectable ECM, either as particles or hydrogels. Preclinical studies with this approach continue to increase, and a variety of clinical studies have been performed mainly using existing 510 (k) products.

Decellularized ECM generated by cells in vitro has also been explored for regenerative medicine, particularly for vascular grafts. Tissue-engineered vascular grafts have been generated by using allogeneic smooth muscle cells seeded onto a polyglycolic acid tubular scaffold. Over time in culture, the cells secrete a new ECM as the synthetic scaffold degrades, and after decellularization, a matrix-only graft remains. This technology was originally tested in a baboon model of arteriovenous access for hemodialysis and a canine model of peripheral and coronary artery bypass. In all models, the grafts were repopulated with host cells and largely remained open, which supported translation into patients. Two completed single-arm phase II trials in hemodialysis patients were promising leading to several ongoing clinical trials. A similar approach using a decellularized graft generated from human dermal fibroblasts grown in a fibrin gel was also recently reported with promising results in a baboon model of hemodialysis access.

A critical mode of action by which biomaterials can promote tissue repair is through influencing the immune response. Although further work is needed, studies have suggested that biomaterials can play a major role in influencing the polarization of both macrophages and T cells, which have considerable cross-talk. This is illustrated by a study showing that T helper 2 cells are necessary for macrophage polarization to an M2 phenotype and pro-regenerative outcomes with decellularized ECM scaffolds. Given the complexity of immune cell cross-talk, it is likely that even synthetic biomaterials and more purified natural materials also influence the phenotype of cells other than macrophages to stimulate tissue repair, although this has yet to be fully studied. It is also known that location of a biomaterial can play a role in the tissue response. For example, PEG-hyaluronic acid (HA) gels, which are thought to be biocompatible, elicit an increased inflammatory response when implanted near adipose tissue. Similarly, thiolated HA hydrogels led to a limited immune response when injected subcutaneously, but led to granuloma (a mass of inflammatory tissue) formation when injected into myocardium. These studies highlight the importance of studying biomaterial scaffolds in the context of a fully functioning immune system, as well as examining biocompatibility and the repair response in the tissue of interest.

Future development is promising given that biomaterial scaffolds have several advantages compared to cell-based regenerative medicine, including reduced costs and fewer translational barriers. Biomaterials that are synthesized or sourced directly from animals or humans can have considerably reduced costs. Although biomaterials that require cell culture in their manufacturing may not avoid high costs, they still obviate concerns regarding the use of a living product and may be amenable to terminal sterilization, which reduces safety concerns. Although more mechanistic studies are needed to better understand how biomaterial scaffolds can recreate the microenvironment and influence the immune system and tissue regeneration, they are already poised to have immediate patient impact and represent an alternative paradigm for regenerative medicine.

(**Selected from:** Christman K L. Biomaterials for Tissue Repair [J]. Science, 2019, 363 (6425): 340-341.)

New Words and Expressions

modulate   *vt*. 调节，（信号）调制
scaffold   *n*. 支架
transplantation   *n*. （器官）移植，迁移
therapeutics   *n*. 疗法，治疗学
extracellular matrix   细胞外基质
degradation   *n*. 退化，降格，降级，堕落
ligand   *n*. 配位体，配基
infiltration   *n*. 渗透，渗透物
cartilage   *n*. 软骨
chondroitin   *n*. 软骨素
detergent   *n*. 清洗剂，洗涤剂
hydrogel   *n*. 水凝胶
decellularization   *n*. 去细胞（化）
smooth muscle   *n*. 平滑肌
arteriovenous   *adj*. 动静脉的

hemodialysis   *n*. 血液透析，血液渗析
macrophage   *n*. 巨噬细胞
hyaluronic acid   透明质酸，玻尿酸
sterilization   *n*. 杀菌
paradigm   *n*. 典范，范例，样板，范式

Notes

(1) In contrast to the cell-only regenerative medicine paradigm, a biomaterial scaffold, if appropriately designed, can create a new microenvironment in diseased tissue that mimics the original healthy ECM and/or provides cues that influence the behavior of infiltrating cells to promote tissue repair or regeneration.

—本句话为全文的中心句，介绍了生物材料支架在促进组织修复的优势。
—参考译文：与单纯细胞再生医学模式不同，如果设计得当，生物材料支架可以在病变组织中创建一个新的微环境，模拟原始的健康ECM和/或提供影响浸润细胞行为的信号，以促进组织修复或再生。

(2) For orthopedics, most synthetic scaffolds for bone regeneration have incorporated hydroxyapatite, a mineral found in bone, to encourage osteogenesis.

—hydroxyapatite 羟基磷灰石，人体和动物骨骼的主要无机成分，能促进缺损组织的修复。
—参考译文：对于整形外科，大多数用于骨再生的合成支架都加入了羟基磷灰石，这是一种在骨骼中发现的能促进成骨的矿物质。

(3) Given the complexity of immune cell cross-talk, it is likely that even synthetic biomaterials and more purified natural materials also influence the phenotype of cells other than macrophages to stimulate tissue repair, although this has yet to be fully studied.

—given 放在句子开头表示在……前提下；考虑到。
—it 代指后面的 even ... repair。英文中常用做法，避免头重脚轻。
—参考译文：考虑到免疫细胞相互作用的复杂性，即使是合成生物材料和更纯净的天然材料也可能影响除巨噬细胞以外的细胞表型以刺激组织修复，尽管这一点还没有得到充分的研究。

(4) Although biomaterials that require cell culture in their manufacturing may not avoid high costs, they still obviate concerns regarding the use of a living product and may be amenable to terminal sterilization, which reduces safety concerns.

—amenable 有责任的，应服从的，在这里可以理解为易于进行。
—句尾为 which 引导的非限制性定语从句，表示生物材料易于进行灭菌这一行为。本句话总结了生物材料应用的优缺点。
—参考译文：尽管在制备过程中需要细胞培养的生物材料可能无法避免高昂的成本，但它们可以消除使用活体产品带来的担忧，并且易于终端灭菌从而降低安全风险。

Chapter 6

Nanoscience, Nanotechnology, and Nanomaterials

【本章导读】

本章介绍纳米科技与纳米材料，包括五个单元，共有五篇课文和五篇阅读材料。五篇课文分别介绍纳米科技的基本概念、纳米材料的结构分类、纳米材料的制备方法和代表性应用、纳米材料独特的物理性质、纳米石墨烯及相应光电器件。五篇阅读材料分别讲解了碳点材料、制备三维纳米材料的卷曲纳米技术、纳米材料的能源应用、天然及人工纳米技术以及催化纳米酶。

Unit 6.1

Text

Introduction to nanoscience and nanotechnology

这篇课文介绍了纳米科技所涉及的尺度范围并给出了纳米科学、纳米技术的定义，说明了两者在概念上的区别。课文还介绍了纳米材料和纳米技术发展的历史进程，重点介绍了现代纳米技术的发展，并选择性简述了几种代表性碳基纳米材料。

Definition of nanoscience and nanotechnology

The prefix "nano" is referred to a Greek prefix meaning "dwarf" or something very small and depicts one thousand millionth of a meter (10^{-9} m). We should distinguish between nanoscience and nanotechnology. Nanoscience is the study of structures and molecules on the scales of nanometers ranging between 1 and 100 nm, and the technology that utilizes it in practical applications such as devices etc. is called nanotechnology. As a comparison, one must realize that a single human hair is 60,000 nm thickness and the DNA double helix has a radius of 1 nm (**Figure 6.1.1**).

Nanotechnology is one of the most promising technologies of the 21st century. It is the ability to convert the nanoscience theory to useful applications by observing, measuring, manipulating, assembling, controlling, and manufacturing matter at the nanometer scale. The

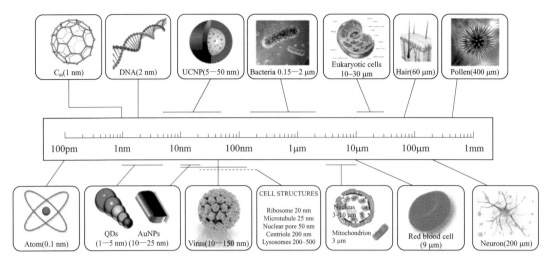

Figure 6.1.1   A comparison of sizes of nanomaterials.

National Nanotechnology Initiative (NNI) in the United States define Nanotechnology as "a science, engineering, and technology conducted at the nanoscale (1 to 100 nm), where unique phenomena enable novel applications in a wide range of fields, from chemistry, physics, and biology, to medicine, engineering, and electronics". This definition suggests the presence of two conditions for nanotechnology. The first is an issue of scale: nanotechnology is concerned to use structures by controlling their shape and size at nanometer scale. The second issue has to do with novelty: nanotechnology must deal with small things in a way that takes advantage of some properties because of the nanoscale.

The history of nanomaterial and nanotechnology

Nanoparticles and structures have been used by humans in fourth century AD, by the Roman, which demonstrated one of the most interesting examples of nanotechnology in the ancient world. The Lycurgus cup, from the British Museum collection, represents one of the most outstanding achievements in ancient glass industry. It is the oldest famous example of dichroic glass. Dichroic glass describes two different types of glass, which change color in certain lighting conditions. This means that the Cup has two different colors: the glass appears green in direct light, and red-purple when light shines through the glass (**Figure 6.1.2**). In 1990, the scientists analyzed the cup using a transmission electron microscopy (TEM) to explain the phenomenon of dichroism. The observed dichroism (two colors) is due to the presence of nanoparticles with 50 – 100 nm in diameter. X-ray analysis showed that these nanoparticles are silver-gold (Ag-Au) alloy, with a ratio of Ag : Au of about 7 : 3, containing in addition about 10% copper (Cu) dispersed in a glass matrix. The Au nanoparticles produce a red color as result of light absorption (~520 nm). The red-purple color is due to the absorption by the bigger particles while the green color is attributed to the light scattering by colloidal dispersions of Ag nanoparticles with a size >40 nm. The Lycurgus cup is recognized as one of the oldest synthetic nanomaterials. A similar effect is seen in late medieval church windows, shining a luminous red and yellow colors due to the fusion of Au and Ag nanoparticles into the glass.

Figure 6.1.2　The Lycurgus cup. The glass appears green in reflected light (a) and red-purple in transmitted light (b).

During the 9th—17th centuries, glowing, glittering "luster" ceramic glazes used in the Islamic world, and later in Europe contained Ag or Cu or other nanoparticles. The Italians employed nanoparticles in creating Renaissance pottery during 16th century. They were influenced by Ottoman techniques: during the 13th—18th centuries, to produce "Damascus" saber blades, cementite nanowires and carbon nanotubes were used to provide strength, resilience, and the ability to hold a keen edge. These colors and material properties were produced intentionally for hundreds of years. Medieval artists and forgers, however, did not know the cause of these surprising effects.

In 1981, the physicists Gerd Binnig and Heinrich Rohrer invented a new type of microscope at IBM Zurich Research Laboratory, the Scanning Tunneling Microscope (STM). The STM uses a sharp tip that moves so close to a conductive surface that the electron wave functions of the atoms in the tip overlap with the surface atom wave functions. When a voltage is applied, electrons "tunnel" through the vacuum gap from the atom of the tip into the surface (or vice versa). In 1983, the group published the first STM image of the Si (111) -7×7 reconstructed surface, which nowadays can be routinely imaged. A few years later, in 1990, Don Eigler of IBM in Almaden and his colleagues used a STM to manipulate 35 individual xenon atoms on a nickel surface and formed the letters of IBM logo. The STM was invented to image surfaces at the atomic scale and has been used as a tool with which atoms and molecules can be manipulated to create structures. The tunneling current can be used to selectively break or induce chemical bonds.

In 1985, Robert Curl, Harold Kroto, and Richard Smalley discovered that carbon can also exist in the form of very stable spheres, the fullerenes or buckyballs. The carbon balls with chemical formula C_{60} or C_{70} are formed when graphite is evaporated in an inert atmosphere. A new carbon chemistry has been developed now, and it is possible to enclose metal atoms and create new organic compounds. A few years later, in 1991, Iijima et al. observed hollow graphitic tubes or carbon nanotubes by TEM which form another member of the fullerene family (**Figure 6.1.3**). The strength and flexibility of carbon nanotubes make them potentially useful in many nanotechnological applications. They also have potential applications as field emitters, energy storage materials, catalysis, and molecular electronic components.

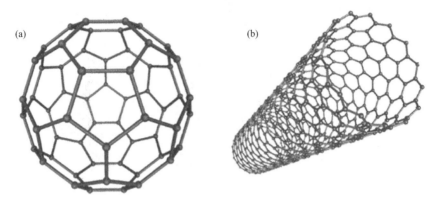

Figure 6.1.3 Schematic of a C_{60} buckyball (Fullerene) (a) and carbon nanotube (b).

In 2004, a new class of carbon nanomaterials called carbon dots (C-dots) with size below 10 nm was discovered accidentally by Xu et al. during the purification of single-walled carbon nanotubes. C-dots with interesting properties have gradually become a rising star as a new nanocarbon member due to their benign, abundant, and inexpensive nature. Possessing such superior properties as low toxicity and good biocompatibility renders C-dots favorable materials for applications in bioimaging, biosensor, and drug delivery. Based on their excellent optical and electronic properties, C-dots can also offer exciting opportunities for catalysis, energy conversion, photovoltaic devices, and nanoprobes for sensitive ion detection. After the discovery of "graphene" in 2004, carbon-based materials became the backbone of almost every field of science and engineering.

Nowadays, nanoscience progressed in other fields of science like in computer science, bio, and engineering. Nanoscience and technology progressed in computer science to decrease the size of a normal computer from a room size to highly efficient moveable laptops. Electrical engineers progressed to design the complex electrical circuits down to nanoscale level. Also, many advances are noticed in smart phone technology and other modern electronic devices for daily uses.

(**Selected from:** Bayda S, Adeel M, Tuccinardi T, et al. The History of Nanoscience and Nanotechnology: From Chemical-Physical Applications to Nanomedicine [J]. Molecules, 2020, 25 (1): 112.)

New Words and Expressions

nanoscience   *n.* 纳米科学
nanotechnology   *n.* 纳米技术
ranging between   范围在……与……之间

nanoparticle　*n*．纳米颗粒
dichroism　*n*．二色性，二向色性
phenomenon　*n*．现象
disperse　*v*．分散，散布
matrix　*n*．基质，母体
absorption　*n*．吸收
scatter　*v*．散射，散布
colloidal　*adj*．胶体的
synthetic　*adj*．合成的，人造的
luminous　*adj*．发亮的
glowing　*adj*．炽热的，发光的
glittering　*adj*．发光的，辉煌的
luster　*n*．光泽，光彩
ceramic　*n*．陶釉，瓷釉
glaze　*n*．釉，光滑面
Renaissance　*n*．复兴，文艺复兴
pottery　*n*．陶器，陶土
blade　*n*．刀片，刀刃
nanowire　*n*．纳米线，纳米丝
nanotube　*n*．纳米管
vice versa　反之亦然
resilience　*n*．恢复力，弹力
intentional　*adj*．故意的，有意的
fullerene　*n*．富勒烯
enclose　*v*．包围，围住
compound　*n*．化合物，混合物
et al.　（源自拉丁语）以及其他人，等人（只能用于人）
biocompatibility　*n*．生物相容性，生物适应性
graphene　*n*．石墨烯

Notes

(1) Nanoscience is the study of structures and molecules on the scales of nanometers ranging between 1 and 100 nm, and the technology that utilizes it in practical applications such as devices etc. is called nanotechnology.

—参考译文：纳米科学是研究1～100纳米尺度上的结构和分子的科学，而将其在器件等实际应用中使用的技术是纳米技术。

(2) Nanoparticles and structures have been used by humans in fourth century AD, by the Roman, which demonstrated one of the most interesting examples of nanotechnology in the ancient world.

—参考译文：罗马人在公元四世纪就使用纳米颗粒和结构，这是古代纳米技术最有趣的例子之一。

(3) X-ray analysis showed that these nanoparticles are silver-gold (Ag-Au) alloy, with a ratio of Ag∶Au of about 7∶3, containing in addition about 10% copper (Cu) dispersed in a glass matrix.

——科技英语中常见定语后置,由 with 开头的介宾短语作为后置定语。翻译时需按照中文习惯改写。

——参考译文:X 射线分析表明这些分布在玻璃基质中的纳米颗粒是银-金(Ag-Au)合金,(其中)Ag 和 Au 含量比值约为 7∶3,(此外还)额外含有大约 10% 的铜(Cu)。

(4) These colors and material properties were produced intentionally for hundreds of years. Medieval artists and forgers, however, did not know the cause of these surprising effects.

——however 表示转折,实际强调转折后的内容。

——参考译文:数百年来,这些颜色和材料性质是有意制造的。然而,中世纪的艺术家和仿制者们并不了解这些奇特效应的成因。

(5) The STM uses a sharp tip that moves so close to a conductive surface that the electron wave functions of the atoms in the tip overlap with the surface atom wave functions.

——sharp tip 锋利的尖端。conductive surface 导电表面。

——参考译文:扫描隧道显微镜(STM)使用一个锋利的尖端,(并将其)移动到非常靠近导电表面,使得尖端中原子的电子波函数能够与表面原子的波函数重叠。

(6) A new carbon chemistry has been now developed, and it is possible to enclose metal atoms and create new organic compounds.

——本句强调新的可能性。

——参考译文:现在已经发展出一种新的碳化学,它可以包封金属原子并产生新的有机化合物。

Exercises

1. Question for discussion

(1) What is the difference between nanoscience and nanotechnology?

(2) Why do we study nanoscience and nanotechnology?

(3) What are two conditions for nanotechnology?

(4) Please explain the mechanism of dichroic glass.

(5) What are advantages of carbon dots?

2. Translate the following into Chinese

(1) The prefix "nano" is referred to a Greek prefix meaning "dwarf" or something very small and depicts one thousand millionth of a meter (10^{-9} m).

(2) It is the ability to convert the nanoscience theory to useful applications by observing, measuring, manipulating, assembling, controlling, and manufacturing matter at the nanometer scale.

(3) A similar effect is seen in late medieval church windows, shining a luminous red and yellow colors due to the fusion of Au and Ag nanoparticles into the glass.

(4) After the discovery of "graphene" in 2004, carbon-based materials became the backbone of almost every field of science and engineering.

3. Translate the following into English

(1) DNA 双螺旋
(2) 新型应用
(3) 胶体分散
(4) 导电表面
(5) 电子波函数
(6) 原子尺度
(7) 化学键
(8) 低毒性
(9) 药物输送
(10) 红色是由较大纳米颗粒的吸收而产生的，而绿色是由>40 nm 的 Ag 纳米颗粒的光散射形成的。
(11) 碳纳米管在场发射、储能、电子器件等领域具有重要的潜在应用。
(12) 碳纳米点具有良好的生物相容性，是生物成像、生物传感器等领域的理想材料。
(13) 纳米科学和纳米技术的进步使得电子器件尺寸不断缩小，因此计算机的尺寸从房间那么大缩小到笔记本一样大。
(14) X 射线分析表明，这种纳米颗粒由金和银构成，其比例为 2∶1，另外还有 10% 的铜。
(15) 研究人员利用 STM 操控了 35 个氙原子，并且在一块 Ni 金属表面排列出了 "IBM"。

Reading Material

Carbon dots

这篇课文介绍了一种新型的碳基纳米材料——碳纳米点。课文阐述了碳点的定义和特点，给出了碳点的主要分类，包括石墨烯量子点、碳量子点和碳化聚合物量子点，并展示了几种碳纳米点材料的特点和应用。

Carbon-based materials play significant roles in the development of material science. From the traditional industrial carbon (e.g., activated carbon, carbon black) to new industrial carbon (e.g., carbon fibers, graphite) and new carbon nanomaterials such as graphene and carbon nanotubes (CNTs), fundamental research and applications of carbon-based materials are always popular in the fields of chemistry, materials, and other interdisciplines due to their environmental friendliness. However, macroscopic carbon material lacks the appropriate band gap, making it difficult to act as an effective fluorescent material. Carbon dots (CDs), a new rising star in the carbon family, have attracted considerable attention due to their excellent and tunable photoluminescence (PL), high quantum yield (QY), low toxicity, small size, appreciable biocompatibility, and abundant low-cost sources, providing important applications in many fields including biomedicine, catalysis, optoelectronic devices, and anticounterfeiting.

CDs can be generally defined as a quasi-0D carbon-based material with a size below 20 nm, and fluorescence is their intrinsic property. In 2004, carbon nanoparticles with fluorescence were first reported, which were accidentally obtained from the purification of single-walled

CNTs. In 2006, Sun and co-workers named nanoscale carbon particles synthesized by laser ablation of carbon target as CDs for the first time, but the QY of these surface passivated CDs was only about 10%. Low QY and complicated preparation procedures limited the development of CDs. Until 2013, Yang's group chose citric acid (CA) and ethylenediamine as precursors and synthesized polymer-like CDs with QY up to 80% through one-step hydrothermal method. This QY was the highest value for carbon-based fluorescent materials. These CDs can be applied both as printing inks and as functional nanocomposites. The facile approach, high QY, low toxicity, and high resistance to photobleaching of CDs caused widespread concern and a research boom. Thereafter, researchers developed different strategies and technologies to pursue CDs with high performance, and lots of significant breakthroughs have taken place in the past few years, including multicolor/deep red/near-infrared (NIR) emission, narrow full width at half maximum (FWHM), two-/multiphoton PL, chirality, room temperature phosphorescence (RTP), and thermally activated delayed fluorescence (TADF) as well as various applications.

At present, CDs are mainly classified into graphene quantum dots (GQDs), carbon quantum dots (CQDs), and carbonized polymer dots (CPDs) according to their different formation mechanism, micro-/nanostructures, and properties, while associations can be built among them by changing the graphene layer and carbonization degree.

GQDs possess one- or multiple-layer graphite structures connected chemical groups on the surface/edge or within the interlayer defect. They have obvious graphene lattices and are usually obtained by "top-down" prepared methods: oxide cutting larger graphitized carbon materials such as graphite powder, carbon rods, carbon fibers, carbon nanotubes, carbon black, or graphene oxide into small pieces. Their optical properties are mainly dominated by the size of π-conjugated domains and the surface/edge structures. GQDs are anisotropic with lateral dimensions larger than their height, but CQDs and CPDs are typically spherical, often produced from small molecules, polymers, or biomass by assembling, polymerization, crosslinking, and carbonization via "bottom-up" methods (e. g., combustion, thermal treatment). CQDs exhibit multiple-layer graphite structures connected surface groups. Intrinsic state luminescence and the quantum confinement effect of size are their PL mechanism.

Notably, CPDs, showing aggregated/crosslinked and carbonized polymer hybrid nanostructures, were first put forward in 2018 according to the formation process, structures, and PL mechanism. They possess special "core-shell" nanostructures, consisting of carbon cores less than 20 nm with highly dehydrated crosslinking polymer frames or slight graphitization and shells of abundant functional groups/polymer chains, which endow CPDs with higher stability, better compatibility, easier modification and functionalization, as well as wider applications. In particular, different from the PL feature and mechanism of GQDs and CQDs, the optical properties of CPDs mainly originate from the molecular state and crosslink enhanced emission effect, which make the relationship between structure and performance of CPDs more controllable.

(**Selected from:** Liu J, Li R, Yang B. Carbon Dots: A New Type of Carbon-Based Nanomaterial with Wide Applications [J]. ACS Central Science, 2020, 6 (12): 2179-2195.)

New Words and Expressions

macroscopic *adj.* 宏观的，肉眼可见的
fluorescent *adj.* 荧光的，发光的
photoluminescence *n.* 光致发光
quantum yield 量子产率
toxicity *n.* 毒性
appreciable *adj.* 可感知的，可观的
anticounterfeiting *n.* 防伪
purification *n.* 净化，提纯
passivated *adj.* 钝化的
ablation *n.* 消融，切除
ethylenediamine *n.* 乙二胺
citric acid 柠檬酸
hydrothermal method 水热法
photobleaching *n.* 光褪色
thereafter *adv.* 其后
full width at half maximum 半高全宽
π-conjugated *adj.* π共轭的
polymerization *n.* 聚合，聚合作用
combustion *n.* 燃烧，氧化
intrinsic *adj.* 内在的，固有的
dehydrated *adj.* 脱水的

Notes

(1) However, macroscopic carbon material lacks the appropriate band gap, making it difficult to act as an effective fluorescent material.

——making 现在分词引导结果状语。
——参考译文：然而，宏观碳材料缺乏合适的带隙，难以作为有效的荧光材料。

(2) CDs can be generally defined as a quasi-0D carbon-based material with a size below 20 nm, and fluorescence is their intrinsic property.

——quasi-0D 准 0 维。
——本句介绍了 CDs 的基本定义。
——参考译文：碳点一般可以定义为 20 nm 以下的准零维碳基材料，荧光是它们的本征性质。

(3) They have obvious graphene lattices and are usually obtained by "top-down" prepared methods: oxide cutting larger graphitized carbon materials such as graphite powder, carbon rods, carbon fibers, carbon nanotubes, carbon black, or graphene oxide into small pieces.

——参考译文：它们具有明显的石墨烯晶格，通常通过"自上而下"的制备方法获得：将较大的石墨化碳材料，如石墨粉、碳棒、碳纤维、碳纳米管、炭黑或氧化石墨烯氧化切割成小块。

(4) In particular, different from the PL feature and mechanism of GQDs and CQDs, the optical properties of CPDs mainly originate from the molecular state and crosslink enhanced emission effect, which make the relationship between structure and performance of CPDs more controllable.

——in particular 特别地，强调特殊的内容。which 引导非限制性定语从句，修饰前面的主句。

——参考译文：特别是，不同于 GQDs 和 CQDs 的发光特性和机理，CPD 的光学性质主要来源于分子态和交联增强发射效应，这使得 CPD 的结构与性能之间的关系更加可控。

(5) Notably, CPDs, showing aggregated/crosslinked and carbonized polymer hybrid nanostructures, were first put forward in 2018 according to the formation process, structures, and PL mechanism.

——showing 现在分词作为插入语修饰 CPDs。

——参考译文：值得注意的是，根据其形成过程、结构和 PL 机理，具有聚集/交联（特征）和碳化聚合物杂化纳米结构的 CPD，于 2018 年被首次提出。

Unit 6.2

Text

Classification of nanomaterials

这篇课文主要按照维度对纳米材料进行分类，并对不同维度的典型材料进行了详细介绍。课文首先介绍了 0 维材料，以量子点、富勒烯及不同形式的纳米颗粒为代表；其次介绍了一维材料，包括纳米管、纳米线等；最后介绍了以二维过渡金属碳化物、石墨烯等为代表的二维材料。

The types differ in the dimensionality of the nanomaterials, which may be zero-dimensional (i. e., isolated nanoparticles), one-dimensional (i. e., consisting of nanotubes or nanorods), or two-dimensional (i. e., existing as stacks or layers). Composites with platelets as the second phase may be thought as two-dimensional.

0D materials

Quantum dots

Quantum dots are a very common and useful type of nanoparticle, where the electrons are confined in all 3 dimensions. Quantum dots are small semiconducting particles that have been greatly used in displays and solar cells. Quantum dots emit a certain wavelength of light when they encounter either light or electricity and many quantum dots can be easily tuned. Quantum dots composed of cadmium, such as cadmium selenide (CdSe), are the widest class of quantum dots that have been studied.

Fullerenes

Fullerenes, also termed Buckminsterfullerenes (because of the geodesic shape they exhibit) come in two forms-pure carbon fullerenes and endohedral fullerenes, with the difference being that endohedral fullerenes contain additional atom (s) inside the carbon fullerene.

Fullerenes come in many shapes and sizes, but the most common is C_{60} as it is the most energetically and structurally stable form. Fullerenes composed of boron have also been predicted. Carbon fullerenes are composed of both single and double bonds, which are arranged into pentagons and hexagons. It is the pentagons which give the fullerenes their curvature. All fullerenes contain 12 pentagons, with a differing number of hexagons.

Nanoparticles

Overall, nanoparticles come in many forms. There are too many to individually discuss, but some of the most common are: single element nanoparticles, such as silver and gold nanoparticles which are used in medical imaging; metal oxide nanoparticles, including titanium dioxide nanoparticles used in white paint formulations; and amphiphilic nanoparticles such as Janus particles, which are used as stabilizers. Janus particles are an interesting class of nanoparticles, as one half of the surface is different to the other, and these two surfaces can differ in their external receptors, hydrophobicity or hydrophilicity, surface charge, and even in their magnetism.

1D materials

Nanotubes

Nanotubes, be it a carbon nanotube or inorganic nanotube, are materials which are elongated in one dimension, with a length-to-diameter ratio of up to 132000000 : 1. Nanotubes direct electrons along the elongated axis and come in many forms, including single-walled nanotubes (SWNT), multi-walled nanotubes (MWNT), chiral nanotubes, armchair nanotubes and zigzag nanotubes.

There has been a lot of hype about how carbon nanotubes could be used for many applications, especially in structural applications. However, issues with dispersing and aligning carbon nanotubes led to them to go out of favour for a while. They have recently been making a resurgence as many of these issues have been negated.

Nanowires

Nanowires, otherwise known as quantum wires, are another well-known 1D material. Again, nanowires are elongated in one direction, albeit with a much lower width to length ratio of 1 : 1000. The most common nanowires are silver nanowires, which are also known to be highly electrically conductive. Nanowires are known for exhibiting many different quantum effects, which alongside their unidirectional electron movement, have made them ideal materials for various electronic applications.

2D materials

Uniatomic 2D materials

There are many types of uniatomic 2D materials, such as germanene (made from germanium atoms), stanene (tin), silicene (silicon), phosphorene (phosphorous) and, of course, graphene (carbon).

Graphene is by far the most useful and the closest material to commercialization within this list, especially as some of these are still theoretical materials. However, graphene and the various other 2D atomic materials possess an excellent array of optical, physical and electrical properties that make them useful for a wide range of applications. Once graphene has been successfully used across many applications at a commercial level, it is expected that many of the other 2D materials will follow suit, although it could take a while.

MXenes

Outside of graphene, the class of MXene 2D materials show some of the best electronic properties. The most common MXene is boron nitride, which exhibits a hexagonal array of alternating boron and nitrogen atoms. Many people think that the MXenes show better properties than graphene, but they are much harder to synthesize. As such, they have not been as widely studied as graphene. However interest in them is significantly growing.

Whilst boron nitride is the most common, and the most widely researched, MXenes come in many forms and are often made from a combination of early transition metals (M), such as Titanium, Vanadium, Chromium and Niobium, alongside carbon or nitrogen (X). Future applications involving the MXenes could include electromagnetic interference shielding, water purification and in energy storage systems.

TMDCs

Transition metal dichalcogenides (TMDCs) are one of the oldest and longest studied class of 2D materials. They are widely used in semiconducting and electronic applications, but do not have as wide a range in properties in each material as other 2D materials. However, there are over 100 TMDCs to date, with the most common being tungsten diselenide (WSe_2), tungsten disulphide (WS_2), molybdenum disulphide (MoS_2) and molybdenum diselenide ($MoSe_2$), although many other transition metals and chalcogen atoms can be used.

One characteristic feature of TMDCs, is that one monolayer is composed of three atomic layers, where a layer of metal atoms is sandwiched between two layers of chalcogen atoms (note that these layers are physically bonded and are not held by van der Waals forces).

(**Selected from:** Critchley L. Nanomaterials: An introduction [J/OL]. AZO Nano, (2018-07-25)[2023-02-22], https://www.azonano.com/article.aspx? ArticleID= 4932.)

New Words and Expressions

composite   *n.* 复合材料

quantum dot　量子点
geodesic shape　测地线形状
endohedral fullerene　内嵌富勒烯
pentagon　*n*．五角形
hexagon　*n*．六角形
amphiphilic　*adj*．两亲的
hydrophobicity　*n*．疏水性
hydrophilicity　*n*．亲水性
magnetism　*n*．磁性，磁力，磁学
elongate　*v*．（使）变长，伸长，拉长
resurgence　*n*．复活，再现，再起
albeit　*conj*．虽然，尽管
unidirectional　*adj*．单向的，单向性的
uniatomic　*adj*．单原子的
transition metal dichalcogenides　过渡金属二硫化物

Notes

(1) Quantum dots emit a certain wavelength of light when they encounter either light or electricity and many quantum dots can be easily tuned.

　　参考译文：量子点在遇到光照或者电流时会散发出一定波长的光，许多量子点很容易被调控。

(2) Nanowires are known for exhibiting many different quantum effects, which alongside their unidirectional electron movement, have made them ideal materials for various electronic applications.

　　参考译文：纳米线以表现出许多不同的量子效应而闻名，这些效应加上它们的单向电子运动，使它们成为各种电子应用的理想材料。

(3) Once graphene has been successfully used across many applications at a commercial level, it is expected that many of the other 2D materials will follow suit, although it could take a while.

　　参考译文：一旦石墨烯在很多方面成功进行了商业层级的应用，预计很多其他二维材料也会效仿，尽管（这样的过程）会花上一段时间。

(4) Whilst boron nitride is the most common, and the most widely researched, MXenes come in many forms and are often made from a combination of early transition metals (M), such as Titanium, Vanadium, Chromium and Niobium, alongside carbon or nitrogen (X).

　　参考译文：虽然氮化硼是最常见并且被研究最多的，但 MXene 有很多的形式，通常是由早期过渡元素金属（M），如钛、钒、铬、铌和碳或氮（X）构成。

Exercises

1. Question for discussion

List one material and its characteristic for each dimension (0, 1, 2D).

2. Translate the following into Chinese

(1) Janus particles are an interesting class of nanoparticle, as one half of the surface is different to the other, and these two surfaces can differ in their external receptors, hydrophobicity or hydrophilicity, surface charge, and even in their magnetism.

(2) However, graphene and the various other 2D atomic materials possess an excellent array of optical, physical and electrical properties that make them useful for a wide range of applications.

3. Translate the following into English

(1) 该技术被大量应用于显示器与太阳能电池上。

(2) 碳纳米管有多种应用，并且已经广为人知。

(3) 迄今为止，石墨烯是最接近商业应用的二维材料。

Reading Material
3D nanomaterials fabricated by rolled-up nanotechnology

这篇课文讲解了利用卷曲纳米技术制备三维纳米材料这一新方法。课文说明了复杂三维纳米材料及结构的研究现状，介绍了主要制备方法及其局限性，并在此基础上阐述了卷曲纳米技术的理念、方法和应用。课文还概述了该技术广阔的应用前景以及发展过程中面临的挑战。

The development of complex 3D micro/nanoarchitectures has sparked enormous research interests due to its potential for advanced electronics, photonics, mechanics, and microelectromechanical systems (MEMS). A number of approaches have been developed to form these 3D architectures, such as 3D printing with polymeric inks, two and multiphoton lithography, large-area projection micro-stereolithography (LAPµSL), and template-assisted deposition (**Figure 6.2.1**). These impressive techniques have offered various manufacturing capabilities with high design freedom and arbitrary critical feature sizes and thus represent an important milestone toward 3D micro/nanoarchitectures. However, most of these techniques fabricate 3D structures from a bottom-up perspective. For instance, two-photon lithography typically begins by fabricating photosensitive polymer-based 3D microstructures, and then covert the materials template to desired materials by exchange growth using atomic layer deposition or chemical vapor deposition (CVD). When it comes to on-chip integrations, these strategies are limited by accessible materials and slow processing rate, and the generated defects in the produced amorphous materials placed restrictions to its further applications in high-performance devices. Therefore, novel approaches and methods of fabricating 3D micro/nanostructures are highly demanded.

Rolled-up nanotechnology is an alternative technique that fabricates 3D nanostructures out of planar films called nanomembranes. Nanomembranes are defined as planar thin films with

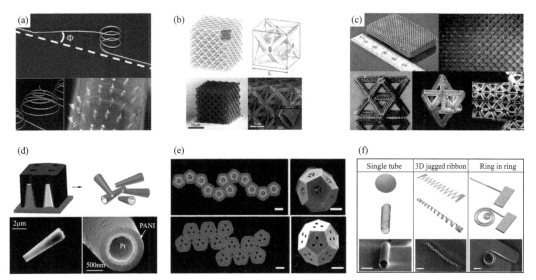

Figure 6.2.1　Fabrication methods of 3D micro/nanoarchitectures. (a) 3D-printed helices and hemispherical spiral array; (b) Architecture and microstructure of nanolattices using two-photon lithography; (c) Hierarchical metamaterial and critical features fabricated by LAPμSL; (d) Template-assisted electrodeposition of microtubes; (e) Self-folded 3D structures (scale bars: 300 μm); (f) Rolled-up 3D structures (scale bars: 10 μm).

thicknesses between one to a few hundred nanometers. These nanomembrane materials behave like soft materials while maintaining excellent properties like their bulk counterpart and therefore is an ideal platform for fabricating 3D nanostructures. Inspired by concepts of origami and kirigami, researchers folded and rolled these nanomembranes into a number of fascinating 3D structures. By releasing strain-engineered planar functional nanomembranes on sacrificial layers, complex rolled-up structures including tubes, rings, and helices can be constructed [for instance, see **Figure 6.2.1 (f)**]. This approach can be applied to engineer a wide range of organic and inorganic materials and their combinations, including metals, insulators, traditional semiconductor families, and recently emerged 2D materials. Given its compatibility to standard CMOS fabrication process and ability to manufacture large periodic arrays with precise geometric control, rolled-up nanotechnology further expands the applications of 3D nanostructures to a number of areas, including optical resonators, photodetectors, micromotors, energy storage, drug delivery, gas detection, and environmental decontamination. The versatility in the materials design and geometric tuning in rolled-up nanotechnology presents tremendous opportunities for further developing sophisticated 3D fine structures.

Despite the rapid development and tremendous progress of rolled-up nanotechnology, there remains some technical challenges before realizing the full potential of multitask systems based on rolled-up platforms. One major challenge lies in the manufacture and control of rolled-up tubular structures. So far, each materials system and corresponding fabrication process has its own limitations. Epitaxial systems produce single-crystalline nanomembranes of highest purity, but the corrosive etchants limit the materials choices. Fabricating nanomembranes on polymers overcame this issue, but the amorphous materials had inferior quality, which is detrimental to its further applications. In order to bring this area forward, it is challenging but rewarding to develop revolutionary approaches to fabricate complex tubular devices comprising multiple materials with excellent performances.

Another challenge is fabricating high-quality micro- and nanotubes based on 2D materials. The emerging family of 2D materials has become a leading topic in materials research. 2D materials as well as their combinations as van der Waals heterostructures have unique physical properties that enable novel designs of nanodevices and flexible electronics. However, to date only limited progress has been made toward experimental realization of rolling up 2D materials into 3D tubular architectures, in which the materials category and quality need to be further improved. The difficulty of rolling up high-quality 2D materials into functional devices and control over its geometry remains a primary barrier for future development due to the fragility and reactive nature of layered materials. A number of 2D materials and their heterostructures are yet to be explored as building blocks of complex 3D nanostructures. Nevertheless, the fascinating properties and promising prospects of rolled-up nanotechnology offer great opportunities for increasingly sophisticated devices and will continue to develop amazing 3D integrated systems in the future.

(**Selected from:** Xu C, Wu X, Huang G, et al. Rolled-up Nanotechnology: Materials Issue and Geometry Capability [J]. Advanced Materials Technologies, 2019, 4 (1): 1800486.)

New Words and Expressions

rolled-up nanotechnology　卷曲纳米技术
spark　*v.* 引发，触发
microelectromechanical system　微机电系统
arbitrary　*adj.* 任意的
critical feature　特征尺寸
milestone　*n.* 里程碑
on-chip integration　片上集成
amorphous　*adj.* 无定形的，非晶的
nanomembrane　*n.* 纳米薄膜
counterpart　*n.* 对应的人（或事物），配对物
origami　*n.* 折纸
kirigami　*n.* 剪纸
versatility　*n.* 多功能性
tremendous　*adj.* 巨大的，精彩的
epitaxial　*adj.* 外延的
inferior　*adj.* 较差的
detrimental　*adj.* 有害的
fragility　*n.* 脆弱
heterostructure　*n.* 异质结构

nevertheless　*adv.* 然而
fascinating　*adj.* 迷人的
prospect　*n.* 前景
sophisticated　*adj.* 先进的，精密的

Notes

(1) A number of approaches have been developed to form these 3D architectures, such as 3D printing with polymeric inks, two and multiphoton lithography, large-area projection microstereolithography (LAPμSL), and template-assisted deposition.

—介绍了目前制备三维结构的一些常规方法。
—参考译文：目前（研究人员）已经开发了许多方法来形成这些 3D 结构，如聚合物油墨 3D 打印、双光子和多光子光刻、大面积投影微立体光刻（LAPμSL）和模板辅助沉积。

(2) By releasing strain-engineered planar functional nanomembranes on sacrificial layers, complex rolled-up structures including tubes, rings, and helices can be constructed.

—sacrificial layers　牺牲层。
—参考译文：通过在牺牲层上释放应变工程的平面功能纳米薄膜，可以构建包括管状、环状和螺旋状在内的复杂卷曲结构。

(3) Given its compatibility to standard CMOS fabrication process and ability to manufacture large periodic arrays with precise geometric control, rolled-up nanotechnology further expands the applications of 3D nanostructures to a number of areas, including optical resonators, photodetectors, micromotors, energy storage, drug delivery, gas detection, and environmental decontamination.

—given　表示由于，鉴于。
—参考译文：考虑到其与标准 CMOS 制造工艺的兼容性，以及制备具有精确几何控制的大型周期阵列的能力，卷曲纳米技术进一步将 3D 纳米结构的应用拓展到许多领域，包括光学谐振腔、光电探测器、微型电机、能量存储、药物传递、气体检测和环境净化。

(4) One major challenge lies in the manufacture and control of rolled-up tubular structures.

—lie in ...　在于……。
—参考译文：一个主要的挑战在于制造和控制卷曲管状结构。

(5) However, to date only limited progress has been made toward experimental realization of rolling up 2D materials into 3D tubular architectures, in which the materials category and quality need to be further improved.

—in which　引导定语从句。
—参考译文：然而，迄今为止，（将）二维材料卷成三维管状结构的实验进展有限，材料种类和质量还有待进一步提高。

(6) A number of 2D materials and their heterostructures are yet to be explored as building blocks of complex 3D nanostructures.

—参考译文：（将）许多二维材料及其异质结构作为复杂三维纳米结构的组成部分仍有待探索。

Unit 6.3

Text

Synthesis and applications of nanomaterials

这篇课文简介了纳米材料的常见合成方法，包括自上而下和自下而上这两类方法，并具体介绍了球磨法和溶胶凝胶法这两种典型合成方法。课文还简要展示了纳米材料在燃料电池、催化剂、能量转换与存储和下一代计算机芯片中的应用。

Synthesis of nanomaterials

We are dealing with very fine structures: a nanometer is a billionth of a meter. This indeed allows us to think in both the "bottom up" or the "top down" approaches to synthesize nanomaterials, i. e. either to assemble atoms together or to dis-assemble (break, or dissociate) bulk solids into finer pieces until they are constituted of only a few atoms. This domain is a pure example of interdisciplinary work encompassing physics, chemistry, and engineering up to medicine.

Methods for creating nanostructures

There are many different ways of creating nanostructures: of course, macromolecules or nanoparticles or buckyballs or nanotubes and so on can be synthesized artificially for certain specific materials. They can also be arranged by methods based on equilibrium or nearequilibrium thermodynamics such as methods of self-organization and self-assembly (sometimes also called bio-mimetic processes). Using these methods, synthesized materials can be arranged into useful shapes so that finally the material can be applied to a certain application.

A broad classification divides method into either those which build from the "bottom up", atom by atom, or those which construct from the "top down" using processes that involve the removal or reformation of atoms to create the desired structure. In the bottom-up approach, atoms, molecules and even nanoparticles themselves can be used as the building blocks for the creation of complex nanostructures; the useful size of the building blocks depends on the properties to be engineered. By altering the size of the building blocks, controlling their surface and internal chemistry, and then controlling their organization and assembly, it is possible to engineer properties and functionalities of the overall nanostructured solid or system. These processes are essentially highly controlled, complex chemical syntheses. On the other hand, top-down approaches are inherently simpler and rely either on the removal or division of bulk material, or on the miniaturization of bulk fabrication processes to produce the desired structure with the appropriate properties. When controlled, both top-down and bottom-up methods may be viewed as essentially different forms of microstructural engineering.

Typical top-down processes: Milling
Mechanical grinding

Mechanical attrition is a typical example of "top down" method of synthesis of nanomaterials, where the material is prepared not by cluster assembly but by the structural decomposition of coarser-grained structures as the result of severe plastic deformation. This has become a popular method to make nanocrystalline materials because of its simplicity, the relatively inexpensive equipment needed, and the applicability to essentially the synthesis of all classes of materials. The major advantage often quoted is the possibility for easily scaling up to tonnage quantities of material for various applications. Similarly, the serious problems that are usually cited are; 1. contamination from milling media and/or atmosphere, and 2. to consolidate the powder product without coarsening the nanocrystalline microstructure.

In fact, the contamination problem is often given as a reason to dismiss the method, at least for some materials. Here we will review the mechanisms presently believed responsible for formation of nanocrystalline structures by mechanical attrition of single phase powders, mechanical alloying of dissimilar powders, and mechanical crystallization of amorphous materials. The two important problems of contamination and powder consolidation will be briefly considered.

Mechanical milling is typically achieved using high energy shaker, planetary ball, or tumbler mills. The energy transferred to the powder from refractory or steel balls depends on the rotational (vibrational) speed, size and number of the balls, ratio of the ball to powder mass, the time of milling and the milling atmosphere. Nanoparticles are produced by the shear action during grinding.

Milling in cryogenic liquids can greatly increase the brittleness of the powders influencing the fracture process. As with any process that produces fine particles, an adequate step to prevent oxidation is necessary. Hence this process is very restrictive to produce non-oxide materials since then it requires that the milling take place in an inert atmosphere and that the powder particles be handled in an appropriate vacuum system or glove box. This method of synthesis is suitable for producing amorphous or nanocrystalline alloy particles, elemental or compound powders. If the mechanical milling imparts sufficient energy to the constituent powders a homogeneous alloy can be formed. Based on the energy of the milling process and thermodynamic properties of the constituents the alloy can be rendered amorphous by this processing.

Typical bottom-up method: Sol-gel process

The sol-gel process, involves the evolution of inorganic networks through the formation of a colloidal suspension (sol) and gelation of the sol to form a network in a continuous liquid phase (gel). The precursors for synthesizing these colloids consist usually of a metal or metalloid element surrounded by various reactive ligands. The starting material is processed to form a dispersible oxide and forms a sol in contact with water or dilute acid. Removal of the liquid from the sol yields the gel, and the sol/gel transition controls the particle size and shape. Calcination of the gel produces the oxide.

Sol-gel processing refers to the hydrolysis and condensation of alkoxide-based precursorssuch as $Si(OEt)_4$ (tetraethyl orthosilicate, or TEOS). The reactions involved in the sol-gel chemistry based on the hydrolysis and condensation of metal alkoxides $M(OR)_z$ can be described as follows:

$$MOR + H_2O \longrightarrow MOH + ROH \text{ (hydrolysis)}$$
$$MOH + ROM \longrightarrow M-O-M + ROH \text{ (condensation)}$$

Sol-gel method of synthesizing nanomaterials is very popular amongst chemists and is widely employed to prepare oxide materials. The interest in this synthesis method arises due to the possibility of synthesizing nonmetallic inorganic materials like glasses, glass ceramics or ceramic materials at very low temperatures compared to the high temperature process required by melting glass or firing ceramics.

The major difficulties to overcome in developing a successful bottom-up approach is controlling the growth of the particles and then stopping the newly formed particles from agglomerating. Other technical issues are ensuring the reactions are complete so that no unwanted reactant is left on the product and completely removing any growth aids that may have been used in the process. Also, production rates of nano powders are very low by this process. The main advantage is one can get monosized nanoparticles by any bottom up approach.

Selected applications of nanomaterials

Since there are almost too many applications of nano to catalog here, this section is not meant to be comprehensive. However, we briefly touch upon some applications that are seen in the current literature.

Fuel cells

A fuel cell is an electrochemical energy conversion device that converts the chemical energy from fuel (on the anode side) and oxidant (on the cathode side) directly into electricity. The heart of fuel cell is the electrodes. The performance of a fuel cell electrode can be optimized in two ways: by improving the physical structure and by using more active electrocatalyst. A good structure of electrode must provide ample surface area, provide maximum contact of catalyst, reactant gas and electrolyte, facilitate gas transport and provide good electronic conductance. In this fashion the structure should be able to minimize losses.

Catalysis

Higher surface area available with the nanomaterial counterparts, nano-catalysts tend to have exceptional surface activity. For example, reaction rate at nano-aluminum can go so high, that it is utilized as a solid-fuel in rocket propulsion, whereas the bulk aluminum is widely used in utensils. Nano-aluminum becomes highly reactive and supplies the required thrust to send off pay loads in space. Similarly, catalysts assisting or retarding the reaction rates are dependent on the surface activity, and can very well be utilized in manipulating the rate-controlling step.

Energy conversion and storage

Harvesting clean energy from abundant resources is one of the most important tasks to be accomplished by modern engineering. As clean energysources (e. g., solar, wind) are not correlated with energy demand in time and non-localized energy sources are needed in transportation, too, efficiency in both energy harvesting and energy storage is necessary. Energy harvesting involves macroscopic equipment, e. g., wind turbines, however in many methods nanomaterials are used to improve efficiency. The two most relevant fields are thermo- and piezoelectric conversion and solar energy conversion by solar cells. Thermoelectric nanomaterials convert thermal energy between two locations at different temperatures. The measure of the efficiency of a material in thermoelectric conversion is the ZT parameter, the ratio of the Seebeck coefficient and the thermal and electrical conductivity of the material. Superlattices are the most efficient conversion devices, however their fabrication is costly and scaling up is complicated. Work has been carried out for efficient use of simpler systems, e. g., silicon nanowires.

Next-generation computer chips

The microelectronics industry has been emphasizing miniaturization, whereby the circuits, such as transistors, resistors, and capacitors, are reduced in size. By achieving a significant reduction in their size, the microprocessors, which contain these components, can run much faster, thereby enabling computations at far greater speeds. However, there are several technological impediments to these advancements, including lack of the ultrafine precursors to manufacture these components; poor dissipation of tremendous amount of heat generated by these microprocessors due to faster speeds; short mean time to failures (poor reliability), etc. Nanomaterials help the industry break these barriers down by providing the manufacturers with nanocrystalline starting materials, ultra-high purity materials, materials with better thermal conductivity, and longer-lasting, durable interconnections (connections between various components in the microprocessors).

(**Selected from:** Khan Z H, Kumar A, Husain S, et al. Introduction to Nanomaterials [M]//Husain M, Khan Z H. Advances in Nanomaterials. New Delhi; Springer India. 2016: 1-23.)

New Words and Expressions

billionth   *n.*/*adj.*   十亿分之一（的）
bottom up   自下而上
top down   自上而下
dissociate   *v.*   解离，分离
encompass   *v.*   包括，包含
macromolecule   *n.*   高分子，大分子
bio-mimetic   *adj.*   仿生的

reformation   *n*. 重整，重组
inherently   *adv*. 内在地，固有地，本质上地
miniaturization   *n*. 小型化，微型化
mechanical attrition   机械研磨，机械球磨
coarser-grained   *adj*. 粗颗粒的
deformation   *n*. 变形
quote   *v*. 引用，引述，举例说明
tonnage   *n*. 吨位，(以吨计的) 总重量，总吨数
consolidate   *v*. 使巩固，使加强，合并，统一
refractory   *n*. 耐火物质
vibrational   *adj*. 振动的，震动性的，摇摆的
cryogenic liquid   低温液体
restrictive   *adj*. 限制(性)的，约束(性)的
impart   *v*. 传授，给予，赋予
gelation   *n*. 凝胶化，凝胶
metalloid   *n*. 类金属，准金属；*adj*. 非金属的，准金属的
dispersible   *adj*. 可分散的
calcination   *n*. 煅烧，焙烧
hydrolysis   *n*. 水解作用，水解
condensation   *n*. 凝聚反应，缩合反应，缩聚反应
metal alkoxide   金属醇盐
nonmetallic   *n*. 非金属物质；*adj*. 非金属的
agglomerating   *adj*. 烧结，黏聚
catalog   *n*. 目录，登记；*v*. 登记，为……编目录
ample   *adj*. 充足的，充裕的，丰富的
conductance   *n*. 电导，电导率，电导系数
propulsion   *n*. 推进，推进剂
utensil   *n*. 餐具，器具
thrust   *n*. 推力
retard   *v*. 阻碍，减缓，减慢
manipulate   *v*. 操纵，操作，控制
non-localized   *adj*. 非定域的，非局限的
macroscopic   *adj*. 宏观的
wind turbine   风力涡轮机
piezoelectric   *adj*. 压电的
thermoelectric   *adj*. 热电的
microelectronics   *n*. 微电子学，微电子技术
microprocessor   *n*. 微型处理器
impediment   *n*. 障碍，阻碍，壁垒
dissipation   *n*. 消散，散热

reliability　　*n*. 可靠性，可信度
interconnection　　*n*. 相互连接，相互作用

Notes

（1）Using these methods, synthesized materials can be arranged into useful shapes so that finally the material can be applied to a certain application.

　　—be arranged into　将……分类，设置为……，分成。
　　—so that ...　以便，为了，引导目的状语从句。
　　—参考译文：利用这些方法，合成的材料可被设计成有用的形状，以便材料最终可应用于特定用途。

（2）A broad classification divides methods into either those which build from the "bottom up", atom by atom, or those which construct from the "top down" using processes that involve the removal or reformation of atoms to create the desired structure.

　　—which 引导定语从句，修饰 those。
　　—that 引导定语从句，修饰 process。
　　—参考译文：对这些合成方法进行一个宽泛的分类，一种是从"自下而上"逐原子构建的方法；另一种是从"自上而下"的构建，采用的过程涉及原子的去除或重组以创建所需的结构。

（3）Mechanical attrition is a typical example of "top down" method of synthesis of nanomaterials, where the material is prepared not by cluster assembly but by the structural decomposition of coarser-grained structures as the result of severe plastic deformation.

　　—not ... but ...　不是……而是……，所连接的前后内容在意义上表示转折，在结构上并列。
　　—as the result of ...　由于……的结果，作为……的结果。
　　—参考译文：机械研磨是"自上而下"法合成纳米材料的一个经典示例。在这种方法中，材料不是通过团簇组装，而是通过严重塑性形变导致的粗晶粒结构分解而制备的。

（4）The interest in this synthesis method arises due to the possibility of synthesizing nonmetallic inorganic materials like glasses, glass ceramics or ceramic materials at very low temperatures compared to the high temperature process required by melting glass or firing ceramics.

　　—参考译文：与熔融玻璃或烧制陶瓷所需的高温工艺相比，（这种合成方法具有）在极低温下合成玻璃、玻璃陶瓷或陶瓷材料等无机非金属材料的可能性，（因此）这种合成方法引起了（人们的）兴趣。

（5）It is evident that nanomaterials split their conventional counterparts because of their superior chemical, physical, and mechanical properties and of their exceptional formability.

　　—It is evident that ...　显然，很明显的是……。这里 it 为形式主语，后面 that 引导的从句为真正的主语。

——参考译文：显然，纳米材料因其优越的化学、物理、机械性能以及独特的可成形性而有别于传统材料。

Exercises

1. Question for discussion

(1) What are the major advantages and disadvantages of mechanical grinding for synthesis of nanomaterials?

(2) What measures need to be taken to prevent oxidation during milling?

(3) Please briefly summarize the properties of nanomaterials.

(4) Why are the researchers interested in sol-gel methods?

(5) What are the technological impediments of the microelectronics industry with reduced size?

2. Translate the following into Chinese

(1) deal with　　　　　　　(2) self-organization

(3) division　　　　　　　(4) essentially

(5) scale up　　　　　　　(6) depend on

(7) Superlattice　　　　　　(8) rate-controlling step

(9) This indeed allows us to think in both the "bottom up" or the "top down" approaches to synthesize nanomaterials, i. e. either to assemble atoms together or to dis-assemble (break, or dissociate) bulk solids into finer pieces until they are constituted of only a few atoms.

(10) In the bottom-up approach, atoms, molecules and even nanoparticles themselves can be used as the building blocks for the creation of complex nanostructures; the useful size of the building blocks depends on the properties to be engineered.

(11) By achieving a significant reduction in their size, the microprocessors, which contain these components, can run much faster, thereby enabling computations at far greater speeds.

(12) This has become a popular method to make nanocrystalline materials because of its simplicity, the relatively inexpensive equipment needed, and the applicability to essentially the synthesis of all classes of materials.

(13) The energy band structure and charge carrier density in the materials can be modified quite differently from their bulk and in turn will modify the electronic and optical properties of the materials.

(14) Energy harvesting involves macroscopic equipment, e. g., wind turbines, however in many methods nanomaterials are used to improve efficiency.

3. Translate the following into English

(1) 研磨过程　　　　　　　(2) 热力学特性

(3) 溶胶-凝胶法　　　　　　(4) 悬浮液

(5) 前驱体　　　　　　　　(6) 陶瓷材料

(7) 量子点　　　　　　　　(8) 灵敏度

(9) 纳米材料的性能与相应块体材料的性能有显著不同。

(10) 由于尺寸小,纳米材料具有极大的表面积/体积比,这使得表面或界面原子占比很大,导致更多"表面"依赖的材料特性。

(11) 碳纳米管优越的力学性能众所周知。

(12) 燃料电池是一种电化学能量转换装置,它将燃料(阳极侧)和氧化剂(阴极侧)的化学能直接转换成电能。

(13) 纳米催化剂通常具有优异的表面活性。

(14) 溶胶凝胶法被广泛应用于制备氧化物材料。

Reading Material
Nanomaterials for energy conversion and storage

这篇课文节选自一篇综述论文的背景介绍部分,内容主要是纳米材料在能源转换与存储领域的应用,包括纳米材料在光伏电池、锂离子电池、超级电容器、储氢等领域的应用。课文还简要介绍了纳米材料在能源领域具有广泛应用的原因。

Nanomaterials and nanostructures play a critical role in the recent advancement of some key technologies. Nanomaterials differ from microsized and bulk materials not only in the scale of their characteristic dimensions, but also in the fact that they may possess new physical properties and offer new possibilities for various technical applications. For example, when the characteristic dimensions of a semiconductor reduce to below a certain size, quantum confinement leads to an increased bandgap. The bandgap can be controlled by simply varying the dimensions of the material, so the optical absorption and emission spectra can be tuned to meet the specific requirements of the desired applications. Gold changes color to pink when the size is reduced to a few tens of nanometers due to the surface plasmon resonant absorption, and finds application in enhanced surface Raman scattering. When gold nanoparticles shrink further to less than 3 nanometers, they exhibit excellent catalytic activity due to the relatively smaller shrinkage of the d-orbitals in comparison to that of the s- and p-orbitals.

The pervasive and wide-spread applications of nanomaterials are not necessarily due to the totally new physical properties uniquely associated with nanomaterials. The enhancement in specific surface area and associated surface energy also renders some very important applications. For example, the vapor pressure and solubility of materials change exponentially with the curvature of the surface. Nanomaterials can have solubility or vapor pressure orders of magnitude higher than their bulk counterparts, so Ostwald ripening can be far more serious than in microsized materials. The melting point of gold particles can be significantly lowered when the particle size reduces to the nanometer scale. Magnetics can become superparamagnetics when the particle size reduces to the nanometer scale, corresponding to particles consisting of less than $\sim 10^5$ atoms; in such a case the surface energy becomes sufficiently large to overcome the ordering force that keeps the magnetic moments aligned.

Smaller size or dimension offers a great deal of advantages and is beneficial to the advancement of existing technologies and to the exploration and development of new technologies. For example, their small size makes nanoparticles viable carriers to deliver drugs to specific targets. The shrinking size in electronic devices has made it possible for mobile phones to serve a plethora of functions. The mechanical strength of nanomaterials is much higher than their bulk counterparts, due to the reduced number of defects. However, the small size and large specific surface area can have adverse impacts on some applications of nanomaterials. For example, the electrical conductivity of nanowires or films with thicknesses of several nanometers can be much lower than that of their bulk counterparts due to a much shortened electron mean free path resulting predominantly from surface scattering. Charge mobility in a polycrystalline semiconductor is lower than that in a single crystal, and is much lower when particles reduce to the nanometer scale.

Nanomaterials offer many advantages in energy conversion and storage applications. Energy conversion and storage involve physical interaction and/or chemical reaction at the surface or interface, so the specific surface area, surface energy, and surface chemistry play a very important role. The surface impacts are not limited to the kinetics and rate only. The surface energy and surface chemistry can have appreciable or significant influences on the thermodynamics of heterogeneous reactions occurring at the interface and the nucleation and subsequent growth when phase transitions are involved. The smaller dimensions of nanomaterials may also offer more favorable mass, heat, charge transfer, as well as accommodate dimensional changes associated with some chemical reactions and phase transitions. Nanomaterials also introduce new challenges in the application of energy conversion and storage. For example, large specific surface area offers more sites for charge recombination in photovoltaics and smaller pores may limit the penetration of electrolyte ions in supercapacitors.

This review article takes a few selected topic fields and focuses on the relatively recent progress in those fields, to highlight the most recent developments and the promise and limitations of nanomaterials in energy conversion and storage applications. Specifically, the topics that will be discussed in this review article include (1) nanostructured inorganic materials for photovoltaics, (2) nanostructured electrodes for lithium ion batteries, (3) nanomaterials for supercapacitors, and (4) nanocomposites for hydrogen storage. These four fields deal with different fundamental and technical challenges, but are connected by the common use of nanostructured materials to form electrodes for electron/mass transport under an electrochemical environment involving solid-liquid interfaces.

Nanostructured inorganic materials for photovoltaic applications such as dye-sensitized solar cells and quantum dot-sensitized solar cells are typically required to have a large specific surface area so that sufficient dye molecules and quantum dots can be adsorbed onto the inorganic materials, serving as an antenna for optical absorption. The surface chemistry must be such that the dyes or quantum dots can be adsorbed favorably to form closely packed conformal monolayers, not only for maximizing photon capture, but also for minimizing the interface charge recombination. The nanostructured inorganic network must possess excellent charge mobility and long lifetime, and possibly possess some light scattering or photon trapping capability. Perfect crystallinity and minimal surface and bulk defects are desired, and the grain boundaries connecting individual nanostructures should be controlled to be as low as possible.

Nanostructured materials as electrodes for lithium ion batteries should offer a set of properties or characteristics including large specific surface area for fast interfacial Faradaic reaction, small distance for mass and charge transport, and small dimensions to accommodate the volume change accompanied with lithium ion insertion and extraction. However, in order to achieve high energy and power density as well as long cyclic life, nanostructured electrodes should possess more open space to accommodate more guest ions and to allow the ions to diffuse effectively; for a given chemical composition, an amorphous material may be better-suited than its crystalline counterpart, and electrodes with perfect crystallinity may be less desirable than poor crystallinity. High energy facets and surface defects may promote and catalyze the interfacial reactions and phase transitions. Bulk defects may enlarge the lattice constants and enhance the electrical conductivity. Nano-carbon coating may not only enhance the electrical conductivity of the electrode, but also introduce surface defects that promote the interfacial *Faradaic* reactions and phase transitions.

Porous nanomaterials have been used and commercialized in electric double layer capacitors (EDLCs), or supercapacitors. The capacitance is directly proportional to the total surface area, so nanopores are desirable to achieve high specific surface area; however, the small pores or apertures may exclude electrolyte ions from penetrating, or at least impose significant diffusion resistance, leading to a low power density. Impurities in porous materials can be detrimental, as they may react with the electrolyte to degrade the cyclic stability; however, other impurities may enhance the surface charge density and thus high capacity can be achieved.

Hydrogen storage in solids has different challenges: high dehydrogenation temperature, reversibility, and thermal management, just to name a few. Nanostructured materials can affect the dehydrogenation temperature and manipulate the reaction mechanisms. Appropriately designed and fabricated nanocomposites can have desired thermal conductivity to mediate the heat released or absorbed during the hydrogenation or dehydrogenation process.

This review summarizes some of the important aspects and latest developments regarding applications of nanostructured materials for energy conversion and storage in the fields mentioned above. It will be shown that the most outstanding advantage of nanostructures is their ability to create architectures with significantly larger internal surface area in view of their nano-scaled size. This feature of nanostructures enables them to be suitable for use in devices such as dye-sensitized solar cells, lithium ion batteries, supercapacitors, and hydrogen storage systems. All of these applications involve a chemical reaction that takes place at either a solid-liquid interface or a solid-gas interface. Therefore, a larger interface results in an active material with a stronger ability to induce the reaction. Besides providing large surface area, it will be shown that nanostructures have other specific merits when used for energy-related devices. For example, one-dimensional nanostructures, including nanowires/nanorods and nanotubes, have demonstrated the ability to serve as antireflection layers in solar cells and give rise to highly efficient electron transport, especially in dye-sensitized solar cells and polymer solar cells. Metal nanoparticles may generate surface plasmon resonance, and thus enhance the optical absorption of all types of solar cells. Quantum dots are a promising type of nanostructure that may potentially lead to solar cells with internal conversion efficiencies over 100% owing to the multiple exciton effect. In regard to lithium ion batteries, it will be shown that recently developed micro/nano-structures and hollow structures exhibit enhanced lithium ion intercalation capability and surface permeability in view of their distinct geometrical characteristics that facilitate electrolyte diffusion and electron/ion transport while offering a large surface area. Mesocrystals, which are a relatively new structure comprised of crystallographically oriented nanoparticles, show great promise in creating high-performance lithium ion batteries because of their more prevalent and uniform pores, compared to traditional nanoparticle films, that can ease lithium ion intercalation by decreasing the diffusion distance. Surface modification and the utilization of defects, which are fairly new concepts to lithium ion batteries and still require further understanding, may controllably create nucleation sites at the electrode-electrolyte interface so as to promote phase transitions between the redox and charge/mass transfer processes, thus enhancing the efficiency and cycling performance of lithium ion batteries. In supercapacitors, besides the similar aforementioned effects on lithium ion batteries to provide high surface area, these nanostructures give rise to easy access for electrolyte to the active material and short diffusion distance, leading to improved energy storage. Hydrogen storage systems benefit from the use of nanostructured materials as a result of the reduced gravimetric and volumetric storage densities and additional binding sites provided by the nanostructures on the surface and within pores.

The motivation of this review is to demonstrate that the performance of devices, for example, the energy-related devices discussed here, is closely related to the structure of the materials, and to prove that attentive design of the structure may enable the materials to have desired function(s) or generate new mechanisms that can enhance the overall device performance.

(**Selected from:** Zhang Q, Uchaker E, Candelaria S L, et al. Nanomaterials for energy conversion and storage [J]. Chemical Society Reviews, 2013, 42 (7): 3127-3171.)

New Words and Expressions

microsized *adj*. 微小尺寸的
semiconductor *n*. 半导体
quantum confinement 量子限域效应
bandgap *n*. 能带间带隙
spectra *n*. 光谱（复数）
plasmon resonant 等离子共振
scattering *n*. 散射
orbital *n*. 轨道
render *v*. 给予
exponentially *adv*. 以指数方式地
curvature *n*. 曲率

Ostwald ripening 奥斯瓦尔德熟化
superparamagnetic *adj.* 超顺磁性的
magnetic moment *n.* 磁矩
plethora *n.* 过多，过剩
electron mean free path 电子平均自由行程
polycrystalline *adj.* 多晶的
thermodynamics *n.* 热力学
heterogeneous reaction 异质反应
recombination *n.* 重组，复合
photovoltaic *n.* 光伏
penetration *n.* 渗透
dye-sensitized solar cells 染料敏化太阳能电池
quantum dot-sensitized solar cells 量子点敏化太阳能电池
antenna *n.* 天线
conformal *adj.* 共形的
lattice constant 晶格常数
aperture *n.* 孔
dehydrogenation *n.* 脱氢
hydrogenation *n.* 加氢
antireflection layer 抗反射层
mesocrystal *n.* 介观晶体，介晶
crystallographically *adv.* 按照晶体学地
redox *n.* 氧化还原作用
gravimetric *adj.* 测定重量的
volumetric *adj.* 测定体积的

Notes

(1) Nanomaterials differ from microsized and bulk materials not only in the scale of their characteristic dimensions, but also in the fact that they may possess new physical properties and offer new possibilities for various technical applications.

—characteristic dimension 特征尺寸。

—参考译文：纳米材料与微米级和块状材料的不同之处不仅在于其特征尺寸的尺度，而且还在于它们可能拥有新的物理特性，以及为各种技术应用提供新的可能性。

(2) Nanostructured inorganic materials for photovoltaic applications such as dye-sensitized solar cells and quantum dot-sensitized solar cells are typically required to have a large specific surface area so that sufficient dye molecules and quantum dots can be adsorbed onto the inorganic materials, serving as an antenna for optical absorption.

—参考译文：（具有）纳米结构的无机材料用于光伏应用，如染料敏化太阳能电池和量子点敏化太阳能电池。它们通常要求具有较大的比表面积，以便有足够的染料分子和量子点作为光吸收的天线可以被吸附在无机材料上。

(3) The surface chemistry must be such that the dyes or quantum dots can be adsorbed favorably to form closely packed conformal monolayers, not only for maximizing photon capture, but also for minimizing the interface charge recombination.

— closely packed conformal monolayers 致密的共形单层。
— 参考译文：表面化学（成分）必须使染料或量子点能够被顺利地吸附，以形成致密的共形单层，这不仅是为了最大限度地捕获光子，也是为了最小化界面载流子复合。

(4) All of these applications involve a chemical reaction that takes place at either a solid-liquid interface or a solid-gas interface.

— that 引导一个定语从句，修饰 chemical reaction。
— 参考译文：所有这些应用都涉及在固液界面或固气界面发生的化学反应。

(5) Mesocrystals, which are a relatively new structure comprised of crystallographically oriented nanoparticles, show great promise in creating high-performance lithium ion batteries because of their more prevalent and uniform pores, compared to traditional nanoparticle films, that can ease lithium ion intercalation by decreasing the diffusion distance.

— crystallographically oriented nanoparticles 晶体学取向的纳米粒子。
— which 引导一个定语从句，修饰 Mesocrystals。that 引导一个定语从句，修饰 pores。
— 参考译文：介观晶体是一种由晶体学取向的纳米粒子组成的相对较新的结构，在开发高性能的锂离子电池时显示出巨大的前景，因为与传统的纳米粒子薄膜相比，它们的孔隙更加普遍和均匀，这可以（通过）减小扩散距离使锂离子易于嵌入。

(6) The motivation of this review is to demonstrate that the performance of devices, for example the energy-related devices discussed here, is closely related to the structure of the materials, and to prove that attentive design of the structure may enable the materials to have desired function(s) or generate new mechanisms that can enhance the overall device performance.

— 本句中前两个 that 各引导一个宾语从句，第三个 that 引导一个定语从句。
— 参考译文：本综述的动机是展示器件的性能，比如此处讨论的能源设备，它与材料的结构密切相关，并证明对结构的精心设计可以使材料具有预期的功能或产生能够提高整体器件性能的新机制。

Unit 6.4

Text

Intriguing physical properties of nanomaterials

这篇课文主要介绍纳米材料独特的电学、磁学、光学等物理性质。课文说明了纳米材料具有独特物理性质的主要原因——电子波函数受到量子限制效应影响，并分别举例介绍了纳米材料一些独特的性质。

When nanomaterials are studied, with few exceptions, the goal of the investigation is to reveal the values of physical properties similar to those used to characterize bulk phases of materials. In most cases though, the values of these properties are very different from those in the bulk, depending on the size and morphology of the nanomaterials. Nanoscale materials, especially with dimensionality of zero to two, exhibit intriguing electromagnetic behavior, as the most important effects and properties are derived from the quantum confinement of the wave function of the charge enclosed in the particles. In the nanometer-sized dimensions of the material, the wave function has a limited number of solutions. This results in changes in several electrical properties: with the decreasing feature size the bandgap becomes wider, the conductivity decreases and the density of states (DOS) decreases. Optical properties change accordingly, as the absorbed and emitted photons depend on the energy difference between the states among the bands or in their fine structure. Beyond this, quasiparticles are generated due to the confined space. An electron-hole pair, i. e., an exciton, carries energy in nanomaterials without motion of net charge. Plasmons, surface plasmons (SPs) (polaritons), and polarons play a key role in nanomaterial interactions with optical phonons, e. g., causing the color dependence of gold colloids as a function of particle size. Magnetic properties are also governed by the size of the nanomaterials, e. g., superparamagnetic materials exist below the size of the domain size.

Electrical properties

Nanomaterials, especially 1D structures such as nanowires and nanotubes, conduct electrical current differently compared with bulk conductors. The limited size in the direction perpendicular to the axis and the atomically organized structure of the 1D objects—crystals or nanotubes—results in weak phonon-charge carrier interaction and ballistic transport. In the ballistic regime of electrical conduction, the resistance does not depend on the length of the conductor but rather on the number of conductance channels used, following Landauer's law. Often, there is a single channel, and the quantum nature of the conductance as a function of the number of channels can be demonstrated, e. g., for carbon nanotubes. However, in other reports, e. g., on nanotubes used to build field-effect transistors (FETs), the conductance mechanism is considered to be diffusive.

Another important effect in devices made of nanosized conductors and semiconductors is the Coulomb blockade that occurs when new charge carriers cannot enter the conduction channel while another charge carrier occupies it, as demonstrated for single-molecule devices. This phenomenon enables the functioning of single-electron transistors (SETs) even at room temperature. SETs are also useful to investigate spin-spin interactions between localized and mobile electrons, i. e., the Kondo effect, more effectively than in macroscopic systems.

Figure 6.4.1 shows a collection of electrical properties. **Figure 6.4.1(a)** shows a schematic of the dependence of the density of states (DOS) on the dimensionality of the nanomaterials; the larger the number of quantum-confined dimensions, the higher the number of discrete states in the DOS. **Figure 6.4.1(b)** shows how the resistance of a carbon nanotube increases as it loses carbon layers from its wall; the resistance, and accordingly the conductivity, also change in quanta, illustrating the independence of the conductance channels in each layer. Results of a similar experiment are shown in **Figure 6.4.1(c)**. Here, the conductance of a nanotube bundle was measured when one, two, three, and four nanotubes were included sequentially in a circuit including a contact made of mercury. In both cases, the size of the quantized steps of the conductance change—following the Landauer law—is equal to G_0. By immersing the nanotube bundle further into the mercury electrode, the measured conductance is as high as $\approx 1000\ G_0$.

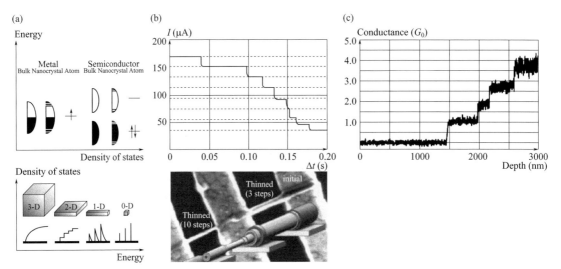

Figure 6.4.1 Examples for electrical properties of nanomaterials. (a) Schematics to represent changes in the DOS of nanomaterials as a function of dimensionality; (b) Quantized decrease of the current when shells of a multi-walled nanotube burned down consecutively and the schematics of the nanotube with partially removed walls; (c) Conductance of carbon nanotubes when a parallel circuit was built from one to four individual tubes. As the voltage was low (100 mV), each nanotube added only one channel to the conductance and the measured values were nG_0, where n is the number of nanotubes/channels. A larger number of nanotubes resulted in conductance up to $1000\ G_0$.

Magnetic properties

The magnetic properties of nanomaterials depend strongly on the characteristic size, i. e., the diameter of nanoparticles or the grain size of nanostructured material, which is normally small in comparison with the magnetic domain size of the material. **Figure 6.4.2(a)** presents these configurations and the energy density for the various configurations in thin films, showing that (for small size) the single-domain system is energetically favorable. Nanomaterials can be classified according to the type of interaction among the magnetic particles, extending from no interaction in a well-distributed nanoparticle system to strongly interacting nanostructured materials [**Figure 6.4.2(b)**].

Superparamagnetic materials behave similarly to paramagnetic materials. However, the magnetization value is considerably higher and the particles keep their magnetization values for a measurable timescale, e. g., several minutes. The magnetization curve of superparamagnetic materials does not show hysteresis, and the exact shape depends on the particle size or its distribution. By introducing defects into the lattice of carbon, e. g., implanting nitrogen or carbon atoms into nanodiamond, ferromagnetic behavior is observable. Also, ferromagnetism is measured on monoatomic cobalt structures at low temperature (10 K), showing the interaction between neighboring atoms. During Mackay transformation and Bain transformation, the crystal structure is changing, and the magnetic properties follow the crystal symmetry, as shown in **Figure 6.4.2(c)**. In all cases, the changes are significant below feature size of 10 nm.

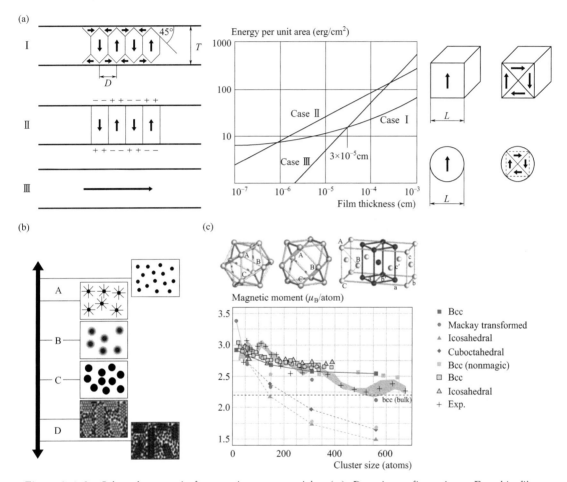

Figure 6.4.2 Selected magnetic features in nanomaterials. (a) Domain configurations. For thin films the single-domain configuration has lowest energy for thickness below 300 nm. Several domain configurations for nanoparticles also were suggested; (b) Schematics for the main types of nanostructures for magnetic behavior, including independent ultrafine particles, core-shell particles with magnetic core, particles as fillers in a matrix, and small crystallites in a noncrystalline matrix; (c) Mackay transformation shows the magnetization changes as a function of crystal structure. Schematics of icosahedron, cuboctahedron, and fcc with inscribed body-centered tetragonal Bain cell and the magnetization as a function of cluster size are displayed.

Optical properties

Optical properties — similarly to the electrical ones — are governed by quantum confinement: the lower dimensionality and smaller size result in a larger energy difference between neighboring discrete energy levels in the DOS, and accordingly higher excitation energy (**Figure 6.4.3**). Inasmuch, the smaller the particle, the shorter the wavelength in the absorption spectrum, and the color shifts from red to blue. This interaction with photons can be interpreted by introducing a quasiparticle known as the SP to represent the oscillations in the confined space. The absorption spectrum has a maximum when the SP-photon interaction is strongest, at the surface plasmon resonance (SPR) wavelength. The SPR for different nanoparticles depends—beyond the size—also on the material and shape of the nanoparticles. **Figure 6.4.3(a)** demonstrates visible-range colors caused by SPRs of different energies for several basic shapes of gold, silver, and alloy nanoparticles. **Figure 6.4.3(b)** shows a particular case, where the resonance wavelength shifts when the silver nanoshell particles are coated with a gold layer of increasing thickness. **Figure 6.4.3(c)** presents a silver nanoprism preparation method where the size distribution and accordingly the absorption spectrum are well controlled by the wavelength of light used in the photo-induced reaction. Another interesting phenomenon is the shift in the absorption of gold nanotriangles from 800 nm to 600 nm with the extent of the truncation of the tips.

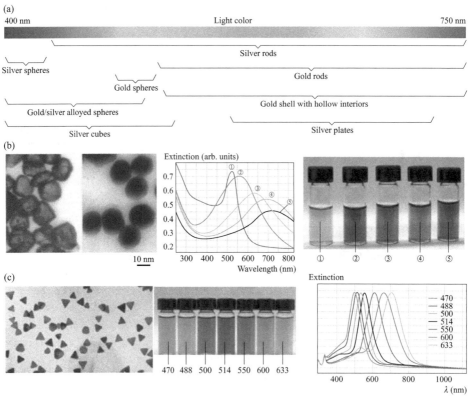

Figure 6.4.3 Selected optical properties of nanomaterials. (a) Typical SPR spectral ranges of silver and gold nanoparticles having various morphologies, compositions, and structures; (b) TEM images of the shell and solid particles; ultraviolet-visible (UV-Vis) extinction spectra and photos of solution of silver nanoshells coated with different thicknesses of gold and solid gold colloids; (c) TEM image of silver nanoprisms and extinction spectra of nanoprisms prepared with illumination by various laser wavelengths.

Along with other microscopy techniques, the optical property corresponding to Raman scattering, provides insightful information into the local bonding system of nanomaterials through phonon-photon interactions. It is used for characterizing a wide range of materials from metals to oxides; one of the characteristic examples is its application in studying carbon nanomaterials. Graphene has two characteristic peaks, the ratio of the intensity of the G and G' peaks depends on the number of layers in the 2D structure. In the case of carbon nanotubes, the D and G peaks dominate the spectra, while for single-walled nanotubes (SWNTs), the radial breathing mode (RBM) is also present and provides information about the diameter and (to some extent) the chirality of the nanotubes. Shifts, shoulders, and new peaks in the spectra can be interpreted as signals caused by defects in the bonds or in the structure.

(**Selected from:** Ajtai R. Science and Engineering of Nanomaterials [M]//VAJTAI R. Springer Handbook of Nanomaterials. Berlin, Heidelberg: Springer Berlin Heidelberg. 2013: 1-36.)

New Words and Expressions

morphology *n.* 形态学

dimensionality *n.* 维度

intriguing *adj.* 有趣的，迷人的

electromagnetic *adj.* 电磁的

quantum confinement 量子限制

emit *v.* 激发

quasiparticle *n.* 准粒子

generate *v.* 产生

net charge 净电荷

plasmon *n.* 等离子体

polariton *n.* 极化激元

polaron *n.* 极化子

colloid *n.* 胶体，胶质

superparamagnetic *adj.* 超顺磁性的

perpendicular to... 垂直于……

ballistic *adj.* 弹道的

field-effect transistor 场效应晶体管

diffusive *adj.* 扩散的

Coulomb blockade 库伦阻塞

spin-spin interaction 自旋相互作用

density of states 态密度

hysteresis *n.* 磁滞回线

ferromagnetism *n.* 铁磁性

nanoprism *n.* 纳米棱镜

to some extent 在某种程度上

Notes

(1) In most cases though, the values of these properties are very different from those in the bulk, depending on the size and morphology of the nanomaterials.

——翻译成中文时,语序需要略作调整。
——参考译文:然而,在大多数情况下,取决于纳米材料的尺寸和形貌,这些性质的数值与体材料中的数值有非常大的不同。

(2) Optical properties change accordingly, as the absorbed and emitted photons depend on the energy difference between the states among the bands or in their fine structure.

——参考译文:光学性质相应地发生变化,因为吸收和发射的光子取决于能带间或它们的精细结构中的状态之间的能量差异。

(3) Another important effect in devices made of nanosized conductors and semiconductors is the Coulomb blockade that occurs when new charge carriers cannot enter the conduction channel while another charge carrier occupies it, as demonstrated for single-molecule devices.

——这是一个含多个从句的复合句,翻译时须根据中文习惯调整。
——参考译文:由纳米尺寸导体和半导体制成的器件的另一个重要效应是库仑阻塞,即当一个载流子占据传导通道时,新的载流子无法进入,如单分子器件所示。

(4) The resistance, and accordingly the conductivity, also change in quanta, illustrating the independence of the conductance channels in each layer.

——分词作状语后置。
——电阻和相应的电导率的变化也量子化了,说明每层中导电通道的独立性。

(5) This interaction with photons can be interpreted by introducing a quasiparticle known as the SP to represent the oscillations in the confined space. The absorption spectrum has a maximum when the SP-photon interaction is strongest, at the surface plasmon resonance (SPR) wavelength.

——参考译文:这种与光子的相互作用可以通过引入一种称为表面等离激元(SP)的准粒子来解释,它代表局域空间中的振荡。在表面等离子体共振(SPR)波长处 SP-光子相互作用最强,吸收光谱具有最大值。

(6) **Figure 6.4.3(b)** shows a particular case, where the resonance wavelength shifts when the silver nanoshell particles are coated with a gold layer of increasing thickness.

——翻译时,语序需要按照中文习惯略作调整。
——参考译文:**Figure 6.4.3(b)** 展示了在一个特殊情况下,共振波长随银纳米壳颗粒表面沉积的金层的厚度增加而移动。

(7) In the case of carbon nanotubes, the D and G peaks dominate the spectra, while for SWNTs, the radial breathing mode (RBM) is also present and provides information about the diameter and (to some extent) the chirality of the nanotubes.

—参考译文：在碳纳米管的例子中，D 和 G 峰在光谱中占主导地位，而对于 SWNT，径向呼吸（振动）模式（RBM）也存在，并提供了有关纳米管直径和（在某种程度上）手性的信息。

Exercises

1. Question for discussion

（1）Why the physical properties of nanomaterials are very different from those of the bulk?

（2）What is Coulomb blockade?

（3）What are the factors that dominate the magnetic properties of nanomaterials?

（4）Why the optical properties of nanomaterials are governed by quantum confinement?

（5）Please present a practical example of nanomaterial/nanostructure's interesting property.

2. Translate the following into Chinese

（1）This results in changes in several electrical properties: with the decreasing feature size the bandgap becomes wider, the conductivity decreases and the density of states decreases.

（2）The limited size in the direction perpendicular to the axis and the atomically organized structure of the 1D objects—crystals or nanotubes—result in weak phonon-charge carrier interaction and ballistic transport.

（3）Nanomaterials can be classified according to the type of interaction among the magnetic particles, extending from no interaction in a well-distributed nanoparticle system to strongly interacting nanostructured materials.

3. Translate the following into English

（1）电磁性能　　　　　　（2）能带

（3）带隙　　　　　　　　（4）电子-空穴对

（5）超顺磁性　　　　　　（6）传导机制

（7）宏观系统　　　　　　（8）铁磁性

（9）根据 Landauer 定律，纳米材料的电阻取决于其导电通道数量。

（10）纳米材料的尺寸通常小于材料磁畴的大小，所以其磁学性质跟体材料相比会发生变化。

（11）随着纳米材料尺寸减小，量子限制效应使得能级间距增大，发光波长向短波长移动。

（12）通过光子-声子相互作用，拉曼光谱可以给出纳米材料的化学键和晶体结构相关的信息。

Reading Material
Natural and artificial nanotechnologies

这篇课文讨论了多种纳米技术及其未来的发展。课文介绍了纳米技术的起源，分析了天然的纳米技术和工业生产中的纳米技术的差异，阐述了发展纳米材料及纳米技术的重要意义。课文在介绍纳米技术研究现状的基础上展望了其未来发展。

Nanotechnology is highly interdisciplinary, involving physics, chemistry, biology, materials science, and the full range of the engineering disciplines. The word nanotechnology is widely used as shorthand to refer to both the science and the technology of this emerging field. Narrowly defined, nanoscience concerns a basic understanding of physical, chemical, and biological properties on atomic and near-atomic scales. Nanotechnology, narrowly defined, employs controlled manipulation of these properties to create materials and functional systems with unique capabilities.

In contrast to recent engineering efforts, nature developed "nanotechnologies" over billions of years, employing enzymes and catalysts to organize with exquisite precision different kinds of atoms and molecules into complex microscopic structures that make life possible. These natural products are built with great efficiency and have impressive capabilities, such as the power to harvest solar energy, to convert minerals and water into living cells, to store and process massive amounts of data using large arrays of nerve cells, and to replicate perfectly billions of bits of information stored in molecules of deoxyribonucleic acid (DNA).

There are two principal reasons for qualitative differences in material behavior at the nanoscale (traditionally defined as less than 100 nanometers). First, quantum mechanical effects come into play at very small dimensions and lead to new physics and chemistry. Second, a defining feature at the nanoscale is the very large surface-to-volume ratio of these structures. This means that no atom is very far from a surface or interface, and the behavior of atoms at these higher-energy sites have a significant influence on the properties of the material. For example, the reactivity of a metal catalyst particle generally increases appreciably as its size is reduced—macroscopic gold is chemically inert, whereas at nanoscales gold becomes extremely reactive and catalytic and even melts at a lower temperature. Thus, at nanoscale dimensions, material properties depend on and change with size, as well as composition and structure.

Using the processes of nanotechnology, basic industrial production may veer dramatically from the course followed by steel plants and chemical factories of the past. Raw materials will come from the atoms of abundant elements—carbon, hydrogen, and silicon—and these will be manipulated into precise configurations to create nanostructured materials that exhibit exactly the right properties for each particular application. For example, carbon atoms can be bonded together in a number of different geometries to create variously a fiber, a tube, a molecular coating, or a wire, all with the superior strength-to-weight ratio of another carbon material—diamond. Additionally, such material processing need not require smokestacks, power-hungry industrial machinery, or intensive human labor. Instead, it may be accomplished either by "growing" new structures through some combination of chemical catalysts and synthetic enzymes or by building them through new techniques based on patterning and self-assembly of nanoscale materials into useful predetermined designs. Nanotechnology ultimately may allow people to fabricate almost any type of material or product allowable under the laws of physics and chemistry. While such possibilities seem remote, even approaching nature's virtuosity in energy-efficient fabrication would be revolutionary.

Even more revolutionary would be the fabrication of nanoscale machines and devices for incorporation into micro- and macroscale systems. Once again, nature has led the way with the fabrication of both linear and rotary molecular motors. These biological machines carry out such tasks as muscle contraction (in organisms ranging from clams to humans) and shuttling little packets of material around within cells while being powered by the recyclable, energy-efficient fuel adenosine triphosphate. Scientists are only beginning to develop the tools to fabricate functioning systems at such small scales, with most advances based on electronic or magnetic information processing and storage systems. The energy-efficient, reconfigurable, and self-repairing aspects of biological systems are just becoming understood.

The potential impact of nanotechnology processes, machines, and products is expected to be far-reaching, affecting nearly every conceivable information technology, energy source, agricultural product, medical device, pharmaceutical, and material used in manufacturing. Meanwhile, the dimensions of electronic circuits on semiconductors continue to shrink, with minimum feature sizes now reaching the nanorealm, under 100 nanometers. Likewise, magnetic memory materials, which form the basis of hard disk drives, have achieved dramatically greater memory density as a result of nanoscale structuring to exploit new magnetic effects at nanodimensions. These latter two areas represent another major trend, the evolution of critical elements of microtechnology into the realm of nanotechnology to enhance performance. They are immense markets driven by the rapid advance of information technology.

(**Selected from:** Picraux S T. Overview of Nanotechnology [J/OL]. Encyclopedia Britannica, (2023-02-03) [2023-03-06], http://www.britannica.com/technology/nanotechnology/Overview-of-nanotechnology.)

New Words and Expressions

interdisciplinary *adj.* 跨学科的
discipline *n.* （尤指大学的）科目，学科
shorthand *n.* 速记（法），简略的表达方式
in contrast to... 与……形成对照
exquisite *adj.* 精致的，细致的
precision *n.* 精确性，严谨
deoxyribonucleic acid 脱氧核糖核酸
quantum mechanical effects 量子力学效应
come into play 开始起作用
appreciably *adv.* 明显地，相当地
veer *v.* （使）转向
configuration *n.* 构造，配置
predetermined *adj.* 预先确定的
ultimately *adv.* 最终，根本上
allowable *adj.* 承认的，容许的
virtuosity *n.* 精湛技巧

macroscale *n*. 宏观尺度
organism *n*. 生物，有机体
far-reaching *adj*. 影响深远的
adenosine triphosphate 三磷酸腺苷
realm *n*. 领域
immense *adj*. 极大的，巨大的

Notes

（1）Nanotechnology is highly interdisciplinary, involving physics, chemistry, biology, materials science, and the full range of the engineering disciplines.

——involving 现在分词，作为后置定语。
——参考译文：纳米技术是高度跨学科的，涉及物理、化学、生物学、材料科学以及工程学科的全部领域。

（2）Nanotechnology, narrowly defined, employs controlled manipulation of these properties to create materials and functional systems with unique capabilities.

——narrowly defined 作为插入语，用于补充含义或解释说明。
——参考译文：狭义的纳米技术，是利用这些特性的可控操作来创造具有独特能力的材料和功能系统。

（3）This means that no atom is very far from a surface or interface, and the behaviour of atoms at these higher-energy sites have a significant influence on the properties of the material.

——参考译文：这意味着没有原子离表面或界面很远，处于这些高能位置的原子的行为对材料的性质有显著的影响。

（4）Using the processes of nanotechnology, basic industrial production may veer dramatically from the course followed by steel plants and chemical factories of the past.

——Using the processes of nanotechnology 现在分词，作为状语。
——参考译文：使用纳米技术的工艺流程（后），基础工业生产可能会相对过去钢铁厂和化工厂所遵循的流程发生戏剧性的转变。

（5）Instead, it may be accomplished either by "growing" new structures through some combination of chemical catalysts and synthetic enzymes or by building them through new techniques based on patterning and self-assembly of nanoscale materials into useful predetermined designs.

——本句介绍纳米材料的制备方法。
——参考译文：相反，（加工）可以这样完成：通过化学催化剂和合成酶的某些组合来"生长"新的结构，或者通过基于纳米尺度材料的图形化和自组装的新技术来构建它们，从而（实现）有用的预先设计。

（6）The potential impact of nanotechnology processes, machines, and products is expected to be far-reaching, affecting nearly every conceivable information technology, energy source, agricultural product, medical device, pharmaceutical, and material used in manufacturing.

—参考译文:预计纳米技术的工艺、机器和产品的潜在影响将是深远的,几乎影响到每一个可以想象到的信息技术、能源、农产品、医疗设备、制药和用于制造业的材料。

Unit 6.5

Text

Nanographenes and optoelectronic devices

这篇课文介绍了纳米石墨烯,如石墨烯量子点、石墨烯纳米带、基于纳米石墨烯的范德华异质结等的制备和光电器件应用。该研究方向将在材料特性和内在机制研究的基础上探索实现光电器件的集成和优化。

Background

The discovery of the outstanding properties of graphene has nurtured the fast development of atomically thin 2D materials for fundamental studies and technological advancements, which has opened a new era in nanoscience, material science, and condensed-matter physics. The superior properties of graphene, including broad optical absorption and response, very high electrical and thermal conductivity, and high specific surface area, render it a fascinating material platform for cutting-edge applications, such as ultra-broadband photodetectors, ultrafast optical modulators, and energy conversion and storage devices. However, the inherent lack of bandgap in its electronic structure limits graphene's applications in field-effect transistors (FETs) related logic devices, limiting its use for the booming semiconductor industry. Thus, opening the bandgap of graphene becomes essential for extending its applications, particularly for electronics and optoelectronics. In principle, bandgap opening in graphene can be achieved by introducing quantum confinement effect, e. g., by "sculpting" graphene into its nanometer-sized structures. Here, these graphene nanostructures are termed as "nanographenes (NGs)", which demonstrate strong electronic confinement effects with at least one of their lateral dimensions being below 10 nm. The emerging NGs include "graphene quantum dots (GQDs)", which are laterally extended polycyclic aromatic hydrocarbons (PAHs), and their 1D extension, i. e., "graphene nanoribbons (GNRs)", as shown in **Figure 6. 5. 1.**

Fabrication

NGs exhibit tunable and finite optical bandgaps (from IR to UV range), making them appealing for next-generation, green electronics, and optoelectronics. The "green" nature and the abundance of the carbon element make NGs appealing for device applications from environmental and economic perspectives. Substantial synthesis effort has been made previously for opening and controlling the bandgap of NGs by tailoring their sizes, dimensions, and edge

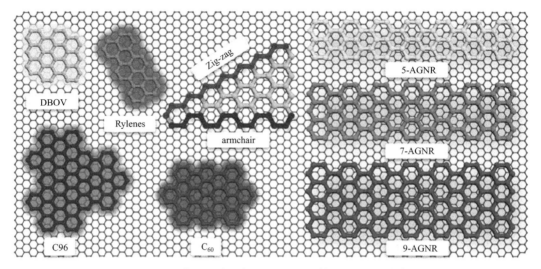

Figure 6.5.1　Exemplary molecular structures of bottom-up synthesized NGs.

structures with ultimate atomic precision. So far, both top-down and bottom-up approaches have been developed. While the former method is based on breaking large carbon allotropes into nanoscale pieces, the latter focuses on assembling the precursor molecules as tiny "LEGO" pieces into the final NGs via organic synthesis, pyrolysis, or other wet chemical methods.

For top-down approaches, previous attempts to fabricate NGs (e.g., GNRs) by hydrothermal cutting of graphene oxide nanosheets, photolithography patterning of graphene, and unzipping of carbon nanotubes have been reported. Yet, despite the successful demonstration of electronic bandgap opening, the use of top-down methods offers poor structural control in sizes below 10 nm, hence limiting the bandgap values below 400 meV. Furthermore, such top-down methods give rise to poorly defined edge structures in the resultant NGs, hindering the precise control of their properties for advanced applications.

In contrast, the bottom-up organic synthesis approach provides atomically precise control of NGs (e.g., width and edges), and consequently, fine-tuning of their electronic properties. The organic synthesis of NGs is typically achieved by the oxidative cyclodehydrogenation of oligophenylene and polyphenylene precursors, either in solution or on a metal surface. Thus, a variety of NGs with different edge structure, size, nonplanarity, and/or chirality have been prepared. Besides the organic synthesis approach, we note that other bottom-up synthesis routes (e.g., for fluorescent GQDs), including pyrolysis and solvothermal methods, have been developed. Despite being successfully applied, these approaches lack precise structural control at the atomic level.

Graphene quantum dot

GQDs are discrete quasi-0D NGs with sizes typically less than 10 nm in all three spatial dimensions, which chemically can be seen as extended polycyclic aromatic hydrocarbons (PAHs). In recent years, GQDs have aroused extensive interest on account of their fascinating optical properties, including high absorption coefficient ($\approx 10^5$ cm^{-1}), near-unity photoluminescence quantum yield (PLQY), and tunable stimulated emission over a wide spectral range (from deep ultraviolet to near-infrared) by engineering their shapes and edge structures. In addition, compared to conventional heavy-metal-based luminescent QDs (e. g., Pb- and Cd-based QDs), GQDs are non-toxic and exhibit high stability and good solution-processability, making them ideal candidates for developing environmentally friendly solid-state light-emitting sources. Along with using GQDs as the active components in LEDs, GQDs have also been considered as optical-gain media for lasing applications due to their low threshold, stable stimulated emission, and wavelength tunability across the visible spectrum. Besides conventional optoelectronic applications, GQDs can also be considered single-photon emitters thanks to their excellent photochemical stability and low cost, holding great promise for next-generation quantum technologies such as quantum cryptography and communication. In addition, GQDs exhibit outstanding photophysical properties for optical imaging applications. The environment-independent blinking properties pave the way to develop GQDs as a new series of fluorophores for optical super-resolution imaging. With the incorporation of nitrogen atoms, the synthesized GQDs can be used as nanoscale sensors for pH or specific metal ion detection.

Graphene nanoribbon

While GQDs with tunable bandgaps are useful for emission and imaging applications thanks to their fascinating optical properties, their applications in electronics are restricted because of their strong spatial confinement in all three directions. Graphene nanoribbons (GNRs), the 1D extension of GQDs, possess not only finite and tunable bandgaps (by controlling the width of the ribbon), but also provide an elongated electrical transport channel for charge carriers with high electrical mobility inherited from graphene. Besides their width dimension, the edge structures substantially affect the electronic and thus electrical properties of GNRs. The most popular edge structures include zig-zag and armchair. Armchair GNRs have been widely investigated due to their high stability and large tuning of bandgaps from IR to UV-vis. Zig-zag GNRs with their fascinating localized edge states have been considered as a promising platform for spintronics, but their synthesis and practical applications are currently impeded by their limited chemical stability.

By uniquely combining high intrinsic mobility and strong exciton effect (featuring strong light absorption), GNRs can be seen as highly conductive organic semiconductors which hold great potential for organic optoelectronics, e. g., organic photovoltaics (OPV). Although the studies illustrate the great potential of bottom-up GNRs for organic optoelectronics, their full potential has not been exploited so far. One of the potential bottlenecks for applications seems to lie in the synthesis, where a large-scale fabrication on the order of gram quantities is challenging. The length of GNRs, despite some exemplary success, is typically limited in the range of sub-100 nm, partially due to the low solubility of longer GNRs. On top of that, the limited processability of GNRs is also a key issue, prohibiting the achievement of highly efficient devices that fully utilize the fascinating intrinsic properties of GNRs. In this regard, besides designing various new GNR structures, further research efforts aimed at large-scale synthesis and improving processing methods (e. g., to avoid aggregation) are key toward implementing GNRs for optoelectronic devices.

NG-based van der Waals heterostructure (vdWH)

An emerging field related to the NG assembly lies in exploiting interfacial vdW interactions, e. g., between NGs and 2D materials, to construct hybrid organic-inorganic vdWHs for exploring new photophysical functionalities. In particular, given the large library of existing NGs and their structural tunability by synthesis, one can envision programming the functionality of NG-2D vdWHs by tuning the molecular structures and compositions. Furthermore, compared to conventional inorganic-inorganic 2D-2D vdWHs, the organic-inorganic nature of NG-2D vdWHs allows readily control of the interfacial coupling between the adjacent layers via solution-processed procedures. This enables fine-tuning of interfacial vdW interactions as ruled by the principles of molecular self-assembly, and thus offers programmable electronic coupling through the selection of the NG dimension, composition, doping level, and deposition methods. Such a unique feature offers a powerful knob to modulate the interfacial electronic coupling and thus charge transport properties of the hybrids for optoelectronics. For instance, a solution-processed approach was used to deposit NGs onto graphene to form strongly coupled NG-graphene vdWHs by taking advantage of the strong interlayer π-π interactions. Such a vdW sensitization scheme combines the strong light absorption in NGs, efficient interfacial charge transfer (CT) between the layers, and excellent electrical transport properties of graphene, making it ideal for photodetection. The resulting photodetector shows an ultrahigh responsivity of 4.5×10^7 A·W^{-1} and a specific detectivity of 4.6×10^{13} Jones. Hole injection from NGs to graphene results in a photoconductivity modulation beyond 1 ns. The long-lived charge separation, together with the efficient hole injection ($\approx 10\%$) across NG-graphene interfaces, rationalizes the superior properties observed in the photodetectors.

Outlook

Future efforts in this field should aim at improving the control over the physical and chemical properties of pristine NGs and at their hybrids when assembled into vdWHs. Endeavors in exploring practical approaches to precisely control vdW interactions in NG assembly are essential to modulate the interfacial charge carrier dynamics for applications. In this aspect, tuning the size and composition (e. g., the relative ratio between the number of hydrogen and carbon atoms) of NG may be a fruitful approach to control the interfacial vdW interactions and thus coupling strength. Interfacing NGs with materials with large lateral sizes and atomically thin thickness would facilitate the formation of strongly coupled NG assembly. In the future, a deep understanding of the underlying mechanisms that determine the fundamental properties of NGs is required for advanced optoelectronic device integration and optimization.

(**Selected from:** Liu Z, Fu S, Liu X, et al. Small Size, Big Impact: Recent Progress in Bottom-Up Synthesized Nanographenes for Optoelectronic and Energy Applications [J]. Advanced Science, 2022, 9 (19): 2106055.)

New Words and Expressions

atomical *adj.* 原子的
nurture *v.* 养育，养护，培养
fundamental *adj.* 基本的，根本的，基础的
era *n.* 时代，年代，纪元
inherent *adj.* 固有的，内在的
ultra-broadband 超宽带
modulator *n.* 调节器，调制器，调幅器
graphene quantum dot 石墨烯量子点
polycyclic *adj.* 多环的，多环麝香
aromatic *adj.* 芳香的，有香味的
hydrocarbon *n.* 烃，碳氢化合物
nanoribbon *n.* 纳米带
exemplary *adj.* 模范的，典型的
tunable *adj.* 可调
finite *adj.* 有限的
tailor *v.* 剪裁，制作
allotrope *n.* 同素异形体
pyrolysis *v.* 热解
nanosheet *n.* 纳米片
unzip *v.* 拉开……的拉锁
demonstration *n.* 示范，证明，展示
hinder *v.* 阻碍，妨碍，阻挡
precise *adj.* 准确的，确切的，精确的
fluorescent *adj.* 发荧光的
discrete *adj.* 离散的，分离的
spatial dimension 空间维度
candidate *n.* 候选者
bottleneck *n.* 瓶颈
sensitization *n.* 敏感化，致敏
endeavor *n.* /*v.* 努力，尽力

Notes

(1) However, the inherent lack of bandgap in its electronic structure limits graphene's applications in field-effect transistors (FETs) related logic devices, limiting its use for the booming semiconductor industry.

一参考译文：然而，其电子结构中固有的带隙缺乏限制了石墨烯在场效应晶体管（FET）相关逻辑器件中的应用，限制了其在蓬勃发展的半导体行业中的应用。

(2) Thus, opening the bandgap of graphene becomes essential for extending its applications, particularly for electronics and optoelectronics.

—become essential 变得至关重要。
—thus 在这里表承接。
—参考译文：因此，打开石墨烯的带隙对扩展其应用至关重要，特别是对于电子和光电子（应用）。

(3) Here, these graphene nanostructures are termed as "nanographenes (NGs)," which demonstrate strong electronic confinement effects with at least one of their lateral dimensions being below 10 nm.

—be termed as... 被称为……。
—which 引导的定语从句修饰前面的纳米石墨烯（NGs）。
—参考译文：在这里，这些石墨烯纳米结构被称为"纳米石墨烯（NG）"，它们至少一个侧向尺寸低于 10 nm，表现出很强的电子限制效应。

(4) Yet, despite the successful demonstration of electronic bandgap opening, the use of top-down methods offers poor structural control in sizes below 10 nm, hence limiting the bandgap values below 400 meV.

—top-down methods 自上而下的方法。
—翻译时须按照中文习惯稍作调整。
—参考译文：然而，尽管成功展示了电子带隙开放，自上而下方法的使用在（制备）小于 10 nm 的尺寸时结构控制较差，因此将带隙值限制在 400 meV 以下。

(5) In contrast, the bottom-up organic synthesis approach provides atomically precise control of NGs (e. g., width and edges), and consequently, fine-tuning of their electronic properties.

—参考译文：与之相反，自下而上的有机合成方法提供了对 NG 原子级的精确控制（例如宽度和边缘），从而（可）对其电子特性进行精细调控。

(6) One of the potential bottlenecks for applications seems to lie in the synthesis, where a large-scale fabrication on the order of gram quantities is challenging.

—large-scale fabrication 大规模制造。
—the order of gram 克量级。
—参考译文：应用的潜在瓶颈之一似乎在于合成，其中克量级的大规模制造具有挑战性。

Exercises

1. Question for discussion

(1) Please detailly illustrate two methods to fabricate graphene.
(2) Please define GQDs.

(3) What are the most popular edge structures of graphene nanoribbons?

(4) How can we control the interfacial vdW interactions?

2. Translate the following into Chinese

(1) The discovery of the outstanding properties of graphene has nurtured the fast development of atomically thin 2D materials for fundamental studies and technological advancements, which has opened a new era in nanoscience, material science, and condensed-matter physics.

(2) Substantial synthesis effort has been made previously for opening and controlling the bandgap of NGs by tailoring their sizes, dimensions, and edge structures with ultimate atomic precision.

(3) In this aspect, tuning the size and composition (e. g., the relative ratio between the number of hydrogen and carbon atoms) of NG may be a fruitful approach to control the interfacial vdW interactions and thus coupling strength.

(4) Interfacing NGs with materials with large lateral sizes and atomically thin thickness would facilitate the formation of strongly coupled NG assembly.

3. Translate the following into English

(1) 凝聚态物质 　　　　　　(2) 光致发光量子产率
(3) 溶剂热 　　　　　　　　(4) 扶手椅
(5) 之字形 　　　　　　　　(6) 场效应晶体管
(7) 石墨烯纳米带 　　　　　(8) 范德华异质结

(9) 扶手椅式石墨烯纳米带因其高稳定性和从 IR 到 UV-vis 的大范围带隙调节能力而受到广泛研究。

(10) 研究人员已经制备了具有不同边缘结构、尺寸、非平面性和/或手性的各种纳米石墨烯。

(11) 具有良好的局部边缘状态的锯齿形石墨烯纳米带被认为是自旋电子学的一个有前途的平台，但它们的合成和实际应用目前因其有限的化学稳定性而受到阻碍。

(12) 除了它们的宽度尺寸外，边缘结构还极大地影响了石墨烯纳米带的电子特性和电特性。

(13) 未来，先进的光电器件集成和优化需要对决定纳米石墨烯基本特性的潜在机制有更深入的了解。

Reading Material
Catalytic nanozymes

这篇课文主要介绍了纳米材料的一种典型应用，即催化纳米酶。课文阐述了纳米酶的定义和基于功能的分类，并针对每一类纳米酶材料举例进行了介绍。课文对比了纳米酶和天然酶的差异，并展望了单原子纳米酶的未来发展。

Definition and types of nanozymes

Enzymes are vital components of biological reactions and living organisms. Enzymes display remarkable specificity and they are capable of enhancing the rate of biochemical reaction, on some occasions nearly by 10^{12} times than that of the uncatalyzed reactions. Nanozymes are catalytic nanomaterials with enzyme-like properties, tunable size, shape, structure, composition, surface area, and selective catalytic reactions responsive to temperature and external stimuli. As alternatives to natural enzymes, highly stable and low-cost "artificial enzymes", have been synthesized using functional nanomaterials. **Figure 6.5.2** demonstrates the classification of nanozymes based on the type of enzymatic reaction.

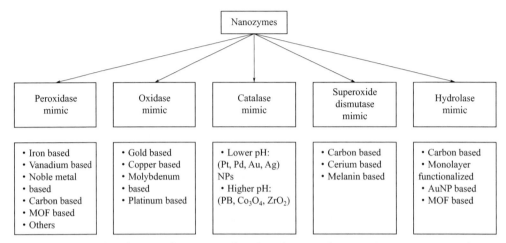

Figure 6.5.2 Classification of nanozymes based on the type of enzymatic reaction: peroxidase, oxidase, catalase, superoxide dismutase, and hydrolase mimic. Each of these types is further categorized depending on the nanomaterials used to replicate the enzymatic functions.

Composition of nanozymes

Nanozymes are composed of nanoparticles of inorganic materials possessing intrinsic enzymatic activities. Nanozymes have been developed with nanoparticle forms of Fe_3O_4, platinum, gold, copper oxide, $CoFe_2O_4$, graphene oxide (GO), carbon, among others. Besides classifying the nanozymes based on the type of enzymatic reactions they mimic, they may also be categorized based on their atomic composition, which largely determine their catalytic activity and specificity in operation. Based on the atomic composition of their core and surface, nanozymes may be distributed into three different categories — metal oxide-based, metal-based, and carbon-based nanozymes. A variety of metal oxide-based nanozymes have been reported to mimic catalytic activities of peroxidase, catalase, and superoxide dismutase (SOD). Since metal oxide nanoparticles are usually chemically and biologically inert, additional surface engineering and conjugation are required to endow these structures with desired functionalities. Metal oxide nanozymes have found wide application in fabricating amperometric and colorimetric detection systems.

Metal-based nanozymes may be further classified as Type Ⅰ and Type Ⅱ, depending on the part of their structure participating in catalysis. The activities of Type Ⅰ nanozymes are entirely from the assembled monolayer onto a metallic core, while for Type Ⅱ, the nanozymes' activities are largely due to the metallic part. Metal based nanozymes have been reported to possess catalytic activities of SOD, peroxidase, and catalase. Composite structures of metal and metal oxide nanozymes have also been found to function as a distinct class of nanozyme having enhanced catalytic activity.

Carbon-based nanozymes are usually found to behave as peroxidase and SOD-mimics, because their unique electrical, optical, thermal, and mechanical properties can offer a variety of multifunctional platforms for biomedical applications. Material selection for nanozymes may also be dependent on their envisaged functionalities or specific application requirements. TiO_2 nanoparticle is a commonly used nanozyme for photocatalysis and environmental remediation. Graphite carbon nitride ($g-C_3N_4$), GO, black phosphorus (BP), carbon quantum dots, metal-organic frameworks (MOFs), and thin chromium thiophosphate ($CrPS_4$) are nanozymes that exhibit remarkable photocatalytic properties. Magnetic nanozymes, such as mussel-inspired nanoflower are considered as antibacterial agents for biosensing applications. Another group reported laccase-like active magnetic nanozyme which can be used for efficient removal of toxic phenolic pollutant like o-phenylenediamine (OPD). Nanozymes can also be endowed with multifunctional capability. Recently, experimentally derived multifunctional nanozyme based on MOFs, which contains catalytic Cu^{2+} and luminescent Tb^{3+} ions, are also reported. This nanozyme not only possessed excellent catalytic activity comparable to horseradish peroxidase but also could indicate the real-time concentration of H_2O_2 through its fluorescence. Recently, biocompatible nanozymes were developed for *in vivo* sensing and drug-delivery applications. Notably, among them was citrate-capped palladium nanozyme that exhibited efficient antioxidant properties within the cellular environment.

Noble metal nanoparticles such as Au, Pd, and Pt demonstrate remarkable catalytic activities and intrinsic antioxidant properties. Functionalizing these nanomaterials with various ligands can greatly improve their biocompatibility and broaden their biomedical application potentials. To this end, efforts have been made to adjust the size and morphology of Pt nanoparticles with a protective layer of sodium alginate to enhance their dispersibility, stability, and possibility of infusing additional surface features. Studies have also been demonstrated with citrate-coated Pd nanoparticles, which displayed enhanced biocompatibility and stability, can be engaged as reactive oxygen species scavengers. Another group used BSA-stabilized Pt nanozyme in developing a rapid, sensitive, and selective Hg^{2+} sensing system. Poly (amidoamine) (PAMAM) encapsulated Pt nanoparticles were used as efficient catalase mimic. Pt nanoparticles anchored MOFs were further shown to display enhanced electrocatalytic property in detecting the activity of telomerase, a ribonucleoprotein. Therefore, the unique physicochemical properties of metal nanoparticles can greatly be enhanced by functionalizing their surfaces with different ligands and functional moieties. This further facilitates development of multifunctional nanocarriers and sensing platforms, where the metal nanoparticles, besides their inherent enzymatic activity can be further endowed with magnetic, fluorescent, and self-propulsion capabilities. With appropriate conjugation and surface charge, hydrodynamic radii of nanozymes may also be optimized for their complete elimination from the body after use and minimum nonspecific tissue/organ uptake during in vivo operations.

Comparison and outlook

Natural enzymes are characterized by a high degree of selectivity and catalytic activity. But, they are often expensive and have limited chemical and biological stability. Functional materials like nanozymes have enzyme-like activity, and possess greater stability and application potentials due to their simpler preparation technologies. Nanozymes offer size and composition-dependent activity, which facilitates design of materials with a broad range of catalytic activity by varying their shape, structure, and composition. Natural enzymes are mostly proteins and can be considered as soft materials, whereas nanozymes are hard with porphyritic nuclei. Compared to natural enzymes, nanozymes have large surface areas, which facilitate their modification and conjugation with other functional groups. The catalytic efficiencies of nanozymes in most cases are lower than that of natural enzymes.

Recently developed single atom nanozymes (SAzymes), owing to their well-defined electronic and geometric structures, have been reported to have maximum atom efficiency, unique quantum size effect, and excellent substrate selectivity. SAzymes could potentially make nanozyme better alternatives to nature enzymes — accomplishing specific enzyme-based applications easily with higher therapeutic efficiency and biosafety. Nanozymes or synthetic catalytic nanomaterials with enzyme-like functions (e. g., peroxidase, catalase, SOD, and oxidase) pave the way toward integration of nano-/micromotors, catalytic reactions, and biological systems.

(**Selected from:** Mujtaba J, Liu J, Dey K K, et al. Micro-Bio-Chemo-Mechanical-Systems: Micromotors, Microfluidics, and Nanozymes for Biomedical Applications [J]. Advanced Materials, 2021, 33 (22), 2007465.)

New Words and Expressions

nanozyme   *n*. 纳米酶
on some occasions   在一些场合
catalytic   *n*. 催化剂；*adj*. 起催化作用的
graphene oxide   氧化石墨烯
atomic   *adj*. 原子的
peroxidase   *n*. 过氧化物酶
amperometric   *adj*. 电流计的
colorimetric   *adj*. 比色的
catalase   *n*. 过氧化氢酶
black phosphorus   黑磷
photocatalysis   *n*. 光催化
antibacterial   *adj*. 抗菌
biosensing   *n*. 生物传感
horseradish   *n*. 芳香胺

fluorescence *n.* 荧光
antioxidant *n.* 抗氧化剂
telomerase *n.* 端粒酶
ligand *n.* 配体
self-propulsion *n.* 自推进
hydrodynamic *n.* 流体动力学的
therapeutic *adj.* 有疗效的，疗法的
owing to 由于

Notes

（1）Nanozymes are catalytic nanomaterials with enzyme-like properties, tunable size, shape, structure, composition, surface area, and selective catalytic reactions responsive to temperature and external stimuli.

——参考译文：纳米酶是一种催化纳米材料，具有类似酶的特性、可调节的尺寸、形状、结构、组成、表面积以及对温度和外部刺激做出响应的选择性催化反应（的能力）。

（2）Besides classifying the nanozymes based on the type of enzymatic reactions they mimic, they may also be categorized based on their atomic composition, which largely determine their catalytic activity and specificity in operation.

——这是一个较长的复合句。最后部分是 which 引导的定语从句。被动语态在翻译时可相应调整。

——参考译文：除了根据它们模拟的酶反应类型对纳米酶进行分类外，还可以根据它们的原子成分对它们进行分类，这在很大程度上决定了它们的催化活性和操作特异性。

（3）The activities of Type Ⅰ nanozymes are entirely from the assembled monolayer onto a metallic core, while for Type Ⅱ, the nanozymes' activities are largely due to the metallic part.

——while 引导的并列句，强调前后句的对比。

——参考译文：第Ⅰ类型的纳米酶的活性完全来自金属核外面的组装单分子层，而对于第Ⅱ类型，纳米酶的活性主要来自金属部分。

（4）Another group reported laccase-like active magnetic nanozyme which can be used for efficient removal of toxic phenolic pollutant like o-phenylenediamine（OPD）.

——参考译文：另一（研究）组报道了类似漆酶的活性磁性纳米酶，它可用于有效去除邻苯二胺（OPD）等有毒酚类污染物。

（5）Therefore, the unique physicochemical properties of metal nanoparticles can greatly be enhanced by functionalizing their surfaces with different ligands and functional moieties.

——参考译文：因此，用不同的配体和功能组件对其表面进行功能化，可以大大增强金属纳米颗粒的独特物理化学性质。

（6）This further facilitates development of multifunctional nanocarriers and sensing platforms, where the metal nanoparticles, besides their inherent enzymatic activity can be further endowed with magnetic, fluorescent, and self-propulsion capabilities.

——句子结构较复杂。where 引导的是定语从句。

——参考译文：这进一步促进了多功能纳米载体和传感平台的发展，其中金属纳米颗粒除了其固有的酶活性外，还可以被进一步赋予磁性、荧光和自我推进能力。

（7）Recently developed single-atom nanozymes (SAzymes), owing to their well-defined electronic and geometric structures, have been reported to have maximum atom efficiency, unique quantum size effect, and excellent substrate selectivity.

——参考译文：据报道，由于具有已知的电子和几何结构，最近开发的单原子纳米酶（SAzymes）具有最大的原子效率、独特的量子尺寸效应和优异的衬底选择性。

Chapter 7
Material Characterization Methods

【本章导读】
　　本章介绍一些常用的材料表征方法，包括四个单元，共有四篇课文和三篇阅读材料。四篇课文分别介绍光谱测试的基础知识、原位谱学、显微学基础知识、现代原位表征技术。三篇阅读材料分别讲解 X 射线吸收谱、显微镜像差、超快透射电镜及电子衍射技术。

Unit 7.1

Text

Introduction to spectroscopic methods

　　这篇课文主要介绍了光谱测试分析技术的种类及其在材料研究领域的应用。谱学分析技术是材料科学领域的基本表征工具，可以在原子、分子和宏观尺度上研究材料的成分、晶体结构和电子结构。课文从光谱学的背景到原理，从其种类到应用都进行了详细的阐述。

What is spectroscopy?

　　Spectroscopy is the study of the interaction between matter and electromagnetic radiation as a function of the wavelength or frequency of the radiation. More recently, the definition has been expanded to include the study of the interactions between particles such as electrons, protons, and ions, as well as their interaction with other particles as a function of their collision energy. Historically, spectroscopy originated as the study of the wavelength dependence of the absorption by gas phase matter of visible light dispersed by a prism. Spectroscopy in the electromagnetic spectrum is a fundamental characterization tool in the fields of materials science, allowing the composition, crystal structure and electronic structure of materials to be investigated at the atomic, molecular and macro scale.

Theory of spectroscopy

　　The theory of spectroscopy is rooted in quantum mechanics. The resonance is a coupling of two quantum mechanical stationary states of one system, such as an atom, via an oscillatory source of energy such as a photon. The coupling of the two states is strongest when the energy of the source matches the energy difference between the two states. The energy E of a photon is related to its frequency ν by $E=h\nu$ where h is Planck's constant, and so a spectrum of the system response vs. photon frequency will peak at the resonant frequency or energy. Particles such as electrons and neutrons have a comparable relationship between their kinetic energy and their wavelength can also excite resonant interactions.

One of the central concepts in spectroscopy is a resonance and its corresponding resonant frequency. Resonances were first characterized in mechanical systems such as pendulums. Mechanical systems that vibrate or oscillate will experience large amplitude oscillations when they are driven at their resonant frequency. A plot of amplitude vs. excitation frequency will have a peak centered at the resonance frequency. This plot is one type of spectrum, with the peak often referred to as a spectral line, and most spectral lines have a similar appearance.

Spectra of atoms and molecules often consist of a series of spectral lines, each one representing a resonance between two different quantum states. The explanation of these series, and the spectral patterns associated with them, were one of the experimental enigmas that drove the development and acceptance of quantum mechanics. The hydrogen spectral series in particular was first successfully explained by the Rutherford-Bohr quantum model of the hydrogen atom. In some cases, spectral lines are well separated and distinguishable, but spectral lines can also overlap and appear to be a single transition if the density of energy states is high enough. Named series of lines include the principal, sharp, diffuse and fundamental series. By comparing spectroscopic measurements to quantum mechanical calculations based on an assumed model of the material, one can use knowledge of a material's electronic structure to determine its physical structure.

Classification of spectroscopic methods

Spectroscopy is a sufficiently broad field that many sub-disciplines exist, each with numerous implementations of specific spectroscopic techniques. Electromagnetic radiation was the first source of energy used for spectroscopic studies. Techniques that employ electromagnetic radiation are typically classified by the wavelength region of the spectrum and include radio wave, microwave, infrared, near-infrared, ultraviolet-visible, X-ray, and gamma spectroscopy. The types of electromagnetic radiation and their wavelength and frequency ranges are listed in **Table 7. 1. 1**. Particles, because of their de Broglie waves, can also be a source of radiative energy. Both electron and neutron spectroscopy are commonly used. For a particle, its kinetic energy determines its wavelength.

Table 7. 1. 1 **The types of electromagnetic radiation and their wavelength and frequency ranges.**

Electromagnetic radiation	Wavelength range (m)	Frequency range (Hz)
radio waves	$10-1000$	$3 \times 10^5 - 3 \times 10^7$
television waves	$1-10$	$3 \times 10^7 - 3 \times 10^8$
microwaves	$1 \times 10^{-3} - 1$	$3 \times 10^8 - 3 \times 10^{11}$
infrared	$8 \times 10^{-7} - 1 \times 10^{-3}$	$3 \times 10^{11} - 4 \times 10^{14}$
visible light	$4 \times 10^{-7} - 7 \times 10^{-7}$	$4 \times 10^{14} - 7 \times 10^{14}$
ultraviolet	$1 \times 10^{-8} - 4 \times 10^{-7}$	$7 \times 10^{14} - 3 \times 10^{16}$
X-rays	$5 \times 10^{-12} - 1 \times 10^{-8}$	$3 \times 10^{16} - 6 \times 10^{19}$
gamma rays	$< 5 \times 10^{-12}$	$> 6 \times 10^{19}$

The types of spectroscopy can be distinguished by the nature of the interaction between the energy and the material. Commonly used spectroscopic methods in materials science are as follows:

X-ray absorption spectroscopy (XAS) is a widely used spectroscopic technique for determining the local geometric and electronic structure of matter. The experiment is usually performed at synchrotron radiation facilities, which provide intense and tunable X-ray beams. XAS can be applied not only to crystals, but also to materials that possess little or no long-range translational order: amorphous systems, glasses, quasicrystals, disordered films, membranes, solutions, liquids, proteins, even molecular gases. This versatility allows it to be used in a wide variety of disciplines. Absorption occurs when energy from the radiative source is absorbed by the material. Absorption coefficient is often determined by measuring the fraction of energy transmitted through the material, with absorption decreasing the transmitted portion. In general, a core electron is excited to an unfilled valence state. This state can then relax via emission of a fluorescence photon, or may be radiationless, leading to the ejection of photoelectrons, Auger electrons and low-energy secondary electrons. As such, XAS probes the unoccupied density of states of the system. By using XAS technique, rich information inside the material can be obtained, including valence state, site symmetry, bond length, bond strength and coordination number.

X-ray diffraction (XRD) is a scattering technique determining the atomic and molecular structure of a crystal, in which the crystalline structure causes a beam of incident X-rays to diffract into many specific directions. By measuring the angles and intensities of these diffracted beams, a crystallographer can produce a three-dimensional picture of the density of electrons within the crystal. From this electron density, the mean positions of the atoms in the crystal can be determined, as well as lattice parameter, chemical bonds, crystallographic disorder, and various other information. Since many materials can form crystals—such as salts, metals, minerals, semiconductors, as well as various inorganic, organic, and biological molecules. XRD has been fundamental in the development of many scientific fields. The method also revealed the structure and function of many biological molecules, including vitamins, drugs, proteins and nucleic acids such as DNA.

Fluorescence spectroscopy is a type of electromagnetic spectroscopy that analyzes fluorescence from a sample. It involves using a beam of X-ray or ultraviolet light, that excites the electrons in molecules of certain materials and causes them to emit fluorescence light. Fluorescence spectroscopy is primarily concerned with electronic and vibrational states. Generally, the species being examined has a ground electronic state (a low energy state) of interest, and an excited electronic state of higher energy. Within each of these electronic states there are various vibrational states. In the field of water research, fluorescence spectroscopy can be used to monitor water quality by detecting organic pollutants. In the special case of single molecule fluorescence spectroscopy, intensity fluctuations from the emitted light are measured from either single fluorophores, or pairs of fluorophores. The main advantage of fluorescence spectroscopy is its high sensitivity. The amount of precious samples used for a measurement can be kept very low because traces of fluorescent species can be detected quantitatively. The sensitivity of fluorescence spectroscopy would be 10 to 100 times as much as that of absorption spectroscopy. This high sensitivity makes fluorescence one of the best available methods for trace analysis.

X-ray photoelectron spectroscopy (XPS) is a surface-sensitive quantitative spectroscopic technique based on the photoelectric effect that can identify the elements that exist within a material or are covering its surface, as well as their chemical state, and the overall electronic structure and density of the electronic states in the material. It involves the collection and measurement of the energies of directly ejected electrons, yielding information about the core- or valence state from which the electrons originated. It is a powerful measurement technique because it not only shows what elements are present, but also what other elements they are bonded to. The technique can be used in line profiling of the elemental composition across the surface, or in depth profiling when paired with ion-beam etching. It is often applied to study chemical processes in the materials in their as-received state or after cleavage, scraping, exposure to heat, reactive gasses or solutions, ultraviolet light, or during ion implantation. XPS belongs to the family of photoemission spectroscopies in which electron population spectra are obtained by irradiating a material with a beam of X-rays. Chemical states are inferred from the measurement of the kinetic energy and the number of the ejected electrons. XPS requires high vacuum (residual gas pressure $P \approx 10^{-6}$ Pa) or ultra-high vacuum ($P < 10^{-7}$ Pa) conditions, although a current area of development is ambient-pressure XPS, in which samples are analyzed at pressures of a few tens of millibar. XPS easily detects all elements except hydrogen and helium. The detection limit is in the parts per thousand range, but parts per million (ppm) are achievable with long collection times and concentration at top surface.

Raman spectroscopy is a spectroscopic technique typically used to determine vibrational modes of molecules, although rotational and other low-frequency modes of systems may also be observed. Raman spectroscopy is commonly used in chemistry to provide a structural fingerprint by which molecules can be identified. Raman spectroscopy relies upon inelastic scattering of photons, known as Raman scattering. A source of monochromatic light, usually from a laser in the visible, near infrared, or near ultraviolet range is used, although X-rays can also be used. The incident light interacts with molecular vibrations, phonons or other excitations in the system, resulting in the energy of the photons being shifted up or down. The shift in energy gives information about the vibrational modes in the system.

Resonant inelastic X-ray scattering (RIXS) is the X-ray spectroscopy technique used to investigate the electronic structure of molecules and materials. RIXS is a fast developing experimental technique in which one scatters high energy X-ray photons inelastically off matter. It is a photon-in/photon-out spectroscopy where one measures both the energy and momentum change of the scattered photon. The energy and momentum lost by the photon are transferred to intrinsic excitations of the material under study and thus RIXS provides information about those excitations. The RIXS event can be thought of as a two-step process. Starting from the initial state, absorption of an incident photon leads to creation of an excited intermediate state, that has a core hole. From this state, emission of a photon leads to the final state. In a simplified picture, the absorption process gives information of the empty electronic states, while the emission gives information about the occupied states. Compared to other scattering techniques, RIXS has a number of unique features: it covers a large scattering phase-space, is polarization dependent, element and orbital specific, bulk sensitive and requires only small sample volumes.

Nuclear magnetic resonance (NMR) spectroscopy is a spectroscopic technique to observe local magnetic fields around atomic nuclei. The sample is placed in a magnetic field and the NMR signal is produced by excitation of the nuclei sample with radio waves into nuclear magnetic resonance, which is detected with sensitive radio receivers. The intramolecular magnetic field around an atom in a molecule changes the resonance frequency, thus giving access to details of the electronic structure of a molecule and its individual functional groups. As the fields are unique or highly characteristic to individual compounds, in modern organic chemistry practice, NMR spectroscopy is the definitive method to identify monomolecular organic compounds. NMR spectra are unique, well-resolved, analytically tractable and often highly predictable for small molecules. Different functional groups are obviously distinguishable, and identical functional groups with differing neighboring substituents still give distinguishable signals. NMR has largely replaced traditional wet chemistry tests such as color reagents or typical chromatography for identification. A disadvantage is that a relatively large amount, 2 — 50 mg, of a purified substance is required. Preferably, the sample should be dissolved in a solvent, because NMR analysis of solids requires a dedicated magic angle spinning machine and may not give equally well-resolved spectra. The timescale of NMR is relatively long, thus it is not suitable for observing fast phenomena, producing only an averaged spectrum.

Infrared (IR) spectroscopy is the measurement of the interaction of infrared radiation with matter by absorption, emission, or reflection. It is used to study and identify chemical substances or functional groups in solid, liquid, or gaseous forms. It can be used to characterize new materials or identify and verify known and unknown samples. The method or technique of infrared spectroscopy is conducted with an instrument called an infrared spectrometer (or spectrophotometer) which produces an infrared spectrum. An IR spectrum can be visualized in a graph of infrared light absorbance (or transmittance) on the vertical axis vs. frequency or wavelength on the horizontal axis. A common laboratory instrument that uses this technique is a Fourier transform infrared (FTIR) spectrometer. The infrared portion of the electromagnetic spectrum is usually divided into three regions; the near-, mid- and far- infrared, named for their relation to the visible spectrum. The higher energy near-IR, approximately $14000 - 4000$ cm^{-1} ($0.8 - 2.5$ μm wavelength) can excite overtone or harmonic vibrations. The mid-infrared, approximately $4000 - 400$ cm^{-1} ($2.5 - 25$ μm) may be used to study the fundamental vibrations and associated rotational-vibrational structure. The far-infrared, approximately $400 - 10$ cm^{-1} ($25 - 1000$ μm), lying adjacent to the microwave region, has low energy and may be used for rotational spectroscopy. The names and classifications of these subregions are conventions, and are only loosely based on the relative molecular or electromagnetic properties.

In summary, spectroscopy techniques are the "eye" of materials scientists. They allow people to look deeply into the materials in atomic scale, including electron configurations of ground and various excited states. Spectroscopy also provides a precise analytical method for finding the constituents in material having unknown chemical composition, and analyzing trace elements with a concentration of a few parts per million in a material.

(**Selected from:** Chu S, Stoner J O, Graybeal J D, et al. Encyclopedia Britannica. Spectroscopy. (2023-06-13). [2023-07-29]. [M]. https://www.britannica.com/science/spectroscopy.)

New Words and Expressions

  resonance *n*. 共振，共鸣，反响
  prism *n*. 棱镜，棱柱
  quantum *n*. 量子，量子论
  coupling *n*. 耦合，结合，联结
  oscillatory *adj*. 振荡的，振动的
  pendulum *n*. 钟摆，摆锤
  ultraviolet *n*. 紫外光，紫外辐射
  quasicrystal *n*. 准晶体
  versatility *n*. 多功能性，用途广泛
  coefficient *n*. 系数
  fluorescence *n*. 荧光，荧光性
  fluctuation *n*. 波动，变动
  fluorophore *n*. 荧光团，荧光分子
  cleavage *n*. 解理，分裂
  inelastic *adj*. 无弹性的，非弹性的
  intramolecular *adj*. 分子内的
  chromatography *n*. 色谱分析法，色谱法
  electromagnetic radiation 电磁辐射，电磁波
  synchrotron radiation 同步辐射
  spectral lines 谱线
  Auger electron 俄歇电子
  density of states 态密度
  coordination number 配位数
  ground electronic state 电子基态
  nuclear magnetic resonance 核磁共振
  trace analysis 痕量分析
  photoelectric effect 光电效应
  vibrational mode 振动模式

Notes

(1) More recently, the definition has been expanded to include the study of the interactions between particles such as electrons, protons, and ions, as well as their interaction with other particles as a function of their collision energy.

——参考译文：最近，该定义已扩大到包括粒子（例如电子、质子和离子）之间的相互作用，以及它们与其他粒子的相互作用随碰撞能量变化的研究。

(2) Spectroscopy in the electromagnetic spectrum is a fundamental characterization tool in the fields of materials science, allowing the composition, crystal structure and electronic structure of materials to be investigated at the atomic, molecular and macro scale.

——参考译文：基于电磁波的谱学技术是材料科学领域的基本表征工具，可以在原子、分子和宏观尺度上研究材料的成分、晶体结构和电子结构。

(3) The theory of spectroscopy is rooted in quantum mechanics.

——be rooted in … 扎根于，植根于……。
——参考译文：光谱学理论植根于量子力学。

(4) The coupling of the two states is strongest when the energy of the source matches the energy difference between the two states.

——参考译文：当（振荡）源的能量与两种（量子）态之间的能量差匹配时，两种（量子）态的耦合最强。

(5) Mechanical systems that vibrate or oscillate will experience large amplitude oscillations when they are driven at their resonant frequency.

——that 引导的定语从句修饰 mechanical systems。
——参考译文：当以共振频率驱动振动或振荡的机械系统时，机械系统会经历大幅度的振荡。

(6) As such, XAS probes the unoccupied density of states of the system.

——as such 因此。
——参考译文：因此，X射线吸收谱（XAS）探测的是体系的未占据态密度。

(7) One of the central concepts in spectroscopy is a resonance and its corresponding resonant frequency.

——参考译文：光谱学的核心概念之一是共振及其相应的共振频率。

Exercises

1. Question for discussion

(1) What information of a material can be obtained by X-ray absorption spectroscopy (XAS)?

(2) Describe the principle of resonant inelastic X-ray scattering (RIXS).

(3) Please list the commonly used spectroscopic methods in materials science.

(4) Why is the fluorescence spectroscopy good at trace analysis?

2. Translate the following into Chinese

(1) Spectroscopy is a sufficiently broad field that many sub-disciplines exist, each with numerous implementations of specific spectroscopic techniques.

(2) The sample is placed in a magnetic field and the NMR signal is produced by excitation of the nuclei sample with radio waves into nuclear magnetic resonance, which is detected with sensitive radio receivers.

(3) An IR spectrum can be visualized in a graph of infrared light absorbance (or transmittance) on the vertical axis vs. frequency or wavelength on the horizontal axis.

3. Translate the following into English

(1) X 射线吸收光谱　　　　　(2) X 射线衍射

(3) 荧光光谱学　　　　　　　(4) X 射线光电子能谱

(5) 拉曼光谱　　　　　　　　(6) 共振非弹性 X 射线散射

(7) 浓度

(8) 当来自辐射源的能量被材料吸收时，就会发生吸收现象。吸收系数通常通过测量能够穿过样品的射线强度占入射光强度的比例来确定。

(9) 光谱学还提供了一种精确的分析方法，用于解析未知材料的组分，并分析材料中含量为 ppm 量级的微量元素。

(10) 红外（IR）光谱是测量红外线与物质吸收、发射或反射的相互作用。它用于研究和识别固态、液态或气态形式的化学物质或官能团。

Reading Material
X-ray absorption spectroscopy

这篇课文介绍了 X 射线吸收谱（XAS）的工作原理及应用。XAS 技术是一种可用于特定元素的检测和分析的谱学表征技术。XAS 谱可以分为两部分：XANES 和 EXAFS。课文讲解了它们能获得的材料内部信息，并介绍了相关技术在电池材料研究中的应用。

X-ray absorption spectroscopy (XAS) is a powerful tool that provides information on a very local scale (4 — 5Å) around a selected atomic species and is well suited for the characterization of not only crystals but also materials that possess little or no long-range translational order. It is based on the absorption: when a sample is exposed to X-rays, it will absorb part of the incoming photon beams, which is mainly generated by the photoelectric effect for energy in the hard X-ray region (3—50 keV). XAS is even selective for the atomic species and also allows us to tune the X-ray beam selectively to a specific atomic core (the absorption energy of next elements are sufficiently spaced), and therefore it probes the local structure around only the selected element that are contained within a material. The element-specific characteristic of XAS, providing both chemical and structural information at the same time, differentiates it from other techniques, such as the X-ray scattering. In this respect, it serves as a unique tool for the investigation of battery materials during charge-discharge cycles.

XAS experiment measures the absorption coefficient μ as a function of energy E: as E increases, μ generally decreases ($\mu - E^{-3}$), that is matter becomes more transparent and X-rays more penetrating, save for some discontinuities, where μ rapidly rises up. These exceptions correspond to particular energies, the so-called absorption edges E_0, which are the characteristic of the material, where the amount of energy exactly matches the core electron binding energy. The edge energies vary with atomic number approximately as a function of Z^2 and both K and L levels can be used in the hard X-ray region (in addition, M edges can be used for heavy elements in the soft X-ray region), which allows most elements to be probed by XAS with X-ray energies between 4 and 35 keV. Because the element of interest is chosen in the experiment, XAFS is element-specific.

Generally, the XAS spectrum can be divided into two parts: X-ray Absorption Near Edge Structure (XANES) and Extended X-ray Absorption Fine Structure (EXAFS)(**Figure 7.1.1**). The XANES part roughly covers the pre-edge and near edge regions from $\sim$30 eV below the edge to 40—50 eV above the edge. In this region, core electrons are excited to bound or quasi-bound states. For bound states, the core electrons are excited to un-occupied orbitals and remain in the vicinity of the absorbing atom. Due to the relatively long life time, peaks are sharp in this region (uncertainty principle). As the energy of incoming photons increases, core electrons can be excited to the quasi-bound states. These correspond to states that are trapped in a potential barrier and the electrons will eventually tunnel out of the barrier into the continuum. Multiple scattering events dominate the scattering process because of the low kinetic energy these excited electrons carry. Peaks are more broad due to shorter core hole life time. In the EXAFS region which is roughly from 50 to 1000 eV, the excited electrons gain sufficient kinetic energy to escape into the continuum. Single scattering pathways dominate because of the high kinetic energy of excited electrons. The XANES spectra reports electronic structure and symmetry of the metal site, and the EXAFS reports numbers, types, and distances to ligands and neighboring atoms from the absorbing element.

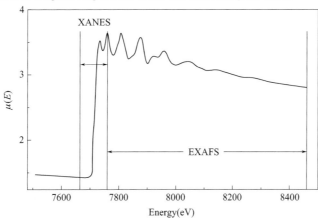

Figure 7.1.1 XANES and EXAFS of a typical Co K-edge XAS of metallic Co.

One of the typical application areas for XAS is battery materials. The main important characteristics of XAS used for the research of battery materials are: (i) its element specificity, which allows the study of a particular element by concentrating on its K (or in some cases L) absorption edge; (ii) the possibility of tuning it to different sites (for instance Fe and P in $LiFePO_4$), thus providing sources of complementary information on the same compound; (iii) the physicochemical information contained in the near-edge structure of the XAS signals, which can be used to reveal the formal oxidation state and the local symmetry of the probed atom; (iv) the possibility of doing operando measurements by collecting XAS spectra during electrochemical cycling using specifically developed *in situ* cells. In this case, the physicochemical properties and the local structure of the studied element can be monitored at all moments during the charge and discharge processes.

(**Selected from:** Mehdi Khodaei, Luca Petaccia. X-ray Characterization of Nanostructured Energy Materials by Synchrotron Radiation [M/CD]. IntechOpen. 2017.)

New Words and Expressions

absorption *n*. 吸收

expose *v*. 暴露

photon beam 光子束

region *n*. 区域，范围

differentiate *v*. 区分，表明……间的差别，构成……间差别的特征

investigation *n*. 调查，研究

discontinuity *n*. 间断，不连贯，不连续

pre-edge *n*. 预边，边前（能量靠近但低于吸收边的部分）

quasi-bound *n*. 准束缚态

trap *v*. 使陷于

dominate *v*. 支配，在……中具有最重要（或明显）的特色

escape *v*. 逃跑

continuum *n*. 连续态

symmetry *n*. 对称性

ligand *n*. 配体

physicochemical *adj*. 物理化学的

excite *v*. 激发，使激动，使兴奋

element *n*. 元素，要素

discharge *v*. 放电

Notes

(1) X-ray absorption spectroscopy (XAS) is a powerful tool that provides information on a very local scale (4 − 5 Å) around a selected atomic species and is well suited for the characterization of not only crystals but also materials that possess little or no long-range

translational order.

—参考译文：X 射线吸收谱（XAS）是一种强大的工具，它能提供围绕选定原子周围非常局部尺度（4~5 Å）上的信息，不仅适用于晶体的表征，也适用于具有很少或没有长程平移序的材料的表征。

(2) In this respect, it serves as a unique tool for the investigation of battery materials during charge-discharge cycles.

—参考译文：就这点而言，它是一种能够在充放电循环过程中研究电池材料的独特工具。

(3) Generally, the XAS spectrum can be divided into two parts: X-ray Absorption Near Edge Structure (XANES) and Extended X-ray Absorption Fine Structure (EXAFS).

—参考译文：一般来说，X 射线吸收谱可分为两部分，X 射线吸收近边结构（XANES）和扩展 X 射线吸收精细结构（EXAFS）。

(4) In this case, the physico-chemical properties and the local structure of the studied element can be monitored at all moments during the charge and discharge processes.

—参考译文：在这种情况下，可以在充电和放电过程中的任何时刻监测所研究元素的物理化学性质和局域结构。

(5) Due to the relatively long life time, peaks are sharp in this region (uncertainty principle).

—参考译文：由于相对较长的寿命，该区域的峰很尖锐（测不准原理）。

Unit 7.2

Text

Operando spectroscopy

课文阐述了原位谱学技术的原理、应用、局限性和使用方法，并以其在电化学领域的应用为例，讲解了原位谱学技术具有的独特优势：这类技术可以精细地表征电极材料在电化学反应全过程的演化，帮助推断和理解电池的电化学性能和储能机制。

Operando (in situ) spectroscopy

Operando spectroscopy is an analytical methodology wherein the spectroscopic characterization of materials undergoing reaction is coupled simultaneously with measurement of other performance, including catalytic activity, electrochemical performance, heating process or pressure process. The primary concern of this methodology is to establish structure-reactivity/selectivity relationships and thereby yield information about mechanisms.

Operando spectroscopy is a class of methodology, rather than a specific spectroscopic technique. It can be used for various spectroscopic techniques, including X-ray diffraction (XRD), X-ray absorption spectroscopy (XAS), Raman spectroscopy, Infrared (IR) spectroscopy, Nuclear magnetic resonance (NMR) spectroscopy etc. Operando spectroscopy is a logical technological progression in *in situ* studies. Catalyst scientists would ideally like to have a "motion picture" of each catalytic cycle, whereby the precise bond-making or bond-breaking events taking place at the active site are known. This would allow a visual model of the mechanism to be constructed. The ultimate goal is to determine the structure-activity relationship. Having two experiments—the performing of a reaction plus the real-time spectral acquisition of the reaction mixture—on a single reaction facilitates a direct link between the structures of the catalyst and intermediates, and of the catalytic activity/selectivity. Although monitoring a catalytic process *in situ* can provide information relevant to catalytic function, it is difficult to establish a perfect correlation because of the current physical limitations of *in situ* reactor cells. Complications arise, for example, for gas phase reactions which require large void volumes, which make it difficult to homogenize heat and mass within the cell. The crux of a successful operando methodology, therefore, is related to the disparity between laboratory setups and industrial setups, i. e., the limitations of properly simulating the catalytic system as it proceeds in industry.

The purpose of operando spectroscopy in catalyst study is to measure the catalytic changes that occur within the reactor during operation using time-resolved (and sometimes spatially-resolved) spectroscopy. Time-resolved spectroscopy theoretically monitor the formation and disappearance of intermediate species at the active site of the catalyst as bond are made and broken in real time. However, current operando instrumentation often only works in the second time scale and therefore, only relative concentrations of intermediates can be assessed. Spatially resolved spectroscopy combines spectroscopy with microscopy to determine active sites of the catalyst studied and spectator species present in the reaction.

Operando spectroscopy requires measurement of the materials under (ideally) real working conditions, but with a spectroscopic device inserted into the reaction vessel. The parameters of the reaction are then measured continuously during the reaction using the appropriate instrumentation, i. e., electrochemistry stations or battery cyclers (**Figure 7.2.1**). Operando instruments (*in situ* cells) must ideally allow for spectroscopic measurement under optimal reaction conditions. For catalyst study, most industrial catalysis reactions require excessive pressure and temperature conditions which subsequently degrades the quality of the spectra by lowering the resolution of signals. Currently many complications of this technique arise due to the reaction parameters and the cell design. The catalyst may interact with the components of the operando apparatus; open space in the cell can have an effect on the absorption spectra, and the presence of spectator species in the reaction may complicate analysis of the spectra. Continuing development of operando reaction-cell design is in line with working towards minimizing the need for compromise between optimal conditions and spectroscopy. Other requirements considered when designing operando experiments include reagent and product flow rates, catalyst position, beam paths, and window positions and sizes. All of these factors must also be accounted for while designing operando experiments, as the spectroscopic techniques used may alter the reaction conditions.

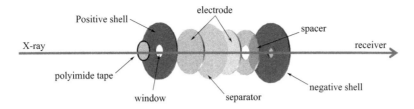

Figure 7.2.1 *In situ* cells for XRD and XAS measurements of battery materials.

Operando spectroscopy plays an exceptional role in battery materials research. Real-time physical and chemical properties of the electrodes and interface, especially their changes during the operation of the batteries, are quite important for the full picture of the electrochemical process. In recent years, operando characterization during battery cycling has been developed very quickly, enabling the capability to capture the real-time information successfully. Compared with the *ex situ* spectroscopic techniques, the operando spectroscopy techniques can deliver more reliable and accurate information by continuously monitoring evolution of the electrode structure and the complicated physical and chemical processes throughout the electrochemical cycling. In addition, it can mimic real battery-operation conditions, and thus eliminate the adverse effects from the external factors. The operando spectroscopy can avoid the contamination or relaxation of samples which might be introduced by *ex situ* sample preparations. (especially, considering the electrode sensitivity to air and moisture). Based on these advantages, operando spectroscopy can provide more in-depth understanding of working mechanisms of a cell, performance degradation process, and origins and so on. **Figure 7.2.2** shows the *in situ* XRD patterns of Mn-based cathode material for sodium-ion batteries in the first cycle. During charge, a new (004) peak at 14.1° appears beside the original (004) peak at the very beginning, suggesting that a two-phase reaction occurs. Along with charging from the open-circuit voltage to 4.2 V, the new peak keeps shifting to lower angle, indicating the interlayer distance (correlated to lattice parameter c) is increasing along with a solid-solution behavior. On the contrary, the (100) peak at 15.8° shifts to higher two-theta angle, indicating the lattice parameters a and b are decreasing during charge. From 4.2 to 4.4 V, the peak shift is not obvious compared with the low-potential region and no new phase can be observed after charging up to 4.4 V. The diffraction peaks at the fully charged state can be assigned to a $P''2$ phase similar to the initial P2 phase. $P''2$ phase has larger lattice parameter c and smaller a (b) than P2 phase. In the discharge process, the (002) and (004) peaks of the $P''2$ phase shift back toward higher angle, indicating a reversible structure recovery. It can be found that peaks of original P2 phase are regenerated when the voltage decreases to 2.5 V, then accompanying with $P''2$ phase till 1.6 V. The *in situ* XRD patterns provide a full picture of the structural evolution of Mn-based cathode material during charge and discharge processes, which is very important for understanding its electrochemical performance and energy storage mechanisms.

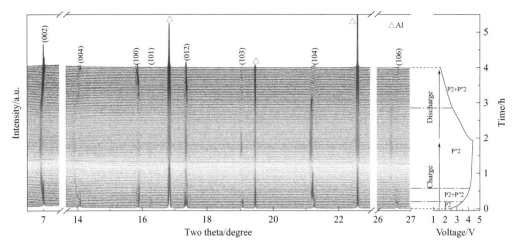

Figure 7.2.2 Time-resolved *in situ* XRD of Mn-based cathode material for sodium-ion batteries during the initial cycle.

(**Selected from:** Yue X Y, Zhou Y N, Fang F, et al. In situ Construction of Lithium Silicide Host with Unhindered Lithium Spread for Dendrite-Free Lithium Metal Anode [J]. Advanced Functional Materials, 2021, 31 (9), 2008786.

Li X L, Zhou Y N, Sun D, et al. Stabilizing Transition Metal Vacancy Induced Oxygen Redox by Co^{2+}/Co^{3+} Redox and Sodium-Site Doping for Layered Cathode Materials [J]. Angewandte Chemie International Edition, 2021, 60 (40): 22026-22034.)

New Words and Expressions

operando spectroscopy   原位光谱

analytical *adj.* 分析的，分析性的，（科学）分析的

methodology *n.* 方法论

wherein *adv.* 其中，在那种情况下

simultaneously *adv.* 同时，联立

catalytic *adj.* （化学物质）起催化作用的，有催化性的，催化性的

primary *adj.* 初级的，主要的

thereby *adv.* 因此，由此

in situ   在线的，原位的

visual *adj.* 视力的，视觉的

mechanism *n.* 机理，机制

acquisition *n.* （知识、技能等的）获得，得到

intermediates *n.* 中间产物

homogenize *v.* 均匀化，（尤指不适宜地）使统一，使单一

crux *n.* 症结，（难题或问题的）关键，最难点

disparity *n.* 差距，不等，差异，悬殊

instrumentation *n.* 仪器，仪表

apparatus *n.* 仪器，器械，装置

adverse *adj.* 不利的，有害的，反面的
reversible *adj.* 可逆的

Notes

(1) The primary concern of this methodology is to establish structure-reactivity/selectivity relationships and thereby yield information about mechanisms.

——参考译文：该方法的主要关注点是建立结构-反应性/选择性之间的关系，从而产生与其机理相关的信息。

(2) Operando spectroscopy is a class of methodology, rather than a specific spectroscopic technique.

——参考译文：原位谱学技术是一类方法，而不是某种特定的光谱技术。

(3) Catalyst scientists would ideally like to have a "motion picture" of each catalytic cycle, whereby the precise bond-making or bond-breaking events taking place at the active site are known.

——参考译文：理想情况下，催化剂科学家希望获得每个催化循环过程的"动态图像"，由此可以知道在活性位点发生的成键或断键事件的确切情况。

(4) Although monitoring a catalytic process *in situ* can provide information relevant to catalytic function, it is difficult to establish a perfect correlation because of the current physical limitations of *in situ* reactor cells.

——参考译文：尽管原位监测催化过程可以提供与催化功能相关的信息，（但）由于现有原位反应池的物理限制，很难建立完美的相关性。

(5) All of these factors must also be accounted for while designing operando experiments, as the spectroscopic techniques used may alter the reaction conditions.

——be accounted for 被考虑进去。
——参考译文：在设计原位实验时，所有这些因素都必须被考虑进去，因为使用的光谱技术可能会改变反应条件。

(6) Along with charging from the open-circuit voltage to 4.2 V, the new peak keeps shifting to lower angle, indicating the interlayer distance (correlated to lattice parameter c) is increasing along with a solid-solution behavior.

——参考译文：随着从开路电压充电到 4.2 V，新的峰不断向低角度移动，表明层间距离（与晶格参数 c 相关）随着固溶行为而增大。

Exercises

1. Question for discussion

(1) What spectroscopic techniques are mainly included in Operando spectroscopy?

(2) What is the purpose of Operando spectroscopy?

(3) Please illustrate the differences between *ex situ* spectroscopic techniques and operando spectroscopy techniques.

2. Translate the following into Chinese

(1) Operando spectroscopy plays an exceptional role in battery materials research. Real-time physical and chemical properties of the electrodes and interface, especially their changes during the operation of the batteries, are quite important for the full picture of the electrochemical process.

(2) The *in situ* XRD patterns provide a full picture of the structural evolution of Mn-based cathode material during charge and discharge processes, which is very important for understanding its electrochemical performance and energy storage mechanisms.

(3) The purpose of operando spectroscopy in catalyst study is to measure the catalytic changes that occur within the reactor during operation using time-resolved (and sometimes spatially-resolved) spectroscopy.

3. Translate the following into English

(1) 红外光谱 (2) 原位
(3) 峰值   (4) 表征
(5) 适时   (6) 夹层
(7) 转换   (8) 参数

(9) 近年来，表征电池循环过程的原位谱学技术得到了非常迅速的发展，从而能够成功获得电池充放电过程的实时信息。

(10) 对于催化剂研究，大多数工业催化反应需要过高的压力和温度条件，这些极限条件会降低信号的分辨率，从而降低原位谱图的质量。

(11) 基于这些优势，原位谱学技术可以更深入地探究电池的工作机制、性能劣化的过程和原因等。

Unit 7.3

Text

Introduction to microscopic methods

这篇课文介绍了光学显微镜、电子显微镜、扫描探针显微镜的背景、主要组成及基本原理。光学显微镜是观察材料微观结构的基本工具，是一切显微测试方法的基础。在电子显微镜方面，课文着重介绍了透射电子显微镜和扫描电子显微镜这两种常见的电子显微镜。课文还介绍了扫描隧道显微镜和原子力显微镜这两种常见的扫描探针显微镜。

Optical microscopy

Optical or light microscopy is the primary means for scientists and engineers to examine the microstructure of materials. Since the 1880s light microscopy has been widely used by metallurgists to examine metallic materials. Light microscopy for metallurgists became a special field named metallography. The basic techniques developed in metallography are not only used for examining metals, but also are used for examining ceramics and polymers. In this chapter, optical microscopy is introduced as a basic tool for microstructural examination of materials including metals, ceramics, and polymers.

Instrumentation

An optical microscope for examining material microstructure can use either transmitted or reflected light for illumination. Reflected-light microscopes are the most commonly used for metallography, while transmitted-light microscopes are typically used to examine transparent or semitransparent materials, such as certain types of polymers. **Figure 7.3.1** illustrates the structure of a light microscope for materials examination. An optical microscope includes the following main components: illumination system; objective lens; eyepiece; specimen stage. The illumination system of a microscope provides visible light by which a specimen is observed. The objective lens generates the primary image of the specimen, and its resolution determines the final resolution of the image, which is the most important optical component of a light microscope. The magnification of the objective lens determines the total magnification of the microscope because eyepieces commonly have a fixed magnification of $10\times$. The eyepiece is used to view the real primary image formed by the objective lens. The specimen stage holds the specimen for microscopic observations.

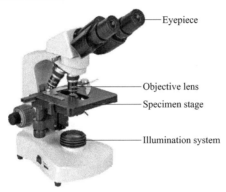

Figure 7.3.1   The structure of a light microscope for materials examination.

Image formation

Image formation can be illustrated by the behavior of a light path in a compound light microscope as shown in **Figure 7.3.2**. A specimen (object) is placed at position A where it is

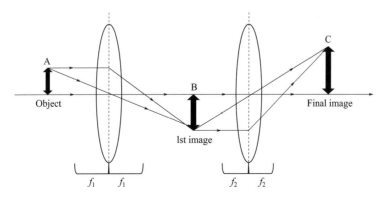

Figure 7.3.2   Principles of magnification in a microscope.

between one and two focal lengths from an objective lens. Light rays from the object first converge at the objective lens and are then focused at position B to form a magnified inverted image. The light rays from the image are further converged by the second lens (projector lens) to form a final magnified image of an object at C.

Magnification

The magnification of a microscope can be calculated by linear optics, which tells us the magnification of a convergent lens M:

$$M = \frac{v-f}{f} \tag{7.3.1}$$

where f is the focal length of the lens and v is the distance between the image and lens. A higher magnification lens has a shorter focal length, as indicated by **Equation (7.3.1)**. The total magnification of a compound microscope as shown in **Figure 7.3.2** should be the magnification of the objective lens multiplied by that of the projector lens.

$$M = M_1 M_2 = \frac{(v_1 - f_1)(v_2 - f_2)}{f_1 f_2} \tag{7.3.2}$$

Resolution

Resolution refers to the minimum distance between two points at which they can be distinguished as two points. The resolution of a microscope is theoretically controlled by the diffraction of light. Light diffraction controlling the resolution of microscope can be illustrated with the images of two self-luminous point objects as shown in **Figure 7.3.3**. When the point object is magnified, its image is a central spot (the Airy disk) surrounded by a series of diffraction rings, not a single spot. To distinguish between two such point objects separated by a short distance, the Airy disks should not severely overlap with each other. Thus, controlling the size of the Airy disk is the key to controlling resolution. The size of the Airy disk (d) is related to the wavelength of light (λ) and the angle of light coming into the lens. The resolution of a microscope (R) is defined as the minimum distance between two Airy disks that can be distinguished.

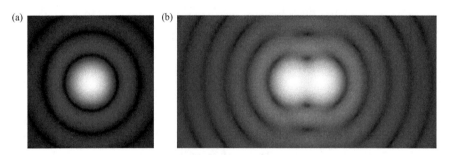

Figure 7.3.3 (a) The Airy disk; (b) Two self-luminous point objects which are overlapping with each other.

Electron microscopy

Electron microscopy utilizes the high-energy electrons as the illumination source for the image formation and the magnification. Electron microscopes generate images of material microstructures with much higher magnification and resolution than light microscopes. The high resolution of electron microscopes results from the short wavelengths of the electrons used for microscope illumination. The wavelength of electrons in electron microscopes is about 10000 times shorter than that of visible light. The resolution of electron microscopes reaches the order of 0.1 nm if lens aberrations can be minimized. Such high resolution makes electron microscopes extremely useful for revealing ultrafine details of material microstructure. There are two main types of electron microscopes: transmission electron microscope (TEM) and scanning electron microscope (SEM).

Transmission electron microscope

The structure of TEM is similar to a transmission light microscope composed of the following components along its optical path: light source, condenser lens, specimen stage, objective lens, and projector lens. The main differences are that, in a TEM, the visible light ray is replaced by an electron beam and glass lenses for visible light are replaced by electromagnetic lens for the electron beam. In a TEM system, an electron gun generates a high-energy electron beam for illumination. Because electron energy determines the wavelength of the electrons and wavelength largely determines the resolution of the microscope, the acceleration voltage determines the resolution of a TEM. The lenses in an electron microscope are electromagnetic. Lenses in light microscopes are made of glass; however, glass lenses cannot be used in electron microscopes because glass does not deflect or focus an electron beam. Noting that electrons have electric charges and that electric charges interact with magnetic fields, we can use an electromagnetic lens to deflect or focus an electron beam.

Scanning electron microscope

SEM is the most widely used type of electron microscope. A SEM consists of an electron gun and a series of electromagnetic lenses and apertures, similar to TEM systems. In an SEM, the electron beam emitted from an electron gun is condensed to a fine probe for surface scanning. The SEM examines microscopic structure by scanning the surface of materials controlled by a scanner. In SEM, elastic scattering produces the back-scattered electrons (BSEs), which are incident electrons scattered by atoms in the specimen. Inelastic scattering produces secondary electrons (SEs), which are electrons ejected from atoms in the specimen. SEs are the primary signals for achieving the topographic contrast corresponding to variation in geometric features on the specimen surface, while BSEs are useful for the formation of elemental composition contrast corresponding to the variation in chemical composition in a specimen.

Scanning probe microscopy

Scanning probe microscopy (SPM) is a technique to examine materials with a solid probe scanning the surfaces. An SPM system consists of several basic components: a probe and its motion sensor, scanner, electric controller, computer, and vibration isolation system. It examines surface features whose dimensions range from atomic spacing to a tenth of a millimeter. SPM started with the scanning tunneling microscope (STM) invented in 1982. An STM uses a tunneling current, a phenomenon of quantum mechanics, to examine material surfaces. The tunneling current flows through an atomic-scale gap between a sharp metallic tip and conducting surface atoms. The atomic force microscope (AFM) was invented in the late 1980s. It uses near-field forces between atoms of the probe tip apex and the surface to generate signals of surface topography. The AFM is more widely used than the STM because it is not restricted to electrically conducting surfaces.

Scanning tunneling microscope

STM relies on a tunneling current across a gap between the probe tip and a surface atom for surface examination. An electric current may penetrate an isolation gap between two conductors if high voltage is applied, but this is not the tunneling current. The tunneling current is a phenomenon of quantum mechanics. It results from electrons penetrating an energy barrier larger than the electron energy. To generate the tunneling current, a bias voltage is applied across a gap between the tip and the sample when the tip is kept near the sample.

Atomic force microscope

The AFM uses a very sharp tip to probe and map sample topography. The AFM detects near-field forces between the tip and sample, instead of detecting the tunneling current. There are several types of near-field forces that are briefly described as follows. The short-range forces refer to atomic forces between atoms when their distance is close to atomic spacing. Overlapping of their electron wave functions, which can be considered as overlapping of their electron clouds, causes either attractive or repulsive forces. The forces will be attractive when the overlap reduces their total energy, and the forces will be repulsive when the Pauli exclusion principle is in play. The attractive short-range forces are in the range of 0.5−1 nN per pair of interactive atoms between a typical tip and a sample. The decay length of the forces is in the range of 0.05−0.2 nm, larger than that of a tunneling current. In any case, the variation of the short-range forces on the atomic scale makes the AFM capable of obtaining topographic images of atoms.

(**Selected from:** Leng Y. Materials Characterization- Introduction to Microscopic and Spectroscopic Methods [M]. 2nd ed. John Wiley &Sons, 2013.)

New Words and Expressions

characterization　*n*. 表征
optical　*adj*. 光学的
microscopy　*n*. 显微术
microscope　*n*. 显微镜

instrumentation   *n*. 仪器
transmission   *n*. 透射
semitransparent   *adj*. 半透明的
illumination   *n*. 照明
objective   *n*. 物镜
eyepiece   *n*. 目镜
photomicrographic   *adj*. 显微照相的
resolution   *n*. 分辨率
magnification   *n*. 放大倍数
focal   *adj*. 焦点的
converge   *vi*. 会聚
diffraction   *n*. 衍射
luminous   *adj*. 发亮的
transmission   *n*. 透射
scanning   *v*. 扫描
wavelength   *n*. 波长
incident   *adj*. 入射的
scattering   *n*. 散射
penetrate   *v*. 穿透
inelastic   *adj*. 非弹性的
condenser   *n*. 聚光（镜）
projector   *n*. 投影
electromagnetic   *adj*. 电磁的
gun   *n*. （电子）枪
probe   *n*. 探针
secondary   *adj*. 二次的
back-scattered   *adj*. 背散射的
contrast   *n*. 衬度
tunneling   *n*. 隧穿
sensor   *n*. 传感器
vibration   *n*. 震动
isolation   *n*. 隔离
attractive   *adj*. 吸引的
repulsive   *adj*. 排斥的

Notes

（1）Light microscopy for metallurgists became a special field named metallography.

—special field 某项研究的专业领域。

—参考译文：对于冶金学家，光学显微术成为一门名为金相学的专业领域。

（2）There are two main types of electron microscopes: transmission electron microscope (TEM) and scanning electron microscope (SEM).

——参考译文：电子显微镜主要有两种，透射电子显微镜（TEM）和扫描电子显微镜（SEM）。

（3）SEM mainly utilizes the secondary electrons and the back-scattered electrons.

——secondary electrons 二次电子。back-scattered electrons 背散射电子。

——参考译文：扫描电子显微镜主要利用二次电子和背散射电子（实现成像）。

（4）The structure of TEM is similar to a transmission light microscope composed of the following components along its optical path: light source, condenser lens, specimen stage, objective lens, and projector lens.

——参考译文：透射电镜的结构类似于透射光学显微镜，沿着光路由以下部件组成，光源、汇聚透镜、样品台、物镜和投影镜。

（5）The forces will be attractive when the overlap reduces their total energy, and the forces will be repulsive when the Pauli exclusion principle is in play.

——Pauli exclusion principle 泡利不相容原理。

——参考译文：当交叠减少了它们的总能量时，这些力是吸引的，而当泡利不相容原理发挥作用时，这些力是排斥的。

Exercises

1. Question for discussion

（1）The components included in a light microscope.

（2）Which component determines the final resolution of the image?

（3）The definition of resolution.

（4）Briefly describe the difference between the optical and electron microscopes.

（5）Briefly describe the difference between the TEM and SEM.

（6）Briefly describe the working principle of scanning probe microscopy.

2. Translate the following into Chinese

（1）materials characterization （2）optical microscope

（3）illumination system （4）objective lens

（5）eyepiece （6）photomicrographic system

（7）specimen stage （8）condenser lens

（9）projector lens

（10）The basic techniques developed in metallography are not only used for examining metals, but also are used for examining ceramics and polymers.

（11）Reflected-light microscopes are the most commonly used for metallography, while transmitted-light microscopes are typically used to examine transparent or semitransparent materials, such as certain types of polymers.

(12) Interaction between electrons and the specimen also produces other signals, such as the secondary electrons with energy less than 50 eV, the back-scattered electrons with energy greater than 50 eV.

(13) The structure of TEM is similar to a transmission light microscope composed of the following components along its optical path: light source, condenser lens, specimen stage, objective lens, and projector lens.

(14) Because electron energy determines the wavelength of the electrons and wavelength largely determines the resolution of the microscope, the acceleration voltage determines the resolution of a TEM.

(15) SEs are the primary signals for achieving the topographic contrast corresponding to variation in geometric features on the specimen surface, while BSEs are useful for the formation of elemental composition contrast corresponding to the variation in chemical composition in a specimen.

3. Translate the following into English

(1) 透射电子显微镜   (2) 扫描电子显微镜
(3) 扫描探针显微镜   (4) 扫描隧道显微镜
(5) 原子力显微镜   (6) 二次电子
(7) 背散射电子   (8) 形貌
(9) 光学显微镜是科学家和工程师用来检测材料微观结构的主要工具。
(10) 电子显微镜利用高能电子作为成像和放大的光源。
(11) 扫描探针显微镜是一项用固体探针扫描表面来检测材料的技术。
(12) 当尖端在样品附近时,在尖端和样品之间的间隙上施加偏置电压以产生隧穿电流。

4. Reading comprehension

(1) Which one is NOT the components in an optical microscope? ____

(A) specimen stage
(B) objective lens
(C) motion sensor
(D) illumination system

(2) Which one is NOT the signals caused by the interaction between electrons and a specimen? ____

(A) back-scattered electrons
(B) incident electrons
(C) elastically scattered electrons
(D) secondary electrons

(3) Which one is NOT the components in a transmission electron microscope? ____

(A) eyepiece
(B) electron gun
(C) condenser lens
(D) projector lens

(4) Which microscope utilizes the elastically scattered electrons? ____

(A) scanning electron microscopy

(B) transmission electron microscope

(C) atomic force microscopy

(D) scanning tunneling microscopy

(5) Which microscope can measure the sample which is not electrically conducting? ____

(A) scanning electron microscopy

(B) atomic force microscopy

(C) transmission electron microscope

(D) scanning tunneling microscopy

(6) Which microscope doesn't need the scanner? ____

(A) scanning tunneling microscopy

(B) transmission electron microscope

(C) atomic force microscopy

(D) scanning electron microscopy

Reading Material
Microscopic aberrations

这篇课文主要介绍了色差、球差、像散和场曲四种导致图像失真的透镜像差。色差和球差会影响图像的整个视场，像散和场曲仅影响图像的离轴点。像差会严重降低显微镜的分辨率。

The aforementioned optical principles are based on the assumptions that all components of the microscope are ideal, and that light rays from any point on an object focus on a correspondingly unique point in the image. However, the lens for a real microscope cannot be prefect, which will cause the distortions of the image. These distortions are called lens aberrations and can be classified into four types: chromatic aberrations, spherical aberrations, astigmatism, and curvature of field. Some aberrations affect the whole field of the image (chromatic and spherical aberrations), while others affect only off-axis points of the image (astigmatism and curvature of field). The lens aberrations can severely diminish the true resolution of a microscopy.

Chromatic aberration

Chromatic aberration is caused by the variation in the refractive index of the lens in the range of light wavelengths (light dispersion). The refractive index of lens glass is greater for shorter wavelengths (for example, blue) than for longer wavelengths (for example, red). Thus, the degree of light deflection by a lens depends on the wavelength of light [**Figure 7.3.4 (a)**]. Because a range of wavelengths is present in ordinary light (white light), light cannot be focused at a single point.

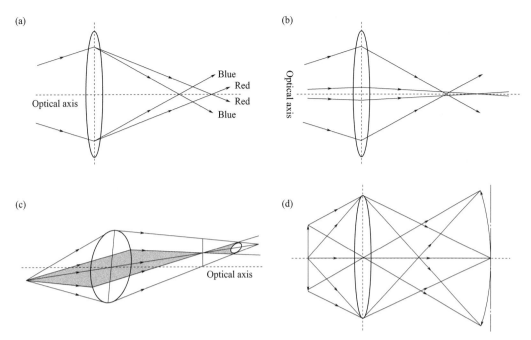

Figure 7.3.4 (a) Paths of rays in white light illustrating chromatic aberration; (b) spherical aberration; (c) astigmatism is an off-axis aberration; (d) curvature of field.

Spherical aberration

Spherical aberration is caused by the spherical curvature of a lens. Light rays from a point on the object on the optical axis enter a lens at different angles and cannot be focused at a single point [**Figure 7.3.4(b)**]. The portion of the lens farthest from the optical axis brings the rays to a focus nearer the lens than does the central portion of the lens.

Astigmatism

Astigmatism results when the rays passing through vertical diameters of the lens are not focused on the same image plane as rays passing through horizontal diameters. In this case, the image of a point becomes an elliptical streak at either side of the best focal plane [**Figure 7.3.4 (c)**]. Astigmatism can be severe in a lens with asymmetric curvature.

Curvature of field

Curvature of field is an off-axis aberration. It occurs because the focal plane of an image is not flat but has a concave spherical surface [**Figure 7.3.4(d)**]. This aberration is especially troublesome with a high magnification lens with a short focal length. It may cause unsatisfactory photography.

Aberration correction

There are a number of ways to reduce the lens aberrations. For example, combining lenses with different shapes and refractive indices corrects chromatic and spherical aberrations.

Selecting single-wavelength illumination by the use of filters helps eliminate chromatic aberrations. We expect that the extent to which lens aberrations have been corrected is reflected in the cost of the lens. It is a reason that we see huge price variation in microscopes.

(**Selected from:** Leng Y. Materials Characterization- Introduction to Microscopic and Spectroscopic Methods [M]. 2nd ed. John Wiley &Sons, 2013.)

New Words and Expressions

  distortion *n*. 畸变
  aberration *n*. 像差
  chromatic *adj*. 颜色的
  spherical *adj*. 球形的
  astigmatism *n*. 像散
  curvature *n*. 弯曲
  off-axis *adj*. 离轴的
  elliptical *adj*. 椭圆的
  streak *n*. 条纹
  asymmetric *adj*. 不对称的
  concave *adj*. 凹面的
  photography *n*. 照相术
  filter *n*. 滤波器

Notes

（1）These distortions are called lens aberrations and can be classified into four types：chromatic aberrations，spherical aberrations，astigmatism，and curvature of field.

  —参考译文：这些畸变被称为透镜像差，可分为四类：色差、球差、像散和场曲。

（2）Chromatic aberration is caused by the variation in the refractive index of the lens in the range of light wavelengths（light dispersion）.

  —参考译文：色差是由透镜的折射率在光波长范围内的变化（光色散）引起的。

（3）In this case，the image of a point becomes an elliptical streak at either side of the best focal plane.

  —参考译文：在该情况下，单点的图像在最佳焦平面的任一侧变成了椭圆条纹。

（4）Selecting single-wavelength illumination by the use of filters helps eliminate chromatic aberrations.

  —参考译文：通过滤波器选择单波长照明可以帮助消除色差。

Unit 7.4

Text

Modern *in situ* characterization techniques

这篇课文介绍了在先进材料分析中起到重要作用的现代原位表征技术。课文讲解了可实时观察微观结构变化的原位透射电子显微镜。课文还介绍了穿透力强、对材料损伤小的原位散射和衍射技术。

In situ characterization techniques has become an increasingly important tool for materials characterization. "*In situ*" is a Latin phrase meaning "in position" or "in place". Therefore, "*In situ* characterization" indicates "observing while exactly in place where things occur". These "things" can be reactions, transformations, alterations, and/or changes to a state of equilibrium of matter. Such changes in equilibrium are unstable and only transient in nature. Consequently, they cannot be isolated, or observed before and after such a reaction or transformation. A thorough examination of these transient stages of dynamic processes is therefore the aim of numerous scientific investigations. The growth, development, and diversification of characterization techniques has led to the utilization of modern *in situ* characterization on the structural dynamics of a material during transformations and the ability to correlate a material's structure and properties with the fourth dimension: time.

In situ transmission electron microscope

Transmission electron microscope (TEM) has hit a significant milestone of sub-angstrom resolution. On one hand, electron microscopists and materials scientists are enjoying the highest TEM spatial resolution ever attainable; on the other hand, study of materials in a steady state is hard to meet the increasing demand in new application fields such as nanocatalysts, nanocrystal growth, nanoelectronics, nanosensors, and nanomechanics in which size effect and structural or property responses to stimuli from the surrounding environment are key information to learn. Special attention is thus paid to *in situ* TEM. Here, *in situ* TEM was defined as "Some form of stimulus is applied to a sample while it is observed in a TEM." According to this definition, *in situ* TEM has two important characters, stimulus and real-time observation. Stimulus, or say an external field, needs to be applied directly to a TEM specimen sitting in an electron microscope column. Typical external fields used for *in situ* TEM include heating, cooling, electric, magnetic fields, as well as mechanical forces (tensile, compressive) and ion beam irradiation. To apply an external field to the specimen area and perform TEM observation simultaneously, one just needs, in most cases, a specially designed TEM specimen holder and a transmission electron microscope, which allows the *in situ* TEM specimen holder to fit in. Also, a fast image recording system (for example, a video recorder or TV rate CCD camera) is important. Sometimes, a modification of the transmission electron microscope is necessary.

Specimen holders

An important point the author would like to make is that to some extent, *in situ* TEM is all about specimen holders. Depending on what external fields or forces that are applied to the specimen in a transmission electron microscope, one can have *in situ* heating TEM, *in situ* cooling TEM, *in situ* environmental TEM (gas or liquid), *in situ* biasing TEM (electric voltage and current), *in situ* straining TEM (tension or compression), *in situ* Lorentz TEM, *in situ* electron holography TEM, and *in situ* ion beam (or laser) irradiation TEM available for study of materials in stimulating environments. Real-time *in situ* TEM observations at high spatial and time resolutions for dynamic structural processes are becoming routine nowadays. In most cases, the *in situ* TEM capability can be realized on any standard transmission electron microscopes, and all that needs to be done is to design a special TEM specimen holder, which supplies a desired stimulus to the TEM specimen. These are holder-based *in situ* TEM technologies. In this chapter, we will briefly introduce the *in situ* heating, *in situ* environmental and in situ nanomechanical TEM.

In situ heating TEM

In situ heating can be realized by placing a heating element at the tip of a TEM specimen holder as shown in **Figure 7.4.1**. Depending on the heating element, heating mechanism, and sample fixing method, there are three typical types of commercially available *in situ* heating TEM holders: furnace heating holders, wire-heating holders, and membrane-heating holders. For the furnace-heating holder, at the tip a heating filament embracing a 3 mm-diameter TEM specimen disk acts like an electric furnace. The thermal radiation heats the specimen, and is therefore an indirect heating mechanism. Cooling water is connected to the holder for use at above 500℃. An embedded thermal couple measures temperature in the furnace cup. Because the 3 mm-diameter heating zone is "huge" and involves many components and supporting materials (e.g. TEM metal grid), the total thermal expansion effect causes a severe problem of sample drifting when changing temperature. It may take a few tens of minutes to seize the drifting, therefore taking high resolution TEM images in a short period of time requires a blessing. Obviously, this type of furnace heating holder is good to perform low resolution *in situ* TEM imaging. For the *in situ* wire-heating holder, a spiral tungsten wire $20-50$ mm in diameter is used to heat the materials attached to the heating wire and is therefore based on a direct heating mechanism. The materials can be heated to as high as 1500℃ although the heating power is maintained at a low level, and the sample area being heated may be millions of times smaller than that of the furnace-heating holder. Cooling water is not required. The sample drifting rate due to the thermal expansion effect is tolerable $\sim$10 min after temperature change, therefore atomic resolution at elevated temperatures is readily achievable. The membrane-heating holder is rather new compared to the furnace-heating and wire-heating TEM holders. The key component of this type of holder is a heating device made from a conductive ceramic membrane suspended on a Si chip. Heating is very local and the heating power is small, resulting in a small sample drift rate at elevated temperatures. A very unique feature of the membrane-heating holder is the fast temperature change rate, as high as 106℃/s is possible. This unique feature makes the holder a good choice for doing *in situ* TEM thermal cycling experiments.

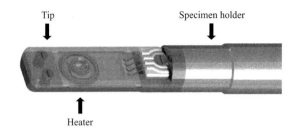

Figure 7.4.1　Illustration of an *in situ* heating TEM holder.

In situ environmental TEM

In our real world, many physical and chemical processes take place in an environment filled with gas and/or liquid. For example, a catalytic process includes catalysts, gas, and a suitable temperature, which is, in most cases, higher than room temperature. It is clearly desirable to have an *in situ* TEM technology that is capable of handling gas or liquid in the specimen chamber of the electron microscope. A solution for the *in situ* environmental TEM is the window-type E-cell concept. The so-called window-type E-cell is built in a TEM specimen holder at the tip area. The TEM specimen sits inside the cell, which is then sealed by two windows above and below the specimen. The window materials must be electron transparent and have a weak interference with the electron beam. This means that the window materials should be amorphous in structure and thin enough. The window membranes must also be strong enough to withstand the pressure difference between the gas cell and the TEM vacuum. Typical window materials are amorphous carbon or silicon nitride with a 15−200 nm thickness depending on the desired gas pressure for the applications. The biggest advantage of the window-type E-cell is the relatively low cost and microscope choice flexibility. Because the E-cell is built into the TEM specimen holder, no modification of the electron microscope is involved (microscope modification is expensive especially when the vacuum system is included) and the holders associated with any electron microscopes can be modified into the E-cell holders.

In situ nanomechanical TEM

Mechanical properties such as strength, hardness, and toughness may be changed significantly when the sizes of materials go down to nanometer scales. With increasingly miniaturized electronic components and the request to correctly understand and predict failure of nanostructures, material size effects on mechanical behaviors become more important and urgent than before. Like all other *in situ* TEM technologies, the development of *in situ* nanomechanical TEM was because of the high resolution that TEM can offer for direct imaging and characterization of internal structures and defects in materials. After decades of development, precisely controlled, quantitative *in situ* nanomechanical TEM holders have been developed to characterize the mechanical behaviors and corresponding structural changes in micro and nanostructure. Obviously, specially designed straining TEM holders are required to load a tensile or compressive mechanical force on TEM specimens while tracking real-time structural changes in order to find out and quantify mechanical property-structure relationships. Various TEM holders with different designs have been reported, including conventional tensile straining holders, microelectromechanical systems (MEMS)-based straining holders, nanoindentation holders, and TEM grid-based straining holders. The basic idea of the conventional straining holder was to make a specimen in a rectangular shape with one end fixed and another end connected to a movable rod, which could slide along the length of the holder. A deficiency of the conventional tensile straining TEM holder described above is its inability to quantify the force load and material deformation displacement. In addition, only tensile force can be applied. To overcome these problems, MEMS-based mechanical testing holders were designed. The key part, a free-standing MEMS chip device integrated into the tip of the holder, is fabricated using the standard silicon-based microfabrication followed by lift-out from the substrate. The force load is applied with a displacement-controlled mechanical, thermal, or electrostatic mechanism and the applied load and straining are measurable by displacement-force sensors or differential capacitive sensors. Both tensile and compressive forces are possible to be applied. The dimension and shape of such a MEMS chip device is similar to those of the specimen.

In situ scattering and diffraction

Many phenomena relate to the atomic order and how this is influenced by external factors. Of particular interest are physical processes, such as phase transitions and crystal growth. These lie at the basis of many important applications in the chemical industry and (electrical) engineering. In order to follow what is happening as a function of the thermodynamic parameters, one needs a probe that does not interfere too much with the process itself, and that is able to reach and leave the area of interest. In many cases, the probe consists of charged particles. However, deeply buried structures and/or high gas pressures do not always allow for such experiments. In those cases, only X-rays and neutrons would qualify.

In situ X-ray diffraction

X-ray diffraction (XRD) is an outstanding tool for structural analyses at the atomic scale, and both the experimental techniques and the theoretical interpretations are well established. X-rays also have the advantage of being highly penetrating, as compared to electrons for instance, allowing for the study of bulk materials, or to study samples in complicated environments. Synchrotron XRD can provide the high photon fluxes which makes it possible to collect full diffraction patterns in relatively short times, and thus to follow time varying processes *in situ*. There are a certain number of advantages of using X-rays as probe rather than, for instances, electrons. These advantages originate from the relatively low scattering cross-section of X-ray photons. One advantage of the low cross section of X-rays is the much larger penetration depth as compared to electrons. Electrons are fully absorbed by a few centimeters of atmospheric air, which means that measurements have to be performed under ultra-high or high vacuum conditions. High energy X-rays (>15 keV) on the other hand have negligible absorption in air. For 22 keV X-rays, even a sample chamber of 0.5 mm aluminum, will only absorbed 50% of the photons. This means that *in situ* experiments under real life conditions in sophisticated sample chambers are possible.

In situ neutron scatterings

Neutrons provide unique possibilities for *in situ* studies of condensed matter due to their ability to penetrate large samples and work pieces, to distinguish between neighboring elements and even between isotopes of the same element, to interact with magnetic moments and to map nuclear and magnetic excitations. The range of applications extends from biological or soft matter systems over basic superconducting and magnetic investigations to all kinds of materials science under a great variety of external conditions. Complex and sophisticated sample environments have been developed in the past to study samples under external perturbations and extreme conditions like high or low temperatures, high pressures and mechanical stresses, electric and magnetic fields and different kinds of chemical environments. Not only equilibrium studies are in the focus of neutron scattering investigations-increasing interest in real-time kinetic studies lead to new techniques that allow even inelastic studies on time scales down to the microsecond regime thus providing most direct information about the changing chemical bonds in materials.

Advantages of X-rays and neutrons scatterings

In this session, we have discussed the advantages of X-rays and Neutrons over charged particles as probe for *in situ* studies. These advantages stem mainly from the low scattering cross section, which makes the theoretical analysis of the scattering intensities much easier, and allows using sophisticated sample chambers, as well real-life experimental conditions, like atmospheric pressure. Both the experimental and the theoretical concepts are therefore well developed and mature. A disadvantage of the low scattering cross-section is the corresponding inefficient use of the available flux, which needs to be compensated by powerful sources, like synchrotrons. This in turn makes these techniques less accessible, especially for students, than laboratory-based techniques.

(**Selected from:** Ziegler A, Graafsma H, Zhang X F, et al. *In-situ Materials Characterization - Across Spatial and Temporal Scales* [M]. 1st ed. Springer, 2014.)

New Words and Expressions

transient *adj.* 转瞬即逝的
dynamic *adj.* 动力学的

dimension *n*. 维度
sub-angstrom *n*. 亚尺度的
stimulus *n*. 刺激
holders *n*. （样品）杆
furnace *n*. 炉子
filament *n*. 灯丝
membrane *n*. 膜
environmental *adj*. 环境的
microelectromechanical *adj*. 微机电
sensor *n*. 传感器
spatial *adj*. 空间的
X-ray *n*. X射线
neutron *n*. 中子
diffraction *n*. 衍射
vacuum *n*. 真空
absorption *n*. 吸收
sophisticated *adj*. 精密的
chamber *n*. 腔
isotopes *n*. 同位素
perturbation *n*. 扰动

Notes

(1) According to this definition, *in situ* TEM has two important characters, stimulus and real-time observation.

——参考译文：根据这个定义，原位 TEM 具有刺激和实时观察这两个重要特征。

(2) Various TEM holders with different designs have been reported, including conventional tensile straining holders, microelectromechanical systems (MEMS)-based straining holders, nanoindentation holders, and TEM grid-based straining holders.

——参考译文：已经报道了很多（基于）不同设计的 TEM 样品杆，包括传统的拉伸应变样品杆、基于微机电系统（MEMS）的应变样品杆、纳米压痕样品杆和基于 TEM 格栅的应变样品杆。

(3) Synchrotron XRD can provide the high photon fluxes which makes it possible to collect full diffraction patterns in relatively short times, and thus to follow time varying processes *in situ*.

——参考译文：同步辐射 X 射线衍射可以提供高光子通量，可以在相对较短的时间内收集完整的衍射花样，因此可以原位跟踪随时间变化的过程。

Exercises

1. Question for discussion

(1) Briefly describe the *in situ* characterization.

(2) The definition of *in situ* TEM.

(3) Classification of *in situ* heating TEM.

(4) Classification of *in situ* nanomechanical TEM.

(5) The advantage of X-rays and neutrons over charged particle.

2. Translate the following into Chinese

(1) *in situ* characterization　　　　(2) specimen holders

(3) *in situ* heating　　　　　　　　(4) window-type

(5) microelectromechanical systems　(6) X-ray diffraction

(7) The growth, development, and diversification of characterization techniques has led to the utilization of modern *in situ* characterization on the structural dynamics of a material during transformations and the ability to correlate a material's structure and properties with the fourth dimension: time.

(8) To apply an external field to the specimen area and perform TEM observation simultaneously, one just needs, in most cases, a specially designed TEM specimen holder and a transmission electron microscope, which allows the *in situ* TEM specimen holder to fit in.

(9) With increasingly miniaturized electronic components and the request to correctly understand and predict failure of nanostructures, material size effects on mechanical behaviors become more important and urgent than before.

(10) In order to follow what is happening as a function of the thermodynamic parameters, one needs a probe that does not interfere too much with the process itself, and that is able to reach and leave the area of interest.

(11) X-ray diffraction (XRD) is an outstanding tool for structural analyses at the atomic scale, and both the experimental techniques and the theoretical interpretations are well established.

(12) These advantages stem mainly from the low scattering cross section, which makes the theoretical analysis of the scattering intensities much easier, and allows using sophisticated sample chambers, as well real-life experimental conditions, like atmospheric pressure.

3. Translate the following into English

(1) 实时观察　　　　　　　　　(2) 原位 X 射线衍射

(3) 加热炉　　　　　　　　　　(4) 纳米力学

(5) 中子散射　　　　　　　　　(6) 同位素

(7) 原位表征技术已经变成材料表征愈发重要的工具。

(8) 根据应用于样品区域的外场类型来设计对应的样品杆。

(9) 可以通过在透射电镜的样品杆的尖端安放一个加热元件来实现原位加热。

(10) 随着材料的尺寸减小到纳米尺度，其强度、硬度和韧性等力学性能可能发生很大的变化。

4. Reading comprehension

(1) Which technology is NOT included in the *in situ* TEM?

(A) environmental

(B) cooling

(C) bias probe

(D) synchrotron

(2) Which one is NOT the heating element for *in situ* heating TEM?

(A) membrane heating

(B) irradiation heating

(C) wire heating

(D) furnace heating

(3) Which process can be observed by in situ environmental TEM?

(A) electron holography

(B) chemical reactions

(C) ion irradiation

(D) nanoindentation

(4) Which one is NOT the holder for *in situ* nanomechanical TEM?

(A) nanoindentation holders

(B) conventional tensile straining holders

(C) TEM grid-based straining holders

(D) window-type E-cell holders

(5) Which one is the disadvantage of X-rays and neutrons scatterings?

(A) inefficient use of the available flux

(B) larger penetration depth

(C) negligible absorption in air

(D) low scattering cross section

Reading Material
Ultrafast transmission electron microscope and diffraction

这篇课文介绍了超快透射电子显微镜和超快电子衍射的基本原理及应用方法。该实验技术可用来研究快速相变、凝固过程、孪晶和冲击传播等。

The development and application of pump-probe instrumentation to study rather complex transient events in the solid state, soft matter and life sciences domains has been an area of enormous interest recently. For many years the emphasis has been on the laser-pump and laser-probe approach, followed by the laser-pump and x-ray probe approach. Lately, another method, the laser-pump electron-probe technique, has been gaining interest. Very early experiments using this technique have used electrons to explore gas-phase diffraction of photo-stimulated chemical reactions, followed by experiments in the solid-state domain, studying for example, rapid phase transformations, solidification processes, twinning, and shock propagation. The session focuses on the emerging area of ultrafast transmission electron microscope, or simply ultrafast electron microscopy (UEM), as well as on ultrafast electron diffraction (UED). The UEM and UED employ energetic electrons or ions as measuring probes, simply because they are charged particles that can (i) be easily accelerated with an electric field (ii) they can be manipulated and focused to form very fine probes—only a few Angstroms in diameter-by means of an electro-magnetic field, (iii) they interact strongly with the material investigated, and (iv) they can be readily detected.

Principle of UEM and UED

The basic principle of many present-day UEM and UED set ups can be described as follows: A reaction is triggered in the sample and the time-delayed probing beam of electrons creates a distinct signal (diffraction pattern, image) that can be collected at specific time delays relative to the beginning of the triggered reaction. The triggering of a reaction is usually done via pulsed laser irradiation. The idea is simple and plausible, however, the technical implementation, the experimental possibilities, instrumental performance and limitations are multiple and often challenging. A few fundamental changes in the way a conventional electron microscope or electron diffraction set up is operated are mandatory to enable these high time-resolved techniques. The major departure is to convert the conventional, continuous electron particle beam into a very short electron pulse. Other alterations evolve around the timing and synchronization of a UEM or UED experiment and the ultrafast electron detection methods need re-development.

Electron emission

There are three principal methods-and a combination thereof-to extract electrons out of a cathode material: (i) thermionic emission, by heating the electron source material to high temperatures, (ii) field emission, by applying high electric fields, and (iii) photoelectron emission, via irradiation of laser light. All three of these methods find application in the UEM and UED. Especially, a combination of them seems to be the most promising way to achieve high electron densities for very short electron pulses.

Electron- and laser-material interactions

Here the word "ultrashort" refers to the time domain ranging from femtoseconds to picoseconds. Accordingly, the word "ultrafast" refers to physical, chemical, and biological events that occur on that time scale, and hence the terms femtophysics, femtochemistry and femtobiology can be found in the recent scientific literature. Femto materials science seems to be less known, possibly because it overlaps with femtophysics and in some aspects with femtochemistry. Even faster processes than "femto" occur on the attosecond time scale. The main driving force for ultrafast science in physics, chemistry, biology and the materials sciences is the development of better, faster, brighter and more stable lasers. Without pico-, femto- and now attosecond laser pulses ultrafast science would hardly be possible.

Forecast

The developments in ultrashort pulsed electron based methods for the pump-probe as well as for related ultrafast types of experiments have reached a point, where pico- and even femtosecond short electron pulses are attainable and controllable to a certain degree to monitor and follow atomic motions during an ultrafast process. The instrumental modifications that need to be accomplished in terms of electron gun design, electron emitter material, pulse propagation, space charge effects in electron pulses, pulse compression and characterization techniques, synchronization and experimental geometric considerations have been presented and discussed. The field of UEM and UED has by far not reached the status of an established characterization technique and multiple innovations and developments are still required to make this a viable and reliable method to assess the dynamics of ultrafast reactions that are moving more and more into the focal point. Scientists have come to realize that static high resolution, and even three-dimensional structural information is not sufficient anymore to satisfy the need to understand structure-property and especially, structure-function relationships.

(**Selected from:** Ziegler A, Graafsma H, Zhang X F, et al. In-situ Materials Characterization- Across Spatial and Temporal Scales [M]. 1st ed. Springer, 2014.)

New Words and Expressions

pump-probe   *n*. 抽运-探测
laser   *n*. 激光
ultrafast   *adj*. 超快的
accelerate   *v*. 加速
trigger   *n*. 触发器
time-delayed   *adj*. 延时的
photoelectron   *n*. 光电子
pulse   *n*. 脉冲
ultrashort   *adj*. 超短的
picosecond   *n*. 皮秒
femtosecond   *n*. 飞秒
attosecond   *n*. 阿秒

Notes

(1) Lately, another method, the laser-pump electron-probe technique, has been gaining interest.

—参考译文：近来，另一种方法，即激光泵浦-电子探针技术，已引起（人们的）兴趣。

(2) The session focuses on the emerging area of ultrafast transmission electron microscope, or simply ultrafast electron microscopy (UEM), as well as on ultrafast electron diffraction (UED).

—参考译文：本章节关注超快透射电子显微镜或简称为超快电子显微技术（UEM）以及超快电子衍射（UED）（这个）新兴的领域。

参考文献

[1] Wikipedia, Materials science, Wikimedia Foundation [DB/OL], (2023-02-17) [2023-02-22]. https://en.wikipedia.org/wiki/Materials_science.

[2] Smith W F, Hashemi J, Presuel-Moreno F. Foundations of Materials Science and Engineering [M]. McGraw-hill, New York, 2006.

[3] Ashby M F, Jones D R H. Engineering Materials 1: An Introduction to Properties, Applications and Design [M]. Elsevier, 2011.

[4] Callister Jr W D, Rethwisch D G. Fundamentals of Materials Science and Engineering: An Integrated Approach [M]. John Wiley & Sons, 2020.

[5] Askeland D R, Wright W J. Essentials of Materials Science and Engineering [M]. Cengage Learning, 2018.

[6] Berns H, Theisen W. Ferrous Materials: Steel and Cast Iron [M]. Springer Science & Business Media, 2008.

[7] China National Space Administration, Chang'e-5 Launch Highlights [DB/OL], (2020-11-24) [2020-11-24]. https://www.cnsa.gov.cn/n6758823/n6758842/c6810575/content.html.

[8] Bouville F, Maire E, Meille S, et al. Strong, Tough and Stiff Bioinspired Ceramics from Brittle Constituents [J]. Nature Materials, 2014, 13 (5): 508-514.

[9] Li S, Yu S, Feng Y. Progress in and Prospects for Electrical Insulating Materials [J]. High Voltage, 2016, 1 (3): 122-129.

[10] Baidu, Advanced Ceramic Parts & Components for Automobiles [DB/OL], (2023-03-02) [2023-03-02]. https://www.ecvv.com/product/4845361.html.

[11] Shi X, Zuo Y, Zhai P, et al. Large-area Display Textiles Integrated with Functional Systems [J]. Nature, 2021, 591 (7849): 240-245.

[12] Rosso M. Ceramic and Metal Matrix Composites: Routes and Properties [J]. Journal of Materials Processing Technology, 2006, 175 (1-3): 364-375.

[13] Department of Materials Science, Fudan University. Research trends [DB/OL], (2021-09-23) [2021-09-23]. https://mse.fudan.edu.cn/49/ec/c22913a412140/page.html.

[14] Zhao T, Zhang X, Lin R, et al. Surface-Confined Winding Assembly of Mesoporous Nanorods [J]. Journal of the American Chemical Society, 2020, 142 (48): 20359-20367.

[15] Wikipedia, Biotic Material, Wikimedia Foundation [DB/OL], (2023-03-02) [2023-03-02]. https://en.wikipedia.org/wiki/Biotic_material.

[16] Hofmann S. On Beyond Uranium: Journey to the End of the Periodic Table [M]. CRC Press, 2018.

[17] Callister W D, Rethwisch D G. Materials Science and Engineering: An Introduction

[M]. John Wiley & Sons, 2007.

[18] Warren M. "Why didn't We Think to do This Earlier?" Chemists Thrilled by Speedy Atomic Structures [J]. Nature, 2018, 563 (7729): 16-18.

[19] Wikipedia, Metal, Wikimedia Foundation [DB/OL]. (2023-02-06) [2023-02-22]. https://en.wikipedia.org/wiki/Metal.

[20] Wikipedia, Alloy, Wikimedia Foundation [DB/OL]. (2023-02-18) [2023-02-22]. https://en.wikipedia.org/wiki/Alloy.

[21] Schweitzer P A, P E. Metallic Materials Physical, Mechanical, and Corrosion Properties vol 19 [M]. CRC press, 2003.

[22] Young K-h, Nei J. The Current Status of Hydrogen Storage Alloy Development for Electrochemical Applications [J]. Materials, 2013, 6 (10): 4574-4608.

[23] Okamoto H, Schlesinger M E, Mueller E M. Introduction to Phase Diagrams vol 3 [M/CD]. ASM International, 2016.

[24] Gong Y, Ma F Q, Xue Y, et al. Failure Analysis on Leaked Titanium Tubes of Seawater Heat Exchangers in Recirculating Cooling Water System of Coastal Nuclear Power Plant [J]. Engineering Failure Analysis, 2019, 101: 172-179.

[25] DOE Fundamentals Handbook: Material Science vol 1 [M/CD]. Washington. DC: USDOE, 1993.

[26] Sun Y, Tan X, Lei LL, et al. Revisiting the Effect of Molybdenum on Pitting Resistance of Stainless Steels [J]. Tungsten, 2021, 3 (3): 329-337.

[27] Callister W D, Rethwisch D G. Materials Science and Engineering - An Introduction: vol 1 [M]. 10th ed. Wiley, 2018.

[28] George E P, Raabe D, Ritchie R O. High-entropy alloys [J]. Nature Reviews Materials, 2019, 4 (8): 515-534.

[29] Mortensen A, Llorca J. Metal Matrix Composites [J]. Annual Review of Materials Science, 2010, 40: 243-270.

[30] Su Y, Ouyang Q, Zhang W, et al. Composite Structure Modeling and Mechanical Behavior of Particle Reinforced Metal Matrix Composites [J]. Materials Science and Engineering A, 2014, 597: 359-369.

[31] Callister W D, Rethwisch D G. Fundamentals of Materials Science and Engineering: An Integrated Approach [M]. 5th ed. Wiley, 2015.

[32] Carter C B, Norton MG. Ceramic Materials: Science and Engineering [M]. 2nd ed. Springer, 2013.

[33] Akrami S, Edalati P, Fuji M, et al. High-Entropy Ceramics: Review of Principles, Production and Applications [J]. Materials Science and Engineering: R: Reports, 2021, 146: 100644.

[34] Wen C. Structural Biomaterials: Properties, Characteristics, and Selection [M]. Woodhead Publishing, (2021-04-06) [2023-02-22] https://www.elsevier.com/books/structural-biomaterials/wen/978-0-12-818831-6.

[35] Baldacchini T. Three-Dimensional Microfabrication Using Two-Photon Polymerization

[M]. 2nd ed. Elsevier, 2020.

[36] Shi Y, Yan C, Zhou Z, et al. Materials for Additive Manufacturing [M/CD]. 1st ed. Elsevier, 2021.

[37] Campanella D, Belanger D, Paolella A. Beyond garnets, Phosphates and Phosphosulfides Solid Electrolytes: New Ceramic Perspectives for All Solid Lithium Metal Batteries [J]. Journal of Power Sources, 2021, 482: 228949.

[38] Liu Y, Xu B, Zhang W, et al. Composition Modulation and Structure Design of Inorganic-in-Polymer Composite Solid Electrolytes for Advanced Lithium Batteries [J]. Small, 2020, 16 (15): 1902813.

[39] Twibanire J K, Grindley T B. Polyester Dendrimers [J]. Polymers, 2012, 4 (1): 794-879.

[40] Zhang B, Chen J, Zhang H, et al. Annealing-Induced Periodic Patterns in Solution Grown Polymer Single Crystals [J]. RSC Advances, 2015, 5 (17): 12974-12980.

[41] Young R J, Lovell P A. Introduction to Polymers [M]. 3rd ed. CRC Press, 2011.

[42] Saldivar-Guerra E, Vivaldo-Lima E. Handbook of Polymer Synthesis, Characterization, and Processing [M]. 1st ed. John Wiley & Sons, 2013.

[43] Tajeddin B, Arabkhedri M. Polymers and Food Packaging [M]. Polymer Science and Innovative Applications, 2020, 525-543.

[44] Mangaraj S, Yadav A, Bal L M, et al. Application of Biodegradable Polymers in Food Packaging Industry: A Comprehensive Review [J/OL], Journal of Packaging Technology and Research, 2019, 3: 77-96.

[45] Lv J, Liu Y, Wei J, et al. Photocontrol of Fluid Slugs in Liquid Crystal Polymer Microactuators [J]. Nature, 2016, 537 (7619): 179-184.

[46] Barbucci R. Integrated Biomaterials Science [M]. Springer Science & Business Media, 2002.

[47] Peppas N A, Khademhosseini A. Make better, Safer Biomaterials [J]. Nature, 2016, 540 (7633): 335-337.

[48] Shen Y, Zhang W, Xie Y, et al. Surface Modification to Enhance Cell Migration on Biomaterials and its Combination with 3D Structural Design of Occluders to Improve Interventional Treatment of Heart Diseases [J]. Biomaterials, 2021, 279: 121208.

[49] Raghavendra G M, Varaprasad K, Jayaramudu T. Biomaterials: design, development and biomedical applications [M]. Nanotechnology Applications for Tissue Engineering. Elsevier, 2015: 21-44.

[50] Wang X, Ma B, Xue J, et al. Defective Black Nano-Titania Thermogels for Cutaneous Tumor-Induced Therapy and Healing [J]. Nano Letters, 2019, 19 (3): 2138-2147.

[51] Ventre T. 3D Printing Metallic Implants: Technologies Available and the Future of the Industry [J/OL] 2021, [02-22 2023]. https://www.azom.com/article.aspx?ArticleID=20542.

[52] Badylak S F. Host Response to Biomaterials: the Impact of Host Response on Biomaterial Selection [M]. Academic Press, 2015.

[53] Attarilar S, Ebrahimi M, Djavanroodi F, et al. 3D Printing Technologies in Metallic Implants: A Thematic Review on the Techniques and Procedures [J]. International Journal of Bioprinting, 2021, 7 (1): 306.

[54] Song C, Zhang X, Wang L, et al. An Injectable Conductive Three-Dimensional Elastic Network by Tangled Surgical-Suture Spring for Heart Repair [J]. ACS Nano, 2019, 13 (12): 14122-14137.

[55] Han W, Zhou B, Yang K, et al. Biofilm-Inspired Adhesive and Antibacterial Hydrogel with Tough Tissue Integration Performance for Sealing Hemostasis and Wound Healing [J]. Bioactive Materials, 2020, 5 (4): 768-778.

[56] Wang W, Wang P, Chen L, et al. Engine-Trailer-Structured Nanotrucks for Efficient Nano-Bio Interactions and Bioimaging-Guided Drug Delivery [J]. Chem, 2020, 6 (5): 1097-1112.

[57] Thomas S, Grohens Y, Ninan N. Nanotechnology Applications for Tissue Engineering [M]. William Andrew, 2015.

[58] Dee K C, Puleo D A, Bizios R. An Introduction to Tissue-Biomaterial Interactions [M]. John Wiley & Sons, 2003.

[59] Boutry C M, Beker L, Kaizawa Y, et al. Biodegradable and Flexible Arterial-Pulse Sensor for the Wireless Monitoring of Blood Flow [J]. Nature Biomedical Engineering, 2019, 3 (1): 47-57.

[60] Christman K L. Biomaterials for Tissue Repair [J]. Science, 2019, 363 (6425): 340-341.

[61] Bayda S, Adeel M, Tuccinardi T, et al. The History of Nanoscience and Nanotechnology: From Chemical-Physical Applications to Nanomedicine [J]. Molecules, 2020, 25 (1): 112.

[62] Liu J, Li R, Yang B. Carbon Dots: A New Type of Carbon-Based Nanomaterial with Wide Applications [J]. ACS Central Science, 2020, 6 (12): 2179-2195.

[63] Critchley L. Nanomaterials: An introduction [J/OL]. AZO Nano, (2018-07-25) [2023-02-22], https://www.azonano.com/article.aspx?ArticleID=4932.

[64] Xu C, Wu X, Huang G, et al. Rolled-up Nanotechnology: Materials Issue and Geometry Capability [J]. Advanced Materials Technologies, 2019, 4 (1): 1800486.

[65] Khan Z H, Kumar A, Husain S, et al. Introduction to Nanomaterials [M] //Husain M, Khan Z H. Advances in Nanomaterials. New Delhi: Springer India. 2016: 1-23.

[66] Zhang Q, Uchaker E, Candelaria S L, et al. Nanomaterials for energy conversion and storage [J]. Chemical Society Reviews, 2013, 42 (7): 3127-3171.

[67] Ajtai R. Science and Engineering of Nanomaterials [M] //VAJTAI R. Springer Handbook of Nanomaterials. Berlin, Heidelberg: Springer Berlin Heidelberg. 2013: 1-36.

[68] Picraux S T. Overview of Nanotechnology [J/OL]. Encyclopedia Britannica, (2023-02-03) [2023-03-06], http://www.britannica.com/technology/nanotechnology/Overview-of-nanotechnology.

[69] Liu Z, Fu S, Liu X, et al. Small Size, Big Impact: Recent Progress in Bottom-Up Synthesized Nanographenes for Optoelectronic and Energy Applications [J]. Advanced Science, 2022, 9 (19): 2106055.

[70] Mujtaba J, Liu J, Dey K K, et al. Micro-Bio-Chemo-Mechanical-Systems: Micromotors, Microfluidics, and Nanozymes for Biomedical Applications [J]. Advanced Materials, 2021, 33 (22), 2007465.

[71] Yue X Y, Zhou Y N, Fang F, et al. In situ Construction of Lithium Silicide Host with Unhindered Lithium Spread for Dendrite-Free Lithium Metal Anode [J]. Advanced Functional Materials, 2021, 31 (9), 2008786.

[72] Li X L, Zhou Y N, Sun D, et al. Stabilizing Transition Metal Vacancy Induced Oxygen Redox by Co^{2+}/Co^{3+} Redox and Sodium-Site Doping for Layered Cathode Materials [J]. Angewandte Chemie International Edition, 2021, 60 (40): 22026-22034.

[73] Leng Y. Materials Characterization — Introduction to Microscopic and Spectroscopic Methods [M]. 2nd ed. John Wiley & Sons, 2013.

[74] Ziegler A, Graafsma H, Zhang X F, et al. In-situ Materials Characterization—Across Spatial and Temporal Scales [M]. 1st ed. Springer, 2014.

[75] Chu S, Stoner J O, Graybeal J D, et al. Encyclopedia Britannica. Spectroscopy. (2023-06-13). [2023-07-29]. [M]. https://www.britannica.com/science/spectroscopy.

[76] Wikipedia. Materials science (2023-06-26) [2023-07-29]. https://en.wikipedia.org/wiki/Materials_science

[77] Mehdi Khodaei, Luca Petaccia. X-ray Characterization of Nanostructured Energy Materials by Synchrotron Radiation [M/CD]. IntechOpen. 2017.